Fungal Biology

Series Editors

Fungal biology has an integral role to play in the development of the biotechnology and biomedical sectors. It has become a subject of increasing importance as new fungi and their associated biomolecules are identified. The interaction between fungi and their environment is central to many natural processes that occur in the biosphere. The hosts and habitats of these eukaryotic microorganisms are very diverse; fungi are present in every ecosystem on Earth. The fungal kingdom is equally diverse, consisting of seven different known phyla. Yet detailed knowledge is limited to relatively few species. The relationship between fungi and humans has been characterized by the juxtaposed viewpoints of fungi as infectious agents of much dread and their exploitation as highly versatile systems for a range of economically important biotechnological applications. Understanding the biology of different fungi in diverse ecosystems as well as their interactions with living and non-living is essential to underpin effective and innovative technological developments. This series will provide a detailed compendium of methods and information used to investigate different aspects of mycology, including fungal biology and biochemistry, genetics, phylogenetics, genomics, proteomics, molecular enzymology, and biotechnological applications in a manner that reflects the many recent developments of relevance to researchers and scientists investigating the Kingdom Fungi. Rapid screening techniques based on screening specific regions in the DNA of fungi have been used in species comparison and identification, and are now being extended across fungal phyla. The majorities of fungi are multicellular eukaryotic systems and therefore may be excellent model systems by which to answer fundamental biological questions. A greater understanding of the cell biology of these versatile eukaryotes will underpin efforts to engineer certain fungal species to provide novel cell factories for production of proteins for pharmaceutical applications. Renewed interest in all aspects of the biology and biotechnology of fungi may also enable the development of "one pot" microbial cell factories to meet consumer energy needs in the 21st century. To realize this potential and to truly understand the diversity and biology of these eukaryotes, continued development of scientific tools and techniques is essential. As a professional reference, this series will be very helpful to all people who work with fungi and should be useful both to academic institutions and research teams, as well as to teachers, and graduate and postgraduate students with its information on the continuous developments in fungal biology with the publication of each volume.

Jorge Poveda • Óscar Santamaría
Jorge Martín-García
Editors

Fungal Metabolites for Agricultural Applications

Biostimulation and Crop Protection by Fungal Biotechnology

Springer

Editors
Jorge Poveda
Higher Technical School of Agricultural Engineering of Palencia, University Institute for Research in Sustainable Forest Management (iuFOR)
University of Valladolid
Palencia, Spain

Óscar Santamaría
Higher Technical School of Agricultural Engineering of Palencia, University Institute for Research in Sustainable Forest Management (iuFOR)
University of Valladolid
Palencia, Spain

Jorge Martín-García
Higher Technical School of Agricultural Engineering of Palencia, University Institute for Research in Sustainable Forest Management (iuFOR)
University of Valladolid
Palencia, Spain

ISSN 2198-7777 ISSN 2198-7785 (electronic)
Fungal Biology
ISBN 978-3-031-76586-5 ISBN 978-3-031-76587-2 (eBook)
https://doi.org/10.1007/978-3-031-76587-2

This Springer imprint is published by the registered company Springer Nature Switzerland AG
The registered company address is: Gewerbestrasse 11, 6330 Cham, Switzerland

Contents

Endophytic Filamentous Fungi as Sources of Metabolites for Agricultural Applications

Everlon Cid Rigobelo

Abstract Filamentous fungi are fungi with structures analogous to roots, playing an important role in nutrient and water absorption named hyphae, as well as extracellular digestion of organic compounds and mineralization of inorganic compounds. Some filamentous fungi have evolved interactions with plants and become endophytic. Endophytic fungi have a closer relationship with plants, improving their efficiency in obtaining energy and nutrients from plant cells. In return, they provide secondary metabolites such as phytohormones, proteins, enzymes, and other substances that play different roles in promoting plant growth, inducing resistance, and enhancing resistance to environmental stresses. Filamentous fungi also produce toxins capable of killing phytophagous insects responsible for significant economic losses in agriculture. This chapter aims to elucidate and discuss aspects related to the role of filamentous fungi as a source of diverse secondary metabolites and their use to face many challenges in agriculture. The use of filamentous fungi is an excellent alternative to face the major challenges of modern agriculture, which include reducing production costs, ensuring food security, and minimizing environmental impacts without compromising productivity. This chapter covers topics such as filamentous and endophytic fungi, biological control promoted by entomopathogenic fungi, the ability of fungi to enhance plant efficiency in soil exploration, and the promotion of plant resistance. All these effects are mediated by the synthesis of secondary metabolites produced by filamentous fungi and can be utilized to obtain various benefits in agricultural production, promoting sustainability and reducing the incidence of diseases and environmental degradation.

Keywords Plant growth · Entomopathogenic fungi · Nematode control · Phytohormone production · Plant protection

E. C. Rigobelo (✉)
Laboratory of Soil Microbiology, Department of Plant Production, UNESP University, Paulo Donatto Castellane, Jaboticabal, Brazil
e-mail: everlon.cid@unesp.br

J. Poveda et al. (eds.), *Fungal Metabolites for Agricultural Applications*, Fungal Biology, https://doi.org/10.1007/978-3-031-76587-2_1

1 Introduction

Fungi are eukaryotic microorganisms that can be unicellular, such as yeasts, or multicellular, such as filamentous fungi, and are widely distributed in nature. They can be found in various environments, such as soil, water, plants, and even the gastrointestinal tract of animals and humans (Bowen et al. 1992; Nasif et al. 2023). Plant associations with microbes benefit plant survival in the environment for various reasons. Plant mutualistic fungi are now recognized as a novel and significant source of bioactive chemicals. They generate a large amount of secondary metabolites, including phytohormones (auxins, gibberellins, cytokinins, and abscisic acid) as well as antifungal and antibacterial chemicals (Rai et al. 2014). Plants treated with a variety of symbiotic fungi are frequently healthier than untreated plants, according to previous studies (Strobel and Daisy 2003; Colla et al. 2015).

These plant–fungus relationships are primarily formed by two types of fungi, mycorrhizal and endophytic fungi (Bonfante and Genre 2010; Bennett and Groten 2022). Endophytic fungi are those that can coexist with plants without creating disease symptoms (Behie and Bidochka 2014; Wen et al. 2022). By releasing antimicrobial compounds, phytohormones, and other bioactive chemicals, they may function as plant growth promoters, boost the germination rate, improve seedling establishment, and increase plant resilience to biotic and abiotic challenges. Endophytic fungi are also responsible for the acquisition of soil nutrients such as phosphorus, nitrogen, potassium, and magnesium, as well as micronutrients such as zinc, iron, and copper (Behie and (Behie and Bidochka 2014; Tariq et al. 2020). Beneficial fungi have the potential to be helpful in agriculture by providing crop plants with a variety of services, such as water status, nutritional enrichment, stress tolerance, protection, weed control, and biocontrol. Fungi are essential in natural agroecosystems because they help with soil organic matter breakdown, nutrient acquisition, organic matter recycling, nutrient recycling, antagonism against plant pests, and crop management (Adeleke et al. 2022). Fungi play an important role in regulating the harmful effects of phenols, HCN, and reactive oxygen species (ROS) in plants through 1-aminocyclopropane-1-carboxylic acid (ACC deaminase) and antioxidant enzymes. Fungi also participate in many physiological processes, such as water absorption, stomatal movement, mineral uptake, photosynthesis, and lignan, auxin, and ethylene production, to increase growth and plant fitness in the face of heat, cold, salt, drought, and heavy metal stress (Baron et al. 2019; Escobar Diaz et al. 2021). We discussed the possible use of related fungi in agriculture, such as entomopathogenic fungi and endophyte fungi, and some of their qualities, such as nutrient acquisition and induced systemic resistance, while emphasizing the advantages of secondary metabolites as a source of many substances with biotechnological importance with great potential to be used to improve agricultural production.

2 Filamentous Fungi

Filamentous fungi are multicellular organisms that differ from yeasts by having visible fruiting bodies and the presence of hyphae. Fungal hyphae are elongated and branched cells forming a network called mycelium, which interconnects and intertwines to form filaments (hence filamentous fungi) (Grimm et al. 2005). Hyphae vary greatly in size and diameter, depending on the fungal species and growth conditions, and they have primary functions in growth and nutrition. Hyphae have a specific region called the growing tip, which is a highly active region capable of recognizing and colonizing the environment in search of nutrients and suitable substrates for fungal growth and nutrition. The growing tip of the hypha is responsible for the expansion and branching of the mycelium. Hyphae can have some morphological differences from one another. For example, some hyphae contain septa and are called septate hyphae (Ruiz-Díez 2002). Septate hyphae have transverse divisions that form separate compartments, while other hyphae do not have these divisions. Hyphae without these divisions, or septa, are called coenocytic hyphae. Hyphae function in the nutrition of filamentous fungi and, therefore, can synthesize various extracellular enzymes that degrade organic matter, releasing immobilized nutrients and making them available for fungal nutrition (White et al. 2002).

Over the years, fungi have acquired great genetic and metabolic diversity, allowing them to colonize virtually all environments on Earth, including the interior of animals, including insects and plants (Pringle and Taylor 2002). With the evolution of fungi, they have become increasingly closer to plants, resulting in tighter and more sophisticated plant–fungus interactions. Interestingly, plants and filamentous fungi have evolved together, promoting changes in their genomes, morphology, and physiology. These changes were made to enhance the interaction between fungi and plants, allowing both an increase in the efficiency and benefits derived from this mutual relationship. Fungi contribute their broad diversity of metabolites, promoting plant growth and development through various services such as water and nutrient absorption, nutrient provision, organic matter decomposition, soil aggregate formation, protection against insect attacks, and synthesis and release of numerous substances known as secondary metabolites, which hold enormous biotechnological potential and are useful in agriculture to address the main and current challenges faced by agricultural production (Li et al. 2022).

3 Endophytic Fungi

Endophytic fungi are microorganisms that live within the internal tissues of plants without causing any visible symptoms of the disease. They are characterized by their ability to colonize living plant tissue without causing any immediate, overt negative effects. The minimal requirement for a fungus to be termed an endophyte is the demonstration of its hyphae in living tissue. Endophytic fungi are prospective producers of an abundant and dependable source of bioactive and chemically novel compounds with the potential for exploitation in a wide variety of fields, such as medicine, agriculture, and industry (Nisa et al. 2015). Endophyte filamentous fungi have been found in every plant species studied thus far. They exhibit high species diversity, with endophyte surveys often finding assemblages consisting of more than 30 fungal species per host plant species. They have attracted considerable attention from natural product chemists and biologists alike, leading to a steady increase in publications devoted to this topic in recent years (Nandy et al. 2020).

Endophytic fungi play an important role in agriculture and plant health. Endophytic fungi are ubiquitous, and their host distribution is very broad, involving all groups of angiosperms, even conifers and algae. Endophytes establish mutualistic coexistence patterns by precisely balancing secondary metabolite production and signaling pathway regulation. They improve the host's biotic and abiotic stress tolerance, increase resistance to herbivore or pathogen attack, affect the niche utilization, diversity and abundance of host plants, and degrade or accumulate organic and nonorganic contaminants at different stages of symbiotic interaction (Lundberg et al. 2012). Overall, endophytic fungi have the potential to improve plant health and productivity in agriculture (Arnold et al. 2001; Verma et al. 2022).

Endophytic fungi are widely distributed and can colonize various types of agriculture crops. Their significance in agriculture stems from their ability to enhance root uptake of nutrients, antagonize phytopathogenic fungi to reduce disease incidence, promote higher root and shoot development through phytohormone production, and produce secondary metabolites that hold potential for biotechnological application (Sornakili et al. 2020).

3.1 *Entomopathogenic Fungi*

Some fungi can infect and cause diseases in insects. These fungi are called entomopathogenic fungi. Entomo refers to insects, and pathogenic refers to the ability to cause diseases. These fungi can utilize insect cells for nutrition. When in contact with insects, whether the fungal spore or hyphae, they grow inside the insect and penetrate their bodies. Fungal penetration of the insect can occur through natural openings or directly through the insect's cuticle, and it begins to grow and spread rapidly. Once the fungus breaches the insect's cuticle and enters its body, it releases various enzymes and toxins that degrade the host's tissue, making the nutrients from the insect's cells available for fungal growth and development (Shah and Pell 2003). This process can lead to the death of the host insect within a period that varies greatly depending on the fungal species, the pathogen–host relationship, and environmental conditions. After the insect's death, the entomopathogenic fungus continues to reproduce and develop inside the insect, forming reproductive structures such as sporangia and fruiting bodies that release reproductive structures called spores. These spores are dispersed by water and wind in the environment, potentially coming into contact with other insects, thus restarting the cycle of infection and fungal propagation. Biological control is called the use of exogenous biological agents in the environment, such as entomopathogenic fungi, to eliminate or reduce the incidence of a pest insect (Sharma and Srivastava 2020).

Biological control, mainly using fungi, is an excellent strategy to reduce the incidence of pest insects and decrease the economic losses caused by them. The fungi used can be parasitoids, parasites, or predators of organisms causing economic losses. Biological control is not only an excellent practice to reduce damage to crops caused by pest insects but also because it contributes to the reduction of dependence on various chemical insecticides, which are used in over two billion tons every year worldwide. These insecticides include fungicides, herbicides, and other chemicals used to eliminate undesirable agents, mainly fungi, invasive plants, and insects, which are considered harmful, to achieve high levels of productivity (Mahmood et al. 2016).

The cost of these practices is not only economically high but also environmentally high. This is due to the excessive use of these agricultural inputs, which contaminate the soil, river water, and groundwater and affect the health of animals and humans due to the toxicity and carcinogenic potential of many of these compounds. Biological control is a relatively old practice, often cheaper than chemical control, and does not have negative impacts on the environment. However, despite the numerous benefits of biological control, it is still underutilized compared to chemical control (Köhl et al. 2019). This is due to various factors. First, it is a cultural issue. Thousands of farmers have been using chemical control with good or average results for years. It is not easy to change behaviors and practices that have been used for generations overnight. Second, biological control is not widely promoted and encouraged. On the other hand, large companies spend millions of dollars on advertising, promotions, and agricultural fairs for the sale of chemical inputs. Third, biological control has some disadvantages that can be easily overcome by chemical control. Biological control has some specificities in pest-insect control. From a certain perspective, this characteristic of biological control can be classified as an advantage, where the biological agent controls a specific insect without harming nontarget insects, such as natural enemies of the pest. On the other hand, specificity becomes a problem when there are two or more pest insects in the agricultural crop, and the biological agent controls only one of them. In this regard, the same chemical product tends to be nonspecific and controls a wide variety of pest insects in a few applications. Another disadvantage of biological control compared to chemical control is that biological control requires proper temperature and humidity conditions for storage, usually at temperatures of 4–8 °C, which requires electricity consumption and low humidity. Meanwhile, chemical products can be stored at room temperature. Last, biological control requires suitable environmental conditions and relatively more time to provide effective pest control, whereas chemical control can be used in almost any environmental condition, except during rain, and its action is very rapid. In general, the use of chemical pest control is more facilitated compared to biological control. The main characteristic of biological control is the advantage of not harming human and animal health or the environment. However, until these characteristics are recognized as advantages through awareness of health and environmental preservation or through governmental practices or even consumer market demands, farmers will continue to choose the easiest option, which is chemical control over biological control (Litwin et al. 2020; Zin and Badaluddin 2020).

Fungi are classified as basal fungi and higher fungi based on certain present characteristics. Basal fungi are considered more primitive in terms of evolution and morphological characteristics, such as the *Zygomycetes* and *Oomycetes* phyla, including the genera *Rhizopus* and *Phytophthora*, respectively. Higher fungi are considered more advanced, such as the *Ascomycetes* and *Basidiomycetes* phyla, including the genera *Saccharomyces* and *Agaricus*, respectively. Both basal and higher fungi are used in biological control. Among basal fungi, there are representatives of the phylum *Blastocladiomycota*, such as *Coelomyces*, and the subphylum *Entomophthoromycota*, such as the genera *Conidiobolus*, *Entomophthora, Erynia*, and *Entomophaga*. Among higher fungi, many species of the Ascomycota phylum, subkingdom Dikarya, have been reported with commercial potential and application in the field, such as *Purpureocillum*, *Metarhizium*, *Beauveria*, *Cordyceps*, *Fusarium*, and *Trichoderma* (Gakuubi et al. 2021).

3.2 Nematode Control

Some filamentous fungi have the ability to control nematodes in the soil, contributing to soil health, reducing the occurrence of diseases, and minimizing productivity losses in various crops. Soil fungi from the genus *Trichoderma* (*Hypocreaceae*) and the former *Paecilomyces*, now called *Purpureocillum*, are two examples of filamentous fungi that can be used to reduce the population levels of the nematode *Meloidogyne* spp. in the soil. The reduction in nematode populations in the soil is an action known as biocontrol, which occurs due to various mechanisms of action of the fungi, such as induced resistance in the host plant, antibiosis (synthesis of metabolites with nematicidal effects directly on the parasite), competition for nutrients and colonization sites, direct parasitism, and enzymatic hydrolysis (Howell 2003). Some species of the *Trichoderma* fungus have been used in tomato production with successful suppression of the reproduction of *Meloidogyne* spp. in tomato roots and other crops (Sharon et al. 2007). The suppression of the fungus in relation to the pathogen occurs due to the fungus's ability to colonize plant roots both endophytically and on the surface. This colonization capability greatly hinders the parasite's access to the root area available for infection (Swarnakumari et al. 2020). Another example of a filamentous fungus with the ability to reduce nematode populations and the damage caused by nematodes is the saprophytic fungus *P. lilacinum*, previously known as *Paecilomyces lilacinum* (Lopez-Lima et al. 2014). A significant advantage of applying this fungus is its control over nematode eggs and females due to its ability to produce enzymes such as proteases and chitinases, which dissolve the vitelline layer of the nematode eggshell, facilitating infection and the penetration of fungal hyphae, thereby destroying the parasite at embryonic stages (Khan et al. 2006). Because of the high capacity of the *P. lilacinum* fungus to control nematodes, some commercial products have been successfully used in crops such as tomatoes, bananas, and beans (Abiola and Abiola 2020)

Another strategy for nematode control is the use of a biopolymer widely distributed in nature with high biocompatibility, biodegradability, and low toxicity, called chitin 1.4 N-acetylglucosamine. Chitin has been employed in agriculture due to its great ability to stimulate the growth of chitinolytic microorganisms in the soil, increasing the biological control activity against nematodes and indirectly promoting plant protection and growth. Although some studies have shown positive results from chitin application in the soil for controlling fungi, bacteria, viruses, and nematodes (Fan et al. 2020), other studies indicate that, due to its low solubility, chitin needs to be applied in large quantities in the soil to exert the expected beneficial effect. This amount must reach five tons per hectare, which is completely unfeasible considering the price of chitin. Moreover, the use of chitin in large quantities, as recommended by some studies, can cause phytotoxicity in plants (Culbreath et al. 1985). With the aim of reducing the required quantities of chitin to obtain some benefits related to soil health, some studies have investigated the combination of chitin with the application of microorganisms used in biological control (Kim, et al. 2019). Certain studies have shown that the combination of chitosan with *Pochonia chlamydosporia* can significantly reduce the occurrence of root-knot nematode diseases. These studies propose that the use of chitosan with *P. chlamydosporia* inoculation favored the spore production of the microorganisms and increased parasitism on nematode eggs (Escudero et al. 2017). These findings suggest that it would be possible to reduce the amount of chitin with methods or practices that increase its solubility and bioavailability.

3.3 Acquisition of Nutrients

Endophytic fungi possess numerous abilities to increase nutrient availability in the soil, providing nutrients to plants. Fungal enzymes produce organic acids that lower the soil pH, and some enzymes, such as phytase and phosphatase, solubilize unavailable phosphorus in the soil (Yan et al. 2019; Manganyi and Ateba 2020). Generally, most soils, due to the large amounts of chemical fertilizers used in the planting of various agricultural crops, have medium to high levels of phosphorus in the soil. Despite these phosphorus levels, the majority of it is unavailable to plants and other microorganisms. This is because phosphorus in the soil is adsorbed and immobilized, usually in the clay fraction of the soil. The organic acids and enzymes synthesized by endophytic fungi solubilize this phosphorus, making it available and readily assimilable for plants. Fungi are capable of not only increasing phosphorus availability in the soil but also enhancing nitrogen availability. Nitrogen is a nutrient required in large quantities by plants and virtually all living organisms. This high demand for nitrogen is due to its involvement in the peptide bonds of amino acids and consequently in proteins, enzymes, and virtually all vital cell organelles. In agricultural crops, the higher the expected productivity, the greater the requirement and need to supplement and provide phosphorus and nitrogen through chemical fertilization. Unlike diazotrophic bacteria, fungi cannot perform biological nitrogen fixation. Fungi lack the required enzyme for this process, which is nitrogenase. However, fungi have an enormous capacity for organic matter decomposition, which releases and makes nitrogen immobilized in organic matter available in the soil, readily assimilable for plants and other microorganisms. Behie et al. 2012 were the pioneers in reporting the endophytic colonization of bean plants (*Phaseolus vulgaris*) and switchgrass (*Panicum virgatum*) by the entomopathogenic fungus *Metarhizium robertsii*. Their study also investigated the fungus's ability to transfer nitrogen (N) from infected and killed *Galleria mellonella* larvae. Since this groundbreaking research, further studies have been conducted on this subject. For instance, Behie and Bidochka (2014) expanded on the endophytic transfer of N from insects, exploring seven fungal species, which included *M. robertsii, M. guizhouense*, *M. brunneum*, *M. flavoviridae*, *M. acridum*, *B. bassiana*, and *Akanthomyces* (=*Lecanicillium*) *lecanii*. These researchers studied four different cultures, comprising two dicots, *Glicyne max* (soybean) and *P. vulgaris* (common bean), and two monocots, *Triticum aestivum* (wheat) and *P. virgatum* (switchgrass). Their findings revealed that five *Metarhizium* species and *B. bassiana* can kill insect larvae and endophytically colonize plants. Moreover, these fungi can transfer N from insects to the plants they inhabit.

3.4 Activation of Induced Systemic Resistance

Filamentous fungi can assist plants in defending themselves against attacks from phytophagous insects, herbivores, and phytopathogenic microorganisms, such as bacteria, fungi, and viruses. This assistance results in an increased efficiency and performance of plants related to two distinct defense mechanisms that share many common points, known as induced systemic resistance (ISR) and acquired systemic resistance (ASR). The activation of systemic resistance begins with recognizing pathogen-associated molecular pattern molecules (MAMPs) present in various pathogenic microorganisms; interestingly, some nonpathogenic microorganisms also possess them. These MAMPs are recognized by receptors in plants called MAMP-triggered immunity (MTI). Examples of MAMPs that induce MTI include substances found in the fungal cell wall, such as chitin and β-glucan, and enzymes such as glucanase, xylanase, and chitinase. Some microorganisms can also produce effector molecules, such as secondary metabolites known as effector-triggered immunity (ETI), which induce the plant's resistance response. The activation of MTI in plants by endophytic fungi is very similar to that caused by pathogenic fungi. However, some endophytic fungi can modify MAMPs to prevent their molecules from being recognized, thus not activating the system. For instance, the MAMP β-glucan present in the cellular wall of *Piriformospora indica* can undergo modification via FGB1 gene expression, which encodes a β-glucan-binding lectin specific to fungi. This modification transforms the composition and characteristics of the endophytic cell wall and proves to be effective in suppressing MTI in various plant hosts (Wawra et al. 2016). The facilitation of colonization is speculated to be enabled by the ability of beneficial microorganisms to surmount this barrier (Yan et al. 2019).

The activation of plant resistance mechanisms through the ISR and ASR pathways by MAMPs and effectors leads to a complex response that involves multiple actions and syntheses, such as ion flux, phosphorylation and dephosphorylation of proteins, production of signaling molecules such as salicylic acid and the hormone ethylene, as ROS and phytoalexins. Interestingly, fungi can only colonize plants due to their ability to neutralize the effects of the plants' defense systems against them. At the same time, fungal colonization activates these defense systems, preparing the plants for future infections caused by phytopathogenic microorganisms, herbivores, or nematodes (Latz et al. 2018; Poveda et al. 2020). The colonization of filamentous endophytic fungi in their host plants only occurs because the fungi lack pathogenic properties. When an endophytic fungus, even if nonpathogenic, possesses elements of pathogenicity, it triggers the virulence mechanisms mentioned above, and its colonization capacity is impaired. On the other hand, when a pathogenic fungus manages to evade the molecules that induce recognition of its pathogenicity, it will not be targeted by the plant's defense mechanism and will have a more effective colonization action, causing the disease to occur with high severity. A good example is the juxtaposition between the endophytic strain of *Colletotrichum tofieldiae* and the pathogenic strain *Colletotrichum incanum* in *Arabidopsis thaliana*. The endophytic strain underwent negative selection of effector protein genes, which directly participate in the pathogenic manifestation during plant colonization, while the pathogenic strain *C. incanum* did not undergo a similar process. In contrast, the *Brassicaceae* plant family exhibits fewer receptors for these proteins. Therefore, the persistence of these genes in pathogenic species may indicate a potential tactic to attack the host. In contrast, species prone to promoting favorable interactions with plants have reduced the variety of these genes in their genome (Hiruma et al. 2016).

4 Metabolites Produced by Endophytic Filamentous Fungi

4.1 *Phytohormone Production*

Filamentous fungi, endophytes of plants, have the ability to produce various phytohormones, including auxins, gibberellins, and cytokinins (Yan et al. 2019). The phytohormones produced by endophytic fungi play important roles in the growth, development, and responses to environmental stresses in plants (Tian et al. 2022). Among these phytohormones, auxins produced by endophytic fungi can promote plant growth by stimulating cell division, elongating plant cells, and the development of vegetative organs. Auxins also assist in the development of the plant's root system, increasing its efficiency in soil exploration by enhancing nutrient and water absorption capacity and enhancing the plant's ability to withstand water stress (Suebrasri et al. 2022). Other phytohormones also produced by filamentous endophytic fungi assist plants in coping with environmental stresses. Phytohormones such as gibberellins and abscisic acid help plants deal with drought, salinity, and extreme temperatures by increasing plant resistance to environmental stresses. Under stress conditions, these phytohormones promote stomatal closure, reduce water loss, and activate protective mechanisms such as the synthesis of antioxidants and stress-responsive proteins that help preserve plant cells even under adverse environmental conditions (Liu et al. 2023).

4.2 Plant Protection

Numerous reports have demonstrated that these fungi and other well-known entomopathogens can provide plant protection through direct interactions with plants. In this scenario, fungi exhibit antagonistic behavior towards plant pathogens, employing diverse mechanisms. These mechanisms include the production of various metabolites, such as antibiotics, volatile compounds, and enzymes. Furthermore, they compete for vital resources such as space, carbon sources, nitrogen, and minerals. Additionally, they can parasitize pathogens and induce systemic resistance in plants, increasing plant growth. Consequently, the activity of the pathogens is significantly reduced (Quesada Moraga 2020; Poveda et al. 2020).

Fungi exert diverse benefits for host plants with several positive implications for agriculture and agricultural production. However, a deeper understanding of plant mycology biology and ecology, as well as the molecular mechanisms that regulate interactions with plants, is necessary to enhance knowledge and improve plant resilience, nutrition, stress tolerance, and defense. This deeper understanding of interactions will practically occur through comprehending gene function, unravelling fungal survival strategies in symbiosis, thereby expanding their biotechnological applications. It is essential to develop molecular tools that allow studies of filamentous fungi in their complex environments and the discovery of new functionalities in plant–fungus interactions (Pozo et al. 2021).

Secondary metabolites produced by filamentous fungi can enhance plant tolerance to grow and develop in contaminated soils. A study investigated the effect of inoculating a fungal consortium in the plant *Alocasia calidora* and its influence on growth and some biochemical responses, such as tolerance, absorption, and translocation of various heavy metals present in the soil. The results showed that inoculation with the fungal consortium substantially increased root shoot growth and the chlorophyll and carotenoid levels in the leaves. Interestingly, several fungal metabolites assisted the plants in tolerance and growth, even in the presence of heavy metals. Further studies are needed to understand the exact mechanism of how fungi help plants grow in the presence of heavy metals in the soil; however, it is suggested that fungal metabolites may alter the osmolarity of fungal cell plasma membranes, leading to reduced metal absorption. They may also facilitate the biotransformation of these metals into less toxic substances, and finally, fungal metabolites may decrease the synthesis of enzymes related to environmental stress, reducing the damage caused to plants due to the stress response (Hassan et al. 2022).

4.3 Diverse Secondary Metabolites Produced by Endophytic Fungi

Furthermore, endophytic filamentous fungi produce a wide range of secondary metabolites that serve crucial ecological functions and have significant biotechnological applications (Ancheeva et al. 2019). Although not essential for microorganism survival, these molecules play adaptive roles, such as defense compounds and signaling molecules, during interactions (Bérdy 2005). Secondary metabolites possess remarkable biological activities, including antibiotic properties, toxin production, ionophores, bioregulators, and intra- and interspecific signaling. Notably, antibiotics produced by microorganisms serve as potent weapons in competitive interactions with other organisms (Challis and Hopwood 2003). Antibiotics are primarily produced by microorganisms when they are experiencing nutritional stress. These findings suggest that microorganisms produce antibiotic molecules to avoid competition for nutrients with other organisms (Ancheeva et al. 2019). Endophytic fungi can also produce antagonistic molecules that target a wide range of pathogenic fungi (Hiruma et al. 2016). These molecules effectively harm pathogenic fungi, reducing the incidence of diseases in plants (Porras-Alfaro and Bayman 2011). Moreover, they have the ability to induce systemic resistance in the plant, making it more resistant to attacks from herbivores and insects (Rodriguez et al. 2009).

Endophytic fungi represent a valuable source of diverse biomolecules with potential applications across multiple disciplines (Alvin et al. 2014). These secondary metabolites are associated with various mechanisms of evolution and contribute to biodiversity (Gouda et al. 2016). Notably, polyketides and terpenoids have emerged as predominant groups of metabolites within this context. Consequently, exploring unique endophytic secondary metabolism has gained increasing significance, offering a promising pathway for sustainable development through a holistic approach to harnessing their full potential (Osemwegie et al. 2020).

Filamentous fungi of the *Epichloe* species can produce a secondary metabolite classified as an alkaloid in different grass species. These alkaloids accumulate in plants and are toxic to various insects and even vertebrates (Lugtenberg et al. 2016). Some studies report the production of molecules to reduce the effect of herbivory caused by insects. One such molecule is the nodulisporic acid produced by the fungus *Nodulisporium* sp. This acid activates glutamate in the chloride channels of nerve and muscle cells of insects, causing flaccid paralysis and consequently loss of movement and death due to respiratory failure in insects (Köhl et al. 2019). Other fungi produce molecules known as VOCs (volatile organic compounds), which have high vapour pressure and are easily diffusible through plant cell membranes, soil, and the atmosphere. This property of VOCs makes them special agents for fungal communication with other organisms, including plants. VOCs also exhibit diverse biological activities against many phytopathogens (Kaddes et al. 2019).

Secondary metabolites produced by fungi can also be used in postharvest technology to preserve fruits, cereals, and tubers (Strobel and Daisy 2003). This practice is known as mycofumigation. One of the fungi that produces a wide variety of VOCs for this purpose is the genus *Muscodor*. This fungus has been used in the postharvest process and as a soil inoculant, inhibiting the growth of pathogenic fungi through VOC production. Additionally, the genus *Nodulisporium* is recognized for producing VOCs with antifungal activity and has been applied for the same purposes as *Muscodor*. Various synthetic chemical compounds used in agriculture have harmful effects on soil, rivers and groundwater, as well as animals and humans. These chemical compounds have been restricted and sometimes even prohibited by regulatory agencies and consumer demand. As filamentous fungi produce a wide diversity of metabolites, many of which have demonstrated extensive potential in agriculture, it is hoped that future research will lead to the development of new products based on secondary metabolites that can be used more sustainably without impacting the environment (Baron et al. 2019, 2020).

New metabolites are needed for treating communicable and noncommunicable diseases and addressing metabolic disorders and autoimmune diseases (Dukare et al. 2018). Microorganisms, particularly endophytic fungi, have an unlimited capacity for synthesizing secondary metabolites that can be used for these various purposes. The secondary metabolites produced by endophytic fungi have been extensively studied, mainly because of the structures of these compounds that allow for the production of several drugs (Gao et al. 2018).

4.4 Biotechnological Potential of Fungal Secondary Metabolites

Many secondary metabolites produced by endophytic fungi have antifungal and antibacterial effects. These compounds are of great interest due to the occurrence of drug-resistant pathogenic microorganisms (Jovetic et al. 2010). The problem of emerging multidrug-resistant bacteria is growing, and its causes are multifactorial. However, one of the causes is attributed to the indiscriminate use of antibiotics and antimicrobials (Di Santo 2010). The discovery of new drugs with antibacterial effects is necessary and urgent to treat various bacterial infections. Endophytic fungi are excellent producers of antibacterial drugs. The agar diffusion test is the most common test to evaluate whether an endophytic fungus produces any antibacterial drug. In this test, a bacterium of interest, which one wants to control, is cultured with the fungus, and the formation of inhibition zones is observed. The formation of the inhibition zone occurs due to the secondary metabolite produced by the fungus, which diffuses in the agar and meets the bacteria, inhibiting their growth and development (Sipiczki 2006).

Some fungi produce phytotoxic secondary metabolites, which are toxic compounds to numerous plants. These compounds are responsible for causing severe symptoms in plants due to phytopathogenic fungi. The symptoms include leaf spots, wilting, chlorosis, necrosis, and growth inhibition. Characterizing these substances and studying their structural activities can aid in a deeper investigation of the importance of these compounds for plant-microorganism interactions. Fungi that produce phytotoxins can be either host-specific or nonspecific. This means that host-specific fungi only produce these toxins in plants they can colonize. Examples of fungi that produce these phytotoxins are *Alternaria* and the fungus *Colletotrichum*. These toxins play a crucial role in the pathogenicity of these fungi. Xu et al. (2021) conducted a survey of over 545 phytotoxic secondary metabolites from fungi that can be used as herbicides. Due to the vast list of phytotoxic secondary metabolites, there are undoubtedly specific compounds that can be utilized for weed control, opening significant potential for reducing herbicide usage in agriculture. Tariq et al. (2020) provide a review of the use of microbial metabolites in plant disease management, highlighting the potential of antibiotics produced by various bacteria and fungi as biopesticides. The present study identifies several prolific producers of secondary metabolites that can be used as substitutes for chemical fungicides, opening the way for their use in sustainable agriculture. The study also provides a list of specific metabolites that have been used in controlling plant diseases, along with their inhibitory effects on various fungal phytopathogens. The metabolites listed in the study can make a significant contribution to reducing the use of agricultural inputs and promoting more sustainable production. Studies conducted for the discovery and development of new drugs with antimicrobial effects are relevant not only to human medicine or veterinary science but also to agriculture. There are numerous diseases in agricultural crops caused by bacteria (Arwiyanto 2014), and the use of antimicrobial substances often proves ineffective and uneconomical. Strategies employed by farmers include the use of disease resistant varieties and a mixture of lime with copper, which helps reduce bacterial and fungal diseases and has been used for many years. Since filamentous fungi have a wide diversity of secondary metabolites, as previously mentioned, many of which have antimicrobial effects, these fungi present enormous biotechnological potential for the discovery of new antimicrobial molecules that could also be used profitably in agriculture in the future. Figure 1 depicts some groups of secondary metabolites and their function and Table 1 shows the examples of some of the secondary metabolites produced by fungi.

5 Future and Perspectives

Despite being a decades-old technique, the usage of fungal species as biological control agents has grown tremendously in recent years. With growing concern about the environmental impacts of agricultural processes in nature and the search for healthier foods free of chemical compounds that are harmful to health, scientific research is moving toward the exploitation of these organisms for this purpose. The

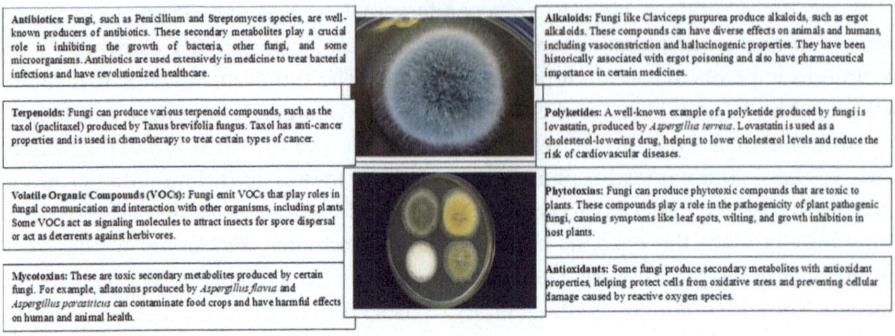

Fig. 1 Some examples of secondary metabolites produced by fungi and their respective functions

Table 1 Secondary metabolites produced by fungi

Secondary metabolite	Fungal source	Biological activity	Reference
Penicillin	*Penicillium spp.*	Antibiotic - inhibits bacterial cell wall synthesis	Koul and Singh (2017)
Aflatoxin	*Aspergillus flavus*	Mycotoxin—potent carcinogen	Kamei et al. (2005)
Lovastatin	*Aspergillus terreus*	Cholesterol-lowering agent	Osman et al. (2011)
Ergotamine	*Claviceps purpurea*	Vasoconstrictor and migraine medication	Osman et al. (2011)
Taxol	*Taxomyces andreanae*	Anticancer agent from Taxus trees	Stierle et al. (1995)
Ciclosporin	*Tolypocladium inflatum*	Immunosuppressive drug for organ transplants	Yang et al. (2018)
Gibberellins	*Gibberella fujikuroi*	Plant growth regulators	Tudzynski (1999)
Mycophenolic acid	*Penicillium brevicompactum*	Immunosuppressive agent	Regueira et al. (2011)
Griseofulvin	*Penicillium griseofulvum*	Antifungal medication	Valente et al. (2020)
Patulin	*Penicillium expansum*	Mycotoxin—found in certain fruits and juices	Andersen et al. (2004)
Citrinin	*Penicillium citrinum*	Nephrotoxic mycotoxin	Geisen et al. (2018)
Gliotoxin	*Aspergillus fumigatus*	Immunosuppressive and cytotoxic properties	Bauer et al. (1989)
Beauvericin	*Beauveria bassiana*	Antibacterial, antifungal, and cytotoxic activity	Xu et al. (2021)
Terrein	*Aspergillus terreus*	Antifungal and anticancer properties	Goutam et al. (2017)
Patulutin	*Penicillium crustosum*	Mycotoxin with antibiotic activity	

(continued)

Table 1 (continued)

Secondary metabolite	Fungal source	Biological activity	Reference
Ergosterol	*Saccharomyces cerevisiae*	Precursor of vitamin D and membrane component	Xu et al. (2021)
Ochratoxin A	*Aspergillus ochraceus*	Nephrotoxic and potentially carcinogenic mycotoxin	Perrone and Gallo (2017)
Fumonisin	*Fusarium verticillioides*	Mycotoxin associated with livestock diseases	Seo et al. (2001)
Resveratrol	*Emericella nidulans*	Antioxidant and anti-inflammatory compound	Antonopoulou et al. (2016)
Xanthomegnin	*Penicillium chrysogenum*	Antibacterial and antiviral properties	Andersen et al. (2004)
Antibiotics	Endophytic filamentous fungi	Potent weapons in competitive interactions with other organisms	Ancheeva et al. (2019)
Antibiotics	Microorganisms	Produced to avoid competition for nutrients with other organisms	Challis and Hopwood (2003)
Antagonistic molecules	Endophytic fungi	Effectively harm pathogenic fungi, reducing the incidence of diseases in plants	Kei Hiruma et al. (2016)
Antagonistic molecules	Endophytic fungi	Effectively harm pathogenic fungi, reducing the incidence of diseases in plants	Porras-Alfaro and Bayman (2011)
VOCs	Endophytic fungi	Increase plant resistance to attacks from herbivores and insects	Rodriguez et al. (2009)
Alkaloids	Epichloe species	Accumulate in plants and are toxic to various insects and even vertebrates	Lugtenberg et al. (2016)
Nodulisporic acid	Nodulisporium sp.	Activates glutamate in the chloride channels of insects, causing paralysis and death	Köhl et al. (2019)
VOCs	Fungi	Diverse biological activities against many phytopathogens	Kaddes et al. (2019)
VOCs	Muscodor spp., Nodulisporium spp.	Inhibit the growth of pathogenic fungi through VOC production	Strobel and Daisy (2003)
Diverse secondary metabolites	Filamentous fungi	Associated with various mechanisms of evolution and contribute to biodiversity	Baron et al. (2019, 2020)
Phytotoxic	Fungi	Toxic compounds to numerous plants, causing severe symptoms and growth inhibition	Xu et al. (2021)
Antibiotics	Bacteria and fungi	Potential of antibiotics produced by various bacteria and fungi as biopesticides	Tariq et al. (2020)

capacity of various fungal species, including those previously utilized for pest management, to colonize crops endophytically has been shown to be a highly promising method for achieving the necessary sustainability in agriculture. Endophytic fungi are helpful because they give various direct and indirect advantages to agricultural plants, as described in this paper, and it is reasonable to suppose that no synthetic chemical can provide such a wide variety of favorable interactions as these microorganisms. Thus, the employment of endophytic fungi is a promising option in biocontrol, biostimulation, and biofertilization, proving that such organisms are a valuable resource for study and business.

References

Abiola OA, Abiola OA (2020) Vegetable production and nematodes infestation: impacts on small-scale farming communities of South. Africa 7:168–177. https://doi.org/10.30574/WJARR.2020.7.2.0211

Adeleke BS, Ayilara MS, Akinola SA, Babalola OO (2022) Biocontrol mechanisms of endophytic fungi. Egypt J Biol Pest Contl 32

Alvin A, Miller KI, Neilan BA (2014) Exploring the potential of endophytes from medicinal plants as sources of antimycobacterial compounds. Microbiol Res 169. https://doi.org/10.1016/j.micres.2013.12.009

Ancheeva E, Daletos G, Proksch P (2019) Bioactive secondary metabolites from endophytic fungi. Curr Med Chem 27:1836–1854. https://doi.org/10.2174/0929867326666190916144709

Andersen B, Smedsgaard J, Frisvad JC (2004) Penicillium expansum: consistent production of patulin, chaetoglobosins, and other secondary metabolites in culture and their natural occurrence in fruit products. J Agric Food Chem 52. https://doi.org/10.1021/jf035406k

Antonopoulou I, Varriale S, Topakas E et al (2016) Enzymatic synthesis of bioactive compounds with high potential for cosmeceutical application. Appl Microbiol Biotechnol 100

Arnold AE, Zuleyka M, Gilbert GS (2001) Fungal endophytes in dicotyledonous neotropical trees: patterns of abundance and diversity. Mycol Res 105. https://doi.org/10.1017/S0953756201004956

Arwiyanto T (2014) Biological control of plant diseases caused by bacteria. Jurnal Perlindungan Tanaman Indonesia 18

Baron NC, Rigobelo EC, Zied DC (2019) Filamentous fungi in biological control: current status and future perspectives. Chil J Agric Res 79. https://doi.org/10.4067/S0718-58392019000200307

Baron NC, de Souza PA, Rigobelo EC (2020) *Purpureocillium lilacinum* and *Metarhizium marquandii* as plant growth-promoting fungi. Peer J. https://doi.org/10.7717/peerj.9005

Bauer J, Gareis M, Bott A, Gedek B (1989) Isolation of a mycotoxin (gliotoxin) from a bovine udder infected with *Aspergillus fumigatus*. Med Mycol 27. https://doi.org/10.1080/02681218980000061

Behie SW, Bidochka MJ (2014) Ubiquity of insect-derived nitrogen transfer to plants by endophytic insect-pathogenic fungi: an additional branch of the soil nitrogen cycle. Appl Environ Microbiol 80. https://doi.org/10.1128/AEM.03338-13

Behie SW, Zelisko PM, Bidochka MJ (2012) Endophytic insect-parasitic fungi translocate nitrogen directly from insects to plants. Science 1979:336. https://doi.org/10.1126/science.1222289

Bennett AE, Groten K (2022) The costs and benefits of plant-arbuscular mycorrhizal fungal interactions. Annu Rev Plant Biol 73

Bérdy J (2005) Bioactive microbial metabolites. J Antibiot (Tokyo) 58. https://doi.org/10.1038/ja.2005.1

Bonfante P, Genre A (2010) Mechanisms underlying beneficial plant - Fungus interactions in mycorrhizal symbiosis. Nat Commun. https://doi.org/10.1038/ncomms1046

Bowen AR, Chen-Wu JL, Momany M et al (1992) Classification of fungal chitin synthases. Proc Natl Acad Sci USA 89. https://doi.org/10.1073/pnas.89.2.519

Challis GL, Hopwood DA (2003) Synergy and contingency as driving forces for the evolution of multiple secondary metabolite production by Streptomyces species. Proc Natl Acad Sci USA 100. https://doi.org/10.1073/pnas.1934677100

Colla G, Rouphael Y, Di Mattia E et al (2015) Co-inoculation of *Glomus intraradices* and *Trichoderma atroviride* acts as a biostimulant to promote growth, yield and nutrient uptake of vegetable crops. J Sci Food Agric 95. https://doi.org/10.1002/jsfa.6875

Culbreath AK, Rodriguez-Kabana R, Morgan-Jones G (1985) The use of hemicellulosic waste matter for reduction of the phytotoxic effects of chitin and control of root-knot nematodes. Nematropica 15

Di Santo R (2010) Natural products as antifungal agents against clinically relevant pathogens. Nat Prod Rep 27

Dukare AS, Paul S, Nambi VE et al (2018) Exploitation of microbial antagonists for the control of postharvest diseases of fruits: a review. Crit Rev Food Sci Nutr 59:1498–1513. https://doi.org/10.1080/10408398.2017.1417235

Escobar Diaz PA, dos Santos RM, Baron NC et al (2021) Effect of *Aspergillus* and *Bacillus* Concentration on Cotton Growth Promotion. Front Microbiol 12. https://doi.org/10.3389/fmicb.2021.737385

Escudero N, Lopez-Moya F, Ghahremani Z et al (2017) Chitosan increases tomato root colonization by *Pochonia chlamydosporia* and their combination reduces root-knot nematode damage. Front Plant Sci 8. https://doi.org/10.3389/fpls.2017.01415

Fan Z, Qin Y, Liu S et al (2020) Chitosan oligosaccharide fluorinated derivative control root-knot nematode (*Meloidogyne incognita*) disease based on the multi-efficacy strategy. Mar Drugs 18. https://doi.org/10.3390/md18050273

Gakuubi MM, Munusamy M, Liang ZX, Ng SB (2021) Fungal endophytes: a promising frontier for discovery of novel bioactive compounds. J Fungi 7:786. https://doi.org/10.3390/JOF7100786

Gao H, Li G, Lou HX (2018) Structural diversity and biological activities of novel secondary metabolites from endophytes. Molecules 23

Geisen R, Schmidt-Heydt M, Touhami N, Himmelsbach A (2018) New aspects of ochratoxin A and citrinin biosynthesis in *Penicillium*. Curr Opin Food Sci 23

Gouda M, Haggag KM, Boraie WE et al (2016) Synthesis and characterization of inorganic pigment nanoparticles for textile coloration using microwave techniques. AATCC J Res 3. https://doi.org/10.14504/ajr.3.3.1

Goutam J, Sharma G, Tiwari VK et al (2017) Isolation and characterization of "Terrein" an antimicrobial and antitumor compound from endophytic fungus *Aspergillus terreus* (JAS-2) associated from *Achyranthus aspera* Varanasi. India Front Microbiol 8. https://doi.org/10.3389/fmicb.2017.01334

Grimm LH, Kelly S, Krull R, Hempel DC (2005) Morphology and productivity of filamentous fungi. Appl Microbiol Biotechnol 69

Hassan A, Pariatamby A, Ossai I, Ahmed A (2022) Synergistic association of endophytic fungi enhances tolerance, growth, and heavy metal uptake of *Alocasia calidora* in landfill contaminated soil. Appl Soil Ecol:104307–104307

Howell CR (2003) Mechanisms employed by trichoderma species in the biological control of plant diseases: the history and evolution of current concepts. Plant Dis 87:4–10. https://doi.org/10.1094/PDIS.2003.87.1.4

Jovetic S, Zhu Y, Marcone GL et al (2010) β-Lactam and glycopeptide antibiotics: First and last line of defense? Trends Biotechnol 28

Kaddes A, Fauconnier ML, Sassi K et al (2019) Endophytic fungal volatile compounds as solution for sustainable agriculture. Molecules 24

Kamei K, mycology AW-M (2005) Undefined Aspergillus mycotoxins and their effect on the host. Kamei, A Watanabe Medical mycology, 2005. academic.oup.com. https://doi.org/10.1080/13693780500051547

Kei Hiruma A, Gerlach N, Sacristá S et al (2016) Root endophyte colletotrichum tofieldiae confers plant fitness benefits that are phosphate status dependent. Cell 165:464–474. https://doi.org/10.1016/j.cell.2016.02.028

Khan A, Williams KL, Nevalainen HKM (2006) Infection of plant-parasitic nematodes by *Paecilomyces lilacinus* and *Monacrosporium lysipagum*. BioControl 51. https://doi.org/10.1007/s10526-005-4242-x

Köhl J, Kolnaar R, Ravensberg WJ (2019) Mode of action of microbial biological control agents against plant diseases: relevance beyond efficacy. Front Plant Sci 10:454982. https://doi.org/10.3389/FPLS.2019.00845/BIBTEX

Koul M, Singh S (2017) Penicillium spp.: prolific producer for harnessing cytotoxic secondary metabolites. Anti-Cancer Drugs 28

Latz MAC, Jensen B, Collinge DB, Jørgensen HJL (2018) Endophytic fungi as biocontrol agents: elucidating mechanisms in disease suppression. Plant Ecol Divers 11

Li L, Guo N, Feng Y et al (2022) Effect of *Piriformospora indica*-induced systemic resistance and basal immunity against rhizoctonia cerealis and *Fusarium graminearum* in wheat. Front Plant Sci 13. https://doi.org/10.3389/fpls.2022.836940

Litwin A, Nowak M, Różalska S (2020) Entomopathogenic fungi: unconventional applications. Rev Environ Sci Biotechnol 19

Liu R, Yang L, Zou Y, Wu Q (2023) Root-associated endophytic fungi modulate endogenous auxin and cytokinin levels to improve plant biomass and root morphology of trifoliate orange. Hortic Plant J 9. https://doi.org/10.1016/j.hpj.2022.08.009

Lopez-Lima D, Carrion G, Núñez-Sánchez ÁE (2014) Isolation of fungi associated with *Criconemoides* sp. and their potential use in the biological control of ectoparasitic and semiendoparasitic nematodes in sugar cane. Aust J Crop Sci 8

Lugtenberg BJJ, Caradus JR, Johnson LJ (2016) Fungal endophytes for sustainable crop production. FEMS Microbiol Ecol 92. https://doi.org/10.1093/femsec/fiw194

Lundberg DS, Lebeis SL, Paredes SH et al (2012) Defining the core Arabidopsis thaliana root microbiome. Nature 488. https://doi.org/10.1038/nature11237

Mahmood I, Imadi SR, Shazadi K et al (2016) Effects of pesticides on environment. Plant Soil Microb 1: Implications in Crop Science:253–269. https://doi.org/10.1007/978-3-319-27455-3_13/COVER

Manganyi MC, Ateba CN (2020) Untapped potentials of endophytic fungi: a review of novel bioactive compounds with biological applications. Microorganisms 8:1934. https://doi.org/10.3390/MICROORGANISMS8121934

Nandy S, Tudu CK, Dk P et al (2020) Fungal endophytes: futuristic tool in recent research area of phytoremediation. South Afr J Bot:285–295. https://doi.org/10.1016/j.sajb.2020.02.015

Nasif SO, Siddique AB, Siddique AB et al (2023) Prospects of endophytic fungi as a natural resource for the sustainability of crop production in the modern era of changing climate. Symbiosis 89

Nisa H, Kamili AN, Nawchoo IA et al (2015) Fungal endophytes as prolific source of phytochemicals and other bioactive natural products: a review. Microb Pathog 82

Osemwegie OO, Adetunji CO, Ayeni EA et al (2020) Exopolysaccharides from bacteria and fungi: current status and perspectives in Africa. Heliyon 6

Osman ME, Khattab OH, Zaghlol GM, Abd El-Hameed RM (2011) Screening for the production of cholesterol lowering drugs (Lovastatin) by some fungi. Aust J Basic Appl Sci 5

Perrone G, Gallo A (2017) *Aspergillus* species and their associated mycotoxins. Methods Mol Biol 1542:33–49. https://doi.org/10.1007/978-1-4939-6707-03_

Porras-Alfaro A, Bayman P (2011) Hidden fungi, emergent properties: endophytes and microbiomes. Annu Rev Phytopathol 49:291–315. https://doi.org/10.1146/ANNUREV-PHYTO-080508-081831

Poveda J, Abril-Urias P, Escobar C (2020) Biological control of plant-parasitic nematodes by filamentous fungi inducers of resistance: trichoderma. Mycorrhizal Endophytic Fungi Front Microbiol 11. https://doi.org/10.3389/FMICB.2020.00992/FULL

Pozo MJ, Zabalgogeazcoa I, Vazquez de Aldana BR, Martinez-Medina A (2021) Untapping the potential of plant mycobiomes for applications in agriculture. Curr Opin Plant Biol 60

Pringle A, Taylor JW (2002) The fitness of filamentous fungi. Trends Microbiol 10

Quesada Moraga E (2020) Entomopathogenic fungi as endophytes: their broader contribution to IPM and crop production. Biocontrol Sci Technol 30

Rai M, Rathod D, Agarkar G et al (2014) Fungal growth promotor endophytes: a pragmatic approach towards sustainable food and agriculture. Symbiosis 62. https://doi.org/10.1007/s13199-014-0273-3

Regueira TB, Kildegaard KR, Hansen BG et al (2011) Molecular basis for mycophenolic acid biosynthesis in *Penicillium brevicompactum*. Appl Environ Microbiol 77. https://doi.org/10.1128/AEM.03015-10

Rodriguez RJ, White JF, Arnold AE, Redman RS (2009) Fungal endophytes: diversity and functional roles: Tansley review. New Phytol 182

Ruiz-Díez (2002) Strategies for the transformation of filamentous fungi. J Appl Microbiol 92:189–195. https://doi.org/10.1046/j.1365-2672.2002.01516.x

Seo JA, Proctor RH, Plattner RD (2001) Characterization of four clustered and coregulated genes associated with fumonisin biosynthesis in *Fusarium verticillioides*. Fungal Genet Biol 34. https://doi.org/10.1006/fgbi.2001.1299

Shah PA, Pell JK (2003) Entomopathogenic fungi as biological control agents. Appl Microbiol Biotechnol 61

Sharma A, Srivastava A (2020) Phytobiome current insights and future. In: Entomopathogenic fungi: a potential source for biological control of insect pests. Springer, pp 225–250. https://doi.org/10.1007/978-981-15-3151-4_9

Sharon E, Chet I, Viterbo A et al (2007) Parasitism of *Trichoderma* on *Meloidogyne javanica* and role of the gelatinous matrix. Eur J Plant Pathol 118. https://doi.org/10.1007/s10658-007-9140-x

Sipiczki M (2006) *Metschnikowia* strains isolated from botrytized grapes antagonize fungal and bacterial growth by iron depletion. Appl Environ Microbiol 72:6716–6724. https://doi.org/10.1128/AEM.01275-06

Sornakili A, Thankappan S, Sridharan A et al (2020) Antagonistic fungal endophytes and their metabolite-mediated interactions against phytopathogens in rice. Physiol Mol Plant Pathol. 101525–101530

Stierle A, Strobel G, Stierle D et al (1995) The search for a taxol-producing microorganism among the endophytic fungi of the pacific yew, taxus brevifolia1. J Nat Prod 58:1315–1324. https://doi.org/10.1021/NP50123A002

Strobel G, Daisy B (2003) Bioprospecting for microbial endophytes and their natural products. Microbiol Mol Biol Rev 67:491–502. https://doi.org/10.1128/MMBR.67.4.491-502.2003

Suebrasri T, Harada H, Jogloy S et al (2022) Auxin-producing fungal endophytes promote growth of sunchoke. Rhizosphere:100271–100271. https://doi.org/10.1016/j.rhisph.2020.100271

Swarnakumari N, Sindhu R, Thiribhuvanamala G, Rajaswaminathan V (2020) Evaluation of oil dispersion formulation of nematophagus fungus, *Pochonia chlamydosporia* against root-knot nematode, *Meloidogyne incognita* in cucumber. J Asia Pac Entomol 23:1283–1287. https://doi.org/10.1016/J.ASPEN.2020.10.008

Tariq M, Khan A, Asif M et al (2020) Biological control: a sustainable and practical approach for plant disease management. Acta Agric Scand Sect B Soil Plant Sci:507–524. https://doi.org/10.1080/09064710.2020.1784262

Tian Y, Fu X, Zhang G et al (2022) Mechanisms in growth-promoting of cucumber by the endophytic Fungus *Chaetomium globosum* Strain ND35. J Fungi 7:180–187

Tudzynski B (1999) Biosynthesis of gibberellins in Gibberella fujikuroi: biomolecular aspects. Appl Microbiol Biotechnol 52:298–310. https://doi.org/10.1007/S002530051524

Valente S, Cometto A, Piombo E, Meloni G (2020) Elaborated regulation of griseofulvin biosynthesis in *Penicillium griseofulvum* and its role on conidiation and virulence. Int Food Microbiol, pp 108687–108687

Verma A, Shameem N, Jatav HS et al (2022) Fungal endophytes to combat biotic and abiotic stresses for climate-smart and sustainable agriculture. Front Plant Sci 13. https://doi.org/10.3389/FPLS.2022.953836/FULL

Wawra S, Fesel P, Widmer H et al (2016) The fungal-specific β-glucan-binding lectin FGB1 alters cell-wall composition and suppresses glucan-triggered immunity in plants. Nature 7:13188–13188

Wen J, Okyere S, Wang S et al (2022) Endophytic fungi: an effective alternative source of plant-derived bioactive compounds for pharmacological studies. J Fungi:205–220

White S, McIntyre M, Berry DR, McNeil B (2002) The autolysis of industrial filamentous fungi. Crit Rev Biotechnol 22:1–14. https://doi.org/10.1080/07388550290789432

Xu Y, Orozco R, Wijeratne E et al (2021) Biosynthesis of the cyclooligomer depsipeptide beauvericin, a virulence factor of the entomopathogenic fungus *Beauveria bassiana*. Chem Biol:898–907

Yan L, Zhu J, Zhao X et al (2019) Beneficial effects of endophytic fungi colonization on plants. Appl Microbiol Biotechnol 103(8):3327–3340. https://doi.org/10.1007/S00253-019-09713-2

Yang X, Feng P, Yin Y et al (2018) Cyclosporine biosynthesis in *Tolypocladium inflatum* benefits fungal adaptation to the environment. mBio:9. https://doi.org/10.1128/MBIO.01211-18

Zin NA, Badaluddin NA (2020) Biological functions of Trichoderma spp. for agriculture applications. Annal Agric Sci 65:168–178. https://doi.org/10.1016/J.AOAS.2020.09.003

Yeast as a Source of Metabolites for Agricultural Applications

Gulam Hassan Mir, Vishal Gupta, Ayushi Sharma, Gayatri Jamwal, Sushil Kumar Gupta, and Vijay K. Razdan

Abstract Microorganisms have emerged as a valuable source of metabolites with significant potential for various agricultural applications. Yeast metabolites encompass a broad spectrum of compounds, including organic acids, enzymes, vitamins, polysaccharides and bioactive secondary metabolites. Yeast-derived biofertilizers and biostimulants have demonstrated efficacy in promoting root development, increasing crop yields, and improving the overall health of plants. Furthermore, yeast-based formulations exhibit biocontrol properties against various plant pathogens, reducing the reliance on synthetic pesticides and contributing to integrated pest management strategies. The production of yeast metabolites can be optimized through biotechnological approaches such as strain improvement, fermentation optimization, and genetic engineering. These advancements enable the customization of yeast strains to maximize the synthesis of specific metabolites tailored for agricultural applications. Among the products available for commercial use, BioGro2 is notable as a multi-strain biofertilizer consisting of a blend of bacteria and the yeast *Candida tropicalis*. Another significant biofertilizer is NutriSmart®, which is made up of *Saccharomyces cerevisiae* yeast strains that are capable of fixing atmospheric nitrogen, breaking down phosphate rock into soluble phosphate and improving the soil health. In the category of biofungicides, Zimevit combines *Bacillus subtilis* strain UYBC38 with *Metschnikowia pulcherrima* strain M26. Other commercial antifungal formulations utilizing yeasts include Noli or Shemer™, Nexy, Botector, and BoniProtect™, each specifically designed to combat particular

G. H. Mir
Advanced Research Station for Saffron and Seed Spices, Sher-e-Kashmir University of Agricultural Sciences & Technology of Kashmir, Pampore, India

V. Gupta (✉) · A. Sharma · G. Jamwal · V. K. Razdan
Division of Plant Pathology, Faculty of Agriculture, Sher-e-Kashmir University of Agricultural Sciences & Technology of Jammu, Chatha, India
e-mail: vishal94gupta@rediffmail.com

S. K. Gupta
Faculty of Horticulture & Forestry, Sher-e-Kashmir University of Agricultural Sciences & Technology of Jammu, Chatha, India

J. Poveda et al. (eds.), *Fungal Metabolites for Agricultural Applications*, Fungal Biology, https://doi.org/10.1007/978-3-031-76587-2_2

phytopathogenic fungi. Therefore, yeast stands out as a promising and sustainable source of metabolites with diverse applications in agriculture. The exploitation of yeast-derived compounds offers a pathway towards reducing the environmental impact of conventional farming practices while simultaneously promoting crop productivity and resilience.

Keywords Yeast · Metabolites · Biocontrol · Biostimulant · Crops

1 Introduction

Yeasts are classified as single-celled fungi that undergo asexual reproduction through budding or cell fission. The term "yeast" is derived from the Dutch word "gist" and the German word "gischt," both signifying boiling or bubbling (Kutty and Philip 2008). These organisms are widespread, inhabiting terrestrial (soil, vegetation) and aquatic (ocean, sea, estuary, mangrove, lakes, and river) ecosystems. The abundance of yeasts in various ecosystems, is facilitated by their extensive nutritional capabilities and the ability to rapidly proliferate. This fungus plays a significant role in enhancing plant growth, providing soluble nutrients, and producing phytohormones and enzymes that act as biostimulants (Falih and Wainwright 1995; Nassar et al. 2005; Cloete et al. 2009). Yeasts also serve as biocontrol agents, employing antagonistic mechanisms against pathogenic microorganisms, primarily functioning as bio-fungicides. In response to environmental stressors such as salinity, high temperature, drought, or metal toxicity affecting plants, yeasts activate defense mechanisms, producing compounds that alleviate these stresses. The diverse mechanisms and intrinsic characteristics of yeasts make them a promising alternative microorganism in various agricultural applications. Inadequate accessibility of vital nutrients in a soluble form may impede plant growth and yeasts tackle this problem through direct mechanisms like nitrogen fixation, phosphorus and potassium solubilization, along with the production of organic acids and plant growth hormones. Indirect mechanisms involve enhancing plant growth by inducing resistance to environmental stress and producing antifungal compounds. Yeast-derived secondary metabolites although not directly associated with their primary metabolic functions, often serve specific ecological roles. These compounds include a diverse range of bioactive molecules such as polyketides, terpenoids, alkaloids, and peptides, which can have various biological activities. Commonly isolated genera of yeasts in soil include *Candida, Cryptococcus, Debaryomyces, Hansenula, Lipomyces, Pichia, Aureobasidium, Rhodotorula, Saccharomyces, Schizoblastosporion, Sporobolomyces, Torulaspora, Torulopsis, Trichosporon, Kluyveromyces,* and *Zygosaccharomyces*species (Nimsi et al. 2023). Exploring and harnessing yeast-derived secondary metabolites in agriculture hold great potential for addressing various challenges while promoting sustainable and environmentally safer farming practices. The diverse mechanisms and intrinsic characteristics of yeasts make them a promising alternative microorganism in various agricultural applications. The use of yeast and their metabolites in agriculture has gained

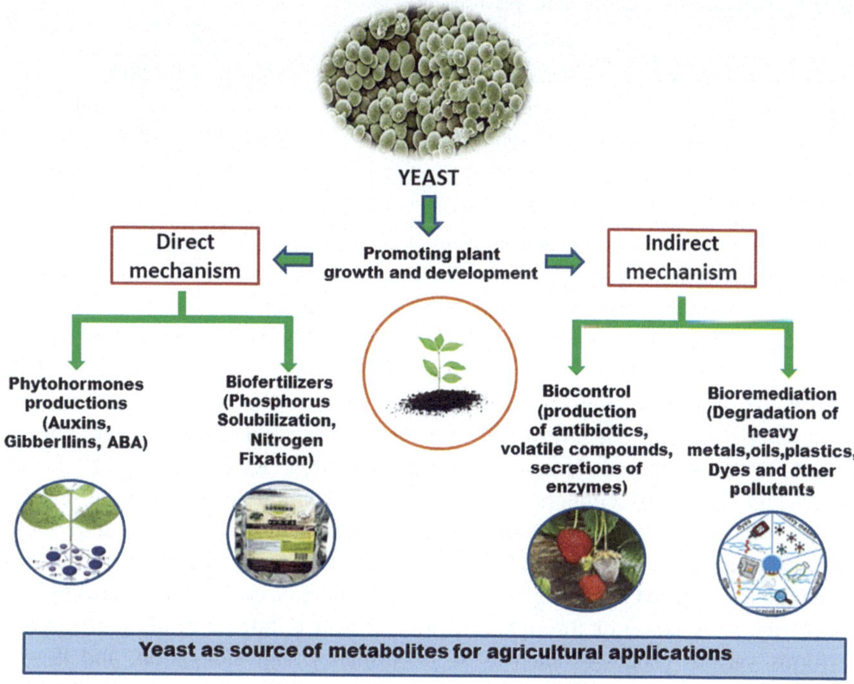

Fig. 1 Yeast as a source of metabolites for agricultural applications

attention for various applications, ranging from enhancing plant growth to improving soil health (Fig. 1).

2 Plant Growth-Promoting Metabolites

2.1 Phytohormones

Plant growth promoting hormones play a pivotal role in facilitating the growth of plants. Phytohormones, including abscisic acid (ABA), auxins, cytokinins, and gibberellins (GA), regulate the growth and development of plant root, shoot, and fruit. The phytohormone auxin, primarily represented by IAA (Indole-3-Acetic Acid), a derivative of an aromatic heterocyclic organic compound with a carboxymethyl substituent, induces a swift and enduring response in plants by fostering the elongation of lateral roots (Cleland 1987). Yeast species are able to produce IAA and significantly influence plant growth and development (El-Tarabily 2004; Nutaratat et al. 2015). Several species of yeast such as *Candida tropicalis, Cryptococcus flavus, Hannaella coprosmaensis, Pseudozyma aphidis, Sporisorium reilianum, Torulaspora globosa, Meyerozyma guilliermondii, and Rhodotorula mucilaginosa*

Table 1 Phytohormones produced by yeast species

	Phytohormones	References
Meyerozyma guilliermondii	Auxins	Oliveira et al. (2019)
Yarrowia lipolytica	Gibberellins (GA)	Kildegaard et al. (2021)
Yarrowia lipolytica	Abscisic acid (ABA)	Arnesen et al. (2022)
Sporobolomyces roseus	Cytokinins	Streletskii et al. (2019)

possess the ability to produce auxins (Sun et al. 2014). Cytokinins, categorized as a type of phytohormone, represent N6-substituted adenine derivatives and play a pivotal role in regulating various aspects of plant growth. This includes cell division, cytokinesis, apical dominance, the growth of axillary buds, photomorphogenic development and the senescence of leaves (Kieber and Schaller 2014). Yeast like *Kluyveromyces lactis, Schizosaccharomyces pombe, Moesziomyces antarcticus, Saccharomyces cerevisiae, Metschnikowia pulcherrima* and *Rodotorula* are known to synthesize cytokinins (Hussain and Hasnain 2009; Kudoyarova et al. 2014; Streletskii et al. 2016). Gibberellic acids (GAs) are slightly acidic plant hormones that play an important role in stem elongation, germination, flowering, and senescence (Hedden and Sponsel 2015). *Yarrowia lipolytica,* an oleaginous yeast has been used to produce GA through genetic modification (specifically, the recombination of the tesA gene) or through cell pretreatment (Ghogare et al. 2020). ABA performs various processes such as seed dormancy, cell elongation, and flower induction (Arnesen et al. 2020). ABA derived from yeast is applied in agriculture to enhance the color of red table grapes, generate hybrid seeds, and induce salt tolerance in citrus, drought tolerance in chickpeas, as well as cold/drought tolerance in wheat (Gomez-Cadenas et al. 2002; Nayyar et al. 2005; Bano et al. 2012; Li et al. 2014; Shi et al. 2017). Some phytohormone producing yeast species are listed in Table 1.

2.2 *Antibiotics*

Certain yeast strains produce antimicrobial compounds that exhibit inhibitory effects against pathogenic bacteria and fungi. By leveraging these yeast-derived antibiotics, agriculture can explore environmentally friendly approaches to control plant diseases, minimizing the reliance on synthetic chemical agents. The antibiotics released by yeast may contribute to biocontrol strategies, protecting crops from harmful pathogens and promoting plant health. The processes by which biocontrol yeasts function to protect plants involve various mechanisms, including competition, enzyme release, toxin production, emission of volatile compounds, mycoparasitism, and induction of resistance. Several yeast species such as *Candida oleophila, Aureobasidium pullulans, Metschnikowia fructicola, Cryptococcus albidus, Saccharomyces cerevisiae* are presently employed as biocontrol products (Freimoser et al. 2019). Antagonistic yeasts belonging to the Hanseniaspora genus

Table 2 Antibiotics produced by yeast species and their potential applications

Species	Antibiotics	Properties	References
Zygosaccharomyces bailii	Zygocin	Antifungal toxin	Wheiler and Schmitt (2003)
Kluyveromyces lactis	Zymotoxin	Antifungal toxin	Strak et al. (1990)
Saccharomyces cerevisiae	Saccharomycin	Antibacterial properties	Branco et al. (2017)

(Ruiz-Moyano et al. 2020) and *A. pullulans* (Di Francesco et al. 2020) have demonstrated inhibitory effects against *B. cinerea* through the production of volatile compounds. Yeasts producing these toxins typically possess resistance to them and others of the same class, while proving lethal to different yeast strains (Mannazzu et al. 2019).

Since toxins often target the cell wall, various components, such as -1,3-D-glucans, -1,6-D-glucans, mannoproteins and chitin, can act as receptors for killer toxins. Some well-known toxins produced by yeasts include: (i) PMKT and PMKT2 from *Pichia membranifaciens*, which exhibits binding to -1,6-D-glucans and mannoproteins of pathogen cell walls, respectively (Belda et al. 2017) (ii) panomycocin from a strain of *P. anomala*, which exerts its activity by hydrolyzing -1,3-glucans (Izgu et al. 2011) (iii) zymocin from *Kluyveromyces lactis*, which hydrolyzes chitins in fungal cell walls, serving as antifungal components (Tzelepis and Karlsson 2019). Few other yeast species producing antibiotics are listed in Table 2. This approach aligns with the growing interest in sustainable and ecologically conscious agricultural practices, as yeast-derived antibiotics have the potential to mitigate the development of antibiotic resistance and reduce environmental impacts associated with conventional pesticide use.

2.3 *Enzymes*

Yeasts have the ability to release enzymes and therefore, assumes a pivotal role in fostering soil health and plant growth. Through the secretion of enzymes, particularly those involved in the breakdown of organic matter, yeast contributes to the decomposition of complex organic compounds present in soil, releasing essential nutrients in more accessible forms for plant uptake. Enzymes like cellulases and hemicellulases released by yeast can aid in the degradation of plant residues, enhancing soil structure and nutrient cycling. This enzymatic activity not only facilitates nutrient availability but also promotes the development of a microbiologically rich soil environment, fostering symbiotic relationships between yeast and other beneficial microorganisms. Harnessing yeast-derived enzymes in agriculture holds promise for sustainable farming practices by promoting soil fertility, optimizing nutrient utilization, and contributing to overall crop resilience in the face of

Table 3 Enzymes produced by yeast species

Species	Enzymes	References
Saccharomyces	Zymase, alcohol dehydrogenase, pyruvate kinase, enolase, proteases, lipases, amylases	Bhatt (2011)
Candida	Proteinases, phospholipases, lipases, Invertase, catalase, NADH oxidase, Xylanases	Kuddus (2018)
Pichia	Alcohol oxidase, catalase, lipases, proteases	Liu et al. (2012)
Kluyveromyces	Invertase, alcohol dehydrogenase, catalase, lipases	Nadaroglu and Polat (2022)
Yarrowia	Lipases, proteinases, phospholipases, Xylanases	Escamilla-Alvarado et al. (2017)
Cryptococcus	Urease, phospholipases, catalase	Almeida et al. (2015)

environmental challenges. Table 3 lists some yeast species secreting enzymes of importance to plants.

3 Nutrient Solubilization and Uptake

3.1 Phosphorus Solubilization

Phosphate-solubilizing microorganisms have the potential to effectively dissolve both fertilizer phosphorus and soil-bound phosphorus in an environmentally friendly and sustainable manner (Khan et al. 2007). Yeast species have the ability to solubilize phosphorus, making it more available to plants which can improve phosphorus uptake and enhance plant development. Several species of yeast that can solubilize phosphate include *Schwanniomyces occidentalis, Oideodendron sp., Pseudonymnoascus sp., Candida tropicalis, Geotrichum candidum, Geotrichum capitatum, Rhodotorula minuta, Rhodotorula rubra, Saccharomyces sp., Hansenula sp., Klockera sp. Pichia norvegensis, Cryptococcus albidus var aerius, Candida etchellsii, Cryptococcus albidus var albidus, Rhodotrula aurantiaca, Cryptococcus luteolus, Cryptococcus albidus var diffluens, Cryptococcus terreus and Debaryomyces spp.* (Al-Fatih 2005; Gizaw et al. 2017; Narsian et al. 2017).

3.2 Nitrogen Fixation

In the process of nitrogen fixation, atmospheric nitrogen is converted into ammonia and other nitrogenous compounds by the metalloenzyme nitrogenase. Nitrogen-fixing ability is not limited to specific bacteria, rather certain groups of yeast genera, including *Torula, Pullularia, Saccharomyces spp., Candida, Rhodotorula, and Lipomyces,* also have the potential for nitrogen fixation (Li and Chen 2020). Some studies propose that yeast may not independently fix nitrogen and they might engage in syntrophic associations with bacteria like *Clostridium* or even stimulate nitrogen fixation by *Azotobacter* and *Beijerinckia* (James 2000).

4 Biocontrol Agents

The antagonistic effects of yeasts against plant pathogens can be attributed to various mechanisms such as colonization, production of antifungal compounds, induced systemic resistance, and mycoparasitism processes (Morath et al. 2012). Yeasts grows faster compared to filamentous fungi thereby creating competition and reducing nutrient availability to other fungi and rapidly colonizing plant niches (Zhang et al. 2010). An advantageous characteristic of beneficial microorganisms is the release of volatile organic compounds (VOCs) like alcohols, esters, aldehydes, ketones, and lactones. These VOCs enhance disease resistance and tolerance to abiotic stresses. The growth of common plant pathogens such as *Mucor* spp., *Penicillium* spp., *Aspergillus* sp., *Fusarium* spp. and *Botrytis cinerea*are is inhibited by yeasts (Choinska et al. 2020). ABA produced by yeast has a potential as an antifungal agent, exerting its inhibitory effect on chorismate mutase and reducing phenyl pyruvate production—an essential metabolite in beta-glucan synthesis required for fungal cell wall formation (Khedr et al. 2018). Yeast strains produce Killer Toxins, which induce damage to the fungal cell wall and cell membrane by triggering apoptosis and inhibiting beta-glucan synthesis.

The yeast *Torulaspora globosa* exhibits antifungal activity against *Colletotrichum graminicola,* the causative agent of anthracnose disease in maize (Rosa-Magri et al. 2011). Yeast is recognized as a potential biocontrol agent for Cacao witches broom disease caused by *Moniliophthora perniciosa* (Ferraz et al. 2019). Various yeasts, including *Pichia kluyveri, P. membranifaciens, P. farinosa, P. acacia, Zygosaccharomyces bailii, Hansenula mrakii, Wickerhamomyces anomalus, Williopsis saturnus, Kluyveromyces phaffii, K. lactis,* and *Wingearo bertsiae,* are capable of triggering antifungal activity. Both Saccharomyces and non-Saccharomyces yeasts, such as, *Pichia membranifaciens, Kloeckera apiculata, Candida guilliermondii, Cryptococcus laurentii, Cryptococcus flavus, Cryptococcus albicus, Candida pyralidae, Pichia kluyveri,* and *P. expansum,* are identified as potential antagonists against phytopathogenic fungi like *Penicillium, Aspergillus* and *Botrytis* (Di Canito et al. 2021) (Table 4).

Table 4 Yeast species as biocontrol agents

Species	Disease target	Target microorganism	Host	References
Cryptococcus albidus	Postharvest decay of pome and citrus	*Botrytis cinerea* and *Penicillium expansum*	Pome and citrus	Elad et al. (1994)
Metschnikowia fructicola	Fruit rot of strawberry	*Botrytis cinerea*	Strawberry	Karabulut et al. (2005)
Candida oleophila	Crown rot of banana	*Colletotrichum musae* and *Fusarium* spp	Banana	Lassois et al. (2008)
Aureobasidium pullulans	Fire blight disease	*Erwinia amylovora*	Pear	Kunz et al. (2023)
Saccharomyces cerevisiae	Fusarium head blight (FHB) or scab	*Fusarium graminearum*	Wneat	Cubaiu et al. (2012)
Pichia anomala	Gray mould of grains	*Botrytis cinerea*	Wheat	Olstorpe et al. (2011)
Saccharomyces cerevisiae	Anthracnose	*Colletotrichum acutatum*	Strawberry	Gramisci et al. (2018)
Saccharomyces paradoxus and*Torulaspora delbrueckii*	Pine canker disease	*Fusarium circinatum* and*Diplodia sapinea.*	Pine	Iturritxa et al. (2023)
Wickerhamomyces anomalus	Gray mould	*Botrytis cinerea*	Tomato	Lanhuang et al. (2022)

5 Soil Health and Bioremediation

5.1 Bioremediation of Pollutants

Some yeast species have the ability to degrade pollutants such as aromatic compounds, plastics and mitigate heavy metal stress contributing to environmental sustainability. Heavy metals accumulating in plants consumed for food or medicine have serious health hazards. Yeast strain *Saccharomyces cerevisiae* is reported for arsenic detoxification in contaminated soils and water (Shah et al. 2010; Todorova et al. 2010; Khullar and Reddy 2020; Mohd et al. 2019; Verma et al. 2019) due to expression of As (V) reductase enzyme involved in the reduction of As (V) to As (III) (Irshad et al. 2021). *Debaryomyces hansenii* strain NBRI-Sh2.11 reduces arsenic transportation in rice (Kaur et al. 2020) whereas other arsenic tolerant yeast strains *Candida tropicalis* (NBRI-B3.4) and *C. dubliniensis* (NBRI-B3.5) have also shown resistance to lead, mercury and cadmium (Rehman and Anjum 2011; Muneer et al. 2016). *Candida tropicalis* strain SDP-1 isolated from coastal soil in Jiangsu, China was reported to degrade phenol in contaminated waste waters (Gong et al. 2021) whereas *Saccharomyces cerevisiae* is reported to remove silver from effluents (Zinicovscaia et al. 2023). *Papiliotrema huenov* yeast is shown to accumulate large quantities of Mn(II) and Cu(II) from aqueous solutions (Nguyen Van et al. 2021). Yeast strains belonging to *Rhodotorula mucilaginosa*, *Clavisporal usitaniae*, and *Wickerhamomyces anomalus* have been reported to harbor resistance against heavy metals like Hg, Cr, Cd, Pb, Cu, Zn, and Fe which can be used for heavy metal removal from environment (Aibeche et al. 2022). A novel yeast, *Geotrichum* sp. CS-67 was shown to accumulate zinc, copper and nickel ions and could be used for heavy metal pollution remediation (He et al. 2022). *Yarrowia lipolytica* has been recently reported to remove copper from chromated copper arsenate treated wood waste (Xing et al. 2023). Another *Rhodotorula mucilaginosa* strain YR29 isolated from rhizosphere of *Prosopis* spp. showed bioremediation potential (Angeles de Paz et al. 2023). Several other yeast strains like *Kluyveromyces marxianus* CMGBP16, *Saccharomyces cerevisiae* S228C, *Saccharomyces cerevisiae* CM6B70, *Saccharomyces cerevisiae* CMGB234 and *Pichia anomala* CMGB88 have also been reported for biotreatment of waste water by removing organic compounds, heavy metals and other contaminants from it (Nicula et al. 2023).

Apart from bioremediation of heavy metals, some yeast like *Rhodotorula mucilaginosa* (Vaksmaa et al. 2023) and new strains of *Exophilia*sp. NS-7 and *Rhodotorula* sp. NS-12 (Giyahchi and Moghimi 2023) are reported to degrade and remove plastics via mineralization and assimilation of carbon in oceanic waters. Polyurethanes can also be degraded by fungal strain *Cladosporium halotolerans* from oceanic waters (Zhang et al. 2022). A yeast consortium consisting of *Sterigmatomyces halophilus, Meyerozyma guilliermondii*, and *Meyerozyma caribbica* was reported to degrade polyethylene (Elsamahy et al. 2023). Oil contaminated soils can be treated with yeast like *Moniliella spathulata* SBUG-Y 2180 (Mikolasch et al. 2021) and *Yarrowia lipolytica* (Louhasakul et al. 2020) having remarkable lipolytic ability. Another yeast *Rhodotorula* sp.CC01 is shown to degrade petroleum hydrocarbons in *in vitro* studies (Lin et al. 2022). Metalaxyl pesticide degradation has been reported for *Candida tropicalis* in aqueous solutions (Derbalah et al. 2020). Species of *Candida* and *Yarrowia* yeast can also remove dyes from textile effluents (Mendes et al. 2022). Reactive Orange 16 has been reported to be degraded by *Candida zeylanoides* (Šlosarčíková et al. 2020) and crystal violet dye by *Saccharomyces cerevisiae*. Other pollutants like ochratoxin A (Wei et al. 2023), Phosphogypsum and titanium gypsum (Chen et al. 2021), aflatoxin B_1 (García-Béjar et al. 2020) are shown to be removed from industrial wastes by several yeast species.

In addition to the natural capability of yeast strains to mitigate heavy metal stress and degrade pollutants, they can also be engineered to enhance this ability. De Oliveira et al. 2020 introduced a mutated version of *Populus trichocarpa* gene (PtMT2b) coding for a metallothionein in *Saccharomyces cerevisiae* and found its ability to extract cadmium from aqueous solutions enhanced than wild-type, having great potential for waste water management. Expression of PETase gene for poly (ethylene terephthalate) PET degradation in *Yarrowia lipolytica* resulted in degradation of PET films (Kosiorowska et al. 2022). Trametestrogii Lac (Ttlcc1) was expressed in *Saccharomyces cerevisiae* strain CEN.PK2-1 C for hydrocarbon tolerance and degradation (Asemoloye and Marchisio 2022) (Table 5). In nutshell, the

Table 5 Yeast species in bioremediation

Yeast species	Pollutants	References
Cryptococcus podzolicus Y3	Ochratoxin A (OTA) pollution in food and environment	Wei et al. (2023)
Rhodotrorula mucilaginosa	Aflatoxin B_1	García-Béjar et al. (2020)
Diutina rugosa	Zn	García-Béjar et al. (2020)
Rhodotorula mucilaginosa	Phosphogypsum (PG) and titanium gypsum (TG), as a by-product (solid waste) in phosphate fertilizer and titanium dioxide industry	Chen et al. (2021)
Rhodotorula mucilaginosa	Plastic degradation	Kitamoto et al. (2011)

bioremediation potential of yeast is huge and can be effectively channelized to mitigate environmental pollution.

6 Fermentation Products

6.1 Biofertilizers

For boosting agriculture food production in a sustainable and cost-effective manner, alternatives to chemical-based fertilizers are being searched. It is well known that plants have a mutualistic association with soil microbiome which includes organisms like filamentous fungi, yeasts, bacteria, algae and protozoa. Although, yeast is present in low proportions in soil as compared to filamentous fungi and bacteria, their plant growth promoting effects are attributed to the release of vitamins, amino acids, phytohormones, ammonia, solubilization of inorganic compounds, enhanced iron uptake and protection from pathogens. Some yeast-based formulations used as biofertilizers have exhibited promising results and newer strains are being reported as potential candidates. Phosphate solubilizing yeast strain *Meyerozyma guilliermondii* CC1 in combination with half dose of chemical fertilizers is reported to improve the dry-weights, and nutrient uptakes of maize under both greenhouse and field conditions (Nakayan et al. 2013). Soil inoculation with maize rhizosphere yeasts *Cryptococcus flavus* and *Candida railenensis,* with or without phosphorus fertilization have shown to strongly modulate maize, rhizosphere yeasts and arbuscular mycorrhizal fungi interactions. Here *C. railenensis* was shown to increase arbuscular mycorrhizal fungi root colonization with P fertilization (Sarabia et al. 2017). Phosphate solubilizing soil yeast *Candida tropicalis* HY along with *Pseudomonas fluorescens* as nitrogen fixing yeast strain are constituents of commercial biofertilizer BioGro and have been shown to stimulate rice seedling growth by increasing the dry weight of inoculated roots and improving grain yield (Hien et al. 2014). Nutria smart is another commercially available biofertilizer containing nitrogen fixing bacteria and some *S. cerevisiae* yeast strains for converting nitrogen for easier uptake by plants (Pang et al. 2003). Another yeast species, *Carlosrosaeavrieseae* UFMG-CM-Y6724 isolated from tank bromeliad *Vrieseaminarum* exhibited inorganic phosphate mobilization, siderophores secretion and IAA and IAA-like compounds synthesizing capability and improved *V. minarum* seedling growth equivalent to that of commercial IAA (Marques et al. 2021). *Debaryomyces hansenii*, *Hansenula polymorpha* (Lucena et al. 2021) and *Fusarium oxysporum* FO12 (Aparicio et al. 2023) have been shown to induce iron deficiency response in cucumber plants in vitro. A marine red yeast *Rhodosporidium paludigenum* JYC100 successfully solubilized calcium phosphate and had growth promoting effects on plants (Chen et al. 2022). Ramos-Garza et al. 2023 have isolated and identified eighty-seven plant growth-promoting yeasts from Mexican maize landraces which produced auxins, siderophores, enzymes and solubilized phosphate and could be used as potential biofertilizer agents. These finding suggest that yeast species harbor great potential as biofertilizers which needs to be harnessed for enhancing agriculture production and reducing dependance on chemical fertilizers.

7 Challenges and Considerations

Utilizing yeast as a source of secondary metabolites in agriculture presents both challenges and considerations. One significant challenge is the intricate regulation of secondary metabolite production in yeast, requiring a thorough understanding of the underlying genetic and environmental factors. Achieving consistent and high yields of specific metabolites can be demanding due to the complex interplay of these factors. Additionally, optimizing fermentation conditions and scaling up production to meet agricultural needs while maintaining cost-effectiveness poses logistical challenges. Furthermore, considerations regarding the potential ecological impact of introducing genetically modified yeast strains into agricultural ecosystems and the necessity for rigorous safety assessments are crucial to address. Balancing the promise of yeast-derived secondary metabolites with these challenges necessitates a multidisciplinary approach, combining genetic engineering, bioprocess optimization, and comprehensive risk assessments to ensure the sustainable and responsible integration of yeast-based formulations in agriculture.

8 Future Prospects

Genetic engineering holds immense potential for tailoring yeast strains to enhance the production of specific metabolites with targeted agronomic benefits, such as improved plant growth promotion, disease resistance, and stress tolerance. The advent of synthetic biology techniques allows for the precise manipulation of yeast genomes, unlocking the ability to design custom metabolic pathways for the synthesis of desired secondary metabolites. This personalized approach could lead to the development of yeast strains capable of producing novel bioactive compounds that address specific agricultural challenges, contributing to sustainable and resilient farming practices.

Moreover, the integration of systems biology and big data analytics is anticipated to play a pivotal role in unraveling the complex regulatory networks governing secondary metabolite production in yeast. By comprehensively understanding the molecular mechanisms involved, researchers can fine-tune fermentation processes, optimize growth conditions, and maximize yields. Additionally, advancements in fermentation technology, such as bioreactor design and process automation, will likely streamline large-scale production, making yeast-derived secondary metabolites more economically viable for widespread agricultural applications. As these technological frontiers converge, yeast-based solutions are poised to become indispensable tools in the pursuit of innovative and sustainable strategies to address the evolving challenges in modern agriculture.

References

Aibeche C, Selami N, Zitouni-Haouar FE, Oeunzar K, Addou A, Kaid-Harche M, Djabeur A (2022) Bioremediation potential and lead removal capacity of heavy metal-tolerant yeasts isolated from Dayet Oum Ghellaz Lake water (northwest of Algeria). Int Microbiol 25:61–73. https://doi.org/10.1007/s10123-021-00191-z. Epub 2021 Jul 6

Al-Fatih AM (2005) Phosphate solubilization in vitro by some soil yeasts. Qatar Univ Sci J25:119–125. http://hdl.handle.net/10576/9742

Almeida P, Barbosa R, Zalar P, Imanishi Y, Shimizu K, Turchetti B, Legras JL, Serra M, Dequin S, Couloux A, Guy J (2015) A population genomics insight into the Mediterranean origins of wine yeast domestication. Mol Ecol 24:5412–5427

Angeles de Paz G, Martínez-Gutierrez H, Ramírez-Granillo A, López-Villegas EO, Medina-Canales MG, Rodríguez-Tovar AV (2023) Rhodotorulamucilaginosa YR29 is able to accumulate Pb^{2+} in vacuoles: a yeast with bioremediation potential. World J Microbiol Biotechnol 39:238. https://doi.org/10.1007/s11274-023-03675-4

Aparicio MA, Lucena C, García MJ, Ruiz-Castilla FJ, Jiménez-Adrián P, López-Berges MS, Prieto P, Alcántara E, Pérez-Vicente R, Ramos J, Romera FJ (2023) The nonpathogenic strain of *Fusarium oxysporum* FO12 induces Fe deficiency responses in cucumber (*Cucumis sativus* L.) plants. Planta 257:50. https://doi.org/10.1007/s00425-023-04079-2. PMID: 36757472; PMCID: PMC9911487

Arnesen JA, Kildegaard KR, Cernuda PastorMet (2020) *Yarrowia lipolytica* strains engineered for the production of terpenoids. Front Bioeng Biotechnol 8:945. https://doi.org/10.3389/fbioe.2020.00945

Arnesen JA, Jacobsen IH, Dyekjær JD, Rago D, Kristensen M, Klitgaard AK, Randelovic M, Martinez JL, Borodina I (2022) Production of abscisic acid in the oleaginous yeast Yarrowia lipolytica. FEMS Yeast Res 22:15. https://doi.org/10.1093/femsyr/foac015

Asemoloye MD, Marchisio MA (2022) Synthetic *Saccharomyces cerevisiae* tolerate and degrade highly pollutant complex hydrocarbon mixture. Ecotoxicol Environ 241:113768. https://doi.org/10.1016/j.ecoenv.2022.113768. Epub Jun 17

Bano A, Ullah F, Nosheen A (2012) A role of abscisic acid and drought stress on the activities of antioxidant enzymes in wheat. Plant Soil Environ 58:181–185. https://doi.org/10.17221/210/2011-PSE

Belda I, Ruiz J, Alonso A, Marquina D, Santos A (2017) The biology of *Pichia membranifaciens* killer toxins. Toxins 9:112. https://doi.org/10.3390/toxins9040112

Bhatt SM (2011) Enzymology and enzyme technology. S. Chand Publishing

Branco P, Kemsawasd V, Santos L, Diniz M, Caldeira J, Almeida MG, Arneborg N, Albergaria H (2017) *Saccharomyces cerevisiae* accumulates GAPDH-derived peptides on its cell surface that induce death of non-saccharomyces yeasts by cell-to-cell contact. FEMS Microb Ecol 93:55. https://doi.org/10.1093/femsec/fix055

Chen H, Lu Y, Zhang C, Min F, Huo Z (2021) Red yeast improves the potential safe utilization of solid waste (phosphogypsum and titanogypsum) through bioleaching. Front Bioeng Biotechnol 9:777957. https://doi.org/10.3389/fbioe.2021.777957. PMID: 35036400; PMCID: PMC8758580

Chen YR, Kuo CY, Fu SF, Chou JY (2022) Plant growth-promoting properties of the phosphate-solubilizing red yeast *Rhodosporidiumpaludigenum*. World J Microbiol Biotechnol 39:54. https://doi.org/10.1007/s11274-022-03498-9. PMID: 36565394; PMCID: PMC9789928

Choinska R, Piasecka-Joz'wiak K, Chabłowska B (2020) Biocontrol ability and volatile organic compounds production as a putative mode of action of yeast strains isolated from organic grapes and rye grains. Antonie Van Leeuwenhoek 113:1135–1146. https://doi.org/10.1007/s10482-020-01420-7

Cleland RE (1987) Auxin and cell elongation. In: Plant hormones and their role in plant growth and development, vol 2. Springer, pp 132–148. https://doi.org/10.1007/978-94-009-3585-3_8

Cloete KJ, Valentine AJ, Stander MA (2009) Evidence of symbiosis between the soil yeast *Cryptococcus laurentii* and a sclerophyllous medicinal shrub, *Agathosma betulina* (Berg.) Pillans. Microb Ecol 57:624–632. https://doi.org/10.1007/s00248-008-9457-9

Cubaiu L, Abbas H, Dobson AD, Budroni M, Migheli Q (2012) A *Saccharomyces cerevisiae* wine strain inhibits growth and decreases ochratoxin a biosynthesis by *Aspergillus carbonarius* and *Aspergillus ochraceus*. Toxins 4:1468–1481. https://doi.org/10.3390/toxins4121468

De Oliveira VH, Ullah I, Dunwell JM, Tibbett M (2020) Bioremediation potential of Cd by transgenic yeast expressing a metallothionein gene from *Populus trichocarpa*. Ecotoxicol Environ Saf 202:110917. https://doi.org/10.1016/j.ecoenv.2020.110917. Epub 2020 Jul 8

Derbalah ASH, El-Banna A, Saad AM (2020) Efficiency of *Candida tropicalis* for Potential Degradation of Metalaxyl in the Aqueous Media. Curr Microbiol 77:2991–2999. https://doi.org/10.1007/s00284-020-02121-0. Epub 2020 Jul 18. PMID: 32683464

Di Canito A, Mateo-Vargas MA, Mazzieri M (2021) The role of yeasts as biocontrol agents for pathogenic fungi on postharvest grapes: a review. Food Secur 10:1650. https://doi.org/10.3390/foods10071650

Di Francesco A, Zajc J, Gunde-Cimerman N, Aprea E, Gasperi F, Placì N, Caruso F, Baraldi E (2020) Bioactivity of volatile organic compounds by *Aureobasidium* species against gray mold of tomato and table grape. World J Microbiol Biotechnol 36:1–11. https://doi.org/10.1007/s11274-020-02947-7

Elad Y, Köhl J, Fokkema NJ (1994) Control of infection and sporulation of *Botrytis cinerea* on bean and tomato by saprophytic yeasts. Phytopathology 84:1193–1200. https://www.apsnet.org/publications/phytopathology/backissues/Documents/1994Articles/Phyto84n10_1193.pdf

Elsamahy T, Sun J, Elsilk SE, Ali SS (2023) Biodegradation of low-density polyethylene plastic waste by a constructed tri-culture yeast consortium from wood-feeding termite: degradation mechanism and pathway. J Hazard Mater 448:130944. https://doi.org/10.1016/j.jhazmat.130944. Epub 2023 Feb 6. PMID: 36860037

El-Tarabily KA (2004) Suppression of *Rhizoctonia solani* diseases of sugar beet by antagonistic and plant growth-promoting yeasts. J Appl Microbiol 96:69–75. https://doi.org/10.1046/j.1365-2672.2003.02043.x

Escamilla-Alvarado C, Pérez-Pimienta JA, Ponce-Noyola T, Poggi-Varaldo HM (2017) An overview of the enzyme potential in bioenergy-producing biorefineries. J Chem Technol Biotechnol 92:906924. https://doi.org/10.1002/jctb.5088

Falih AM, Wainwright M (1995) Nitrification, S-oxidation and Psolubilization by the soil yeast *Williopsis californica* and by *Saccharomyces cerevisiae*. Mycol Res 99:200–204. https://doi.org/10.1016/S0953-7562(09)80886-1

Ferraz RB, Kabeya N, Lopes-Marques MO (2019) A complete enzymatic capacity for long-chain polyunsaturated fatty acid biosynthesis is present in the Amazonian *teleost tambaqui*, *Colossoma macropomum*. Comp Biochem Physiol B Biochem Mol Biol 227:90–97. https://doi.org/10.1016/j.cbpb.2018.09.003

Freimoser FM, Rueda-Mejia MP, Tilocca B, Migheli Q (2019) Biocontrol yeasts: mechanisms and applications. World J Microbiol Biotechnol 35:1–19. https://doi.org/10.1007/s11274-019-2728-4

García-Béjar B, Arévalo-Villena M, Guisantes-Batan E, Rodríguez-Flores J, Briones A (2020) Study of the bioremediatory capacity of wild yeasts. Sci Rep 10:11265. https://doi.org/10.1038/s41598-020-68154-4. PMID: 32647290; PMCID: PMC7347596

Ghogare R, Chen S, Xiong X (2020) Metabolic engineering of oleaginous yeast *Yarrowia lipolytica* for overproduction of fatty acids. Front Microbiol 11:1717. https://doi.org/10.3389/fmicb.2020.01717

Giyahchi M, Moghimi H (2023) Aerobic biodegradation of untreated polyester-polyether urethanes by newly isolated yeast strains *Exophilia* sp. NS-7 and *Rhodotorula* sp. Sci Rep 13:5016. https://doi.org/10.1038/s41598-023-31639-z. PMID: 36977741; PMCID: PMC10050204

Gizaw B, Tsegay Z, Tefera G (2017) Phosphate solubilizing fungi isolated and characterized from teff rhizosphere soil collected from North Showa and Gojam. J Fertil Pestic 8:180. https://doi.org/10.5897/AJMR2017.8525

Gomez-Cadenas A, Arbona V, Jacas J (2002) A bscisic acid reduces leaf abscission and increases salt tolerance in citrus plants. J Plant Growth Regul 21:234–240. https://doi.org/10.1007/s00344-002-0013-4

Gong Y, Ding P, Xu MJ, Zhang CM, Xing K, Qin S (2021) Biodegradation of phenol by a halotolerant versatile yeast *Candida tropicalis* SDP-1 in wastewater and soil under high salinity conditions. J Environ Manag 289:112525. https://doi.org/10.1016/j.jenvman.2021.112525. Epub 2021 Apr 6. PMID: 33836438

Gramisci BR, Lutz MC, Lopes CA, Sangorrín MP (2018) Enhancing the efficacy of yeast biocontrol agents against postharvest pathogens through nutrient profiling and the use of other additives. Biol Control 121:151–158. https://doi.org/10.1016/j.biocontrol.2018.03.001

He M, Xu Y, Qiao Y, Zhang Z, Liang J, Peng Y, Liao J, Qiao Y, Shang C, Guo Z, Chen S (2022) A novel yeast strain Geotrichum sp. CS-67 capable of accumulating heavy metal ions. Ecotoxicol Environ Saf 236:113497. https://doi.org/10.1016/j.ecoenv.2022.113497. Epub Apr 8

Hedden P, Sponsel V (2015) A century of gibberellin research. J Plant Growth Regul 34:740–760. https://doi.org/10.1007/s00344-015-9546-1

Hien NT, van Toan P, Choudhury ATMA, Rose MT, Roughley RJ, Kennedy IR (2014) Field application strategies for the inoculant biofertilizer biogro supplementing fertilizer nitrogen application in rice production. J Plant Nutr 37:1837–1858. https://doi.org/10.1080/01904167.2014.911320

Hussain A, Hasnain S (2009) Cytokinin production by some bacteria: its impact on cell division in cucumber cotyledons. Afr J Microbiol Res 3:704–712. https://academicjournals.org/journal/AJMR/article-full-text-pdf/AC8E36D14631

Irshad S, Xie Z, Mehmood S, Nawaz A, Ditta A, Mahmood Q (2021) Insights into conventional and recent technologies for arsenic bioremediation: a systematic review. Environ Sci Pollut Res Int 28:18870–18892. https://doi.org/10.1007/s11356-021-12487-8. Epub 2021 Feb 14. PMID: 33586109

Iturritxa E, Mesanza N, Torija MJ (2023) The potential of wild yeasts as promising biocontrol agents against pine canker diseases. J Fungi (Basel) 9:840. https://doi.org/10.3390/jof9080840. PMID: 37623611; PMCID: PMC10455916

Izgu DA, Kepekci RA, Izgu F (2011) Inhibition of *Penicillium digitatum* and *Penicillium italicum in vitro* and in planta with Panomycocin, a novel exo-1,3-glucanase isolated from *Pichia anomala* NCYC 434. Antonie Van Leeuwenhoek. Int J Gen Mol Microbiol 99:85–91. https://doi.org/10.1007/s10482-010-9527-0

James EK (2000) Nitrogen fixation in endophytic and associative symbiosis. Field Crop Res 65:197–209. https://doi.org/10.1016/S0378-4290(99)00087-8

Karabulut OA, Arslan U, Ilhan K, Kuruoglu G (2005) Integrated control of postharvest diseases of sweet cherry with yeast antagonists and sodium bicarbonate applications within a hydrocooler. Postharvest Biol Technol 37:135–141. https://doi.org/10.1016/j.postharvbio.2005.03.003

Kaur J, Anand V, Srivastava S, Bist V, Tripathi P, Naseem M, Nand S, Anshu KP, Srivastava PK, Bisht S, Srivastava S (2020) Yeast strain Debaryomyceshansenii for amelioration of arsenic stress in rice. Ecotoxicol Environ Saf 195:110480. https://doi.org/10.1016/j.ecoenv.2020.110480. *Epub Mar* 20

Khan MS, Zaidi A, Wani PA (2007) Role of phosphate solubilizing microorganisms in sustainable agriculture—a review. Agron Sustain Dev 27:29–43. https://doi.org/10.1007/978-90-481-2666-8_34

Khedr MA, Massarotti A, Mohamed ME (2018) Rational discovery of (+) (*S*) abscisic acid as a potential antifungal agent: a repurposing approach. Sci Rep 8:8565. https://doi.org/10.1038/s41598-018-26998-x

Khullar S, Reddy MS (2020) Arsenic toxicity and its mitigation in ectomycorrhizal fungus Hebelomacylindro sporum through glutathione biosynthesis. Chemosphere 240:124914. https://doi.org/10.1016/j.chemosphere.2019.124914

Kieber JJ, Schaller GE (2014) Cytokinins. Arabidopsis Book 12:168. https://doi.org/10.1199/tab.0168. PMID: 24465173; PMCID: PMC3894907

Kildegaard KR, Arnesen JA, Adiego-Perez B (2021) Tailored biosynthesis of gibberellin plant hormones in yeast. Metab Eng 66:1–11. https://doi.org/10.1016/j.ymben.2021.03.010

Kitamoto HK, Shinozaki Y, Hong CX, Morita T, Konishi M, Tago K (2011) Phyllosphere yeasts rapidlybreak down biodegradable plastics. AMB Express 11–11. https://doi.org/10.1186/2191-0855-1-1. PMID: 21906327

Kosiorowska KE, Moreno AD, Iglesias R, Leluk K, Mirończuk AM (2022) Production of PETase by engineered Yarrowialipolytica for efficient poly (ethylene terephthalate) biodegradation. Sci Total Environ 846:157358. https://doi.org/10.1016/j.scitotenv.2022.157358. *Epub* 2022 Jul 16. PMID: 35850328

Kuddus (2018) Enzymes in food biotechnology: production, applications, and future prospects, vol 1. Academic Press, p 883. https://doi.org/10.1016/C2016-0-04555-2

Kudoyarova GR, Melentiev AI, Martynenko EV (2014) Cytokinin producing bacteria stimulate amino acid deposition by wheat roots. Plant Physiol Biochem 83:285–291. https://doi.org/10.1016/j.plaphy.2014.08.015

Kunz S, Zeng Q, Johnson KB (2023) History, efficacy, orchard ecology, and mode of action of *ureobasidium pullulans,* the microbial agent in blossom protect, for suppression of fire blight of pome fruit. Plant Pathol J:1–11. https://doi.org/10.1007/s42161-023-01448-4

Kutty SN, Philip R (2008) Marine yeasts—a review. Yeast 25:465–483. https://doi.org/10.1002/yea.1599A

Lanhuang B, Yang Q, Godana EA, Zhang H (2022) Efficacy of the yeast *Wickerhamomyces anomalus* in biocontrol of gray mold decay of tomatoes and study of the mechanisms involved. Food Secur 11:720. https://doi.org/10.3390/foods11050720. PMID: 35267353; PMCID: PMC8909804

Lassois L, de Bellaire LDL, Jijakli MH (2008) Biological control of crown rot of bananas with *Pichia anomala* strain K and *Candida oleophila* strain O. Biol Control 45:410–418. https://doi.org/10.1016/j.biocontrol.2008.01.013

Li Q, Chen S (2020) Transfer of nitrogen fixation (nif) genes to non-diazotrophic hosts. Chembiochem 21:1717–1722. https://doi.org/10.1002/cbic.201900784

Li X, Cai J, Liu F (2014) Exogenous abscisic acid application during grain filling in winter wheat improves cold tolerance of offspring's seedlings. J Agron Crop Sci 200:467–478. https://doi.org/10.1111/jac.12064

Lin X, Zhou H, Zeng F, Jiang L, Atakpa EO, Chen G, Zhang C, Xie Q (2022) A biosurfactant-producing yeast *Rhodotorula* sp.CC01 utilizing landfill leachate as nitrogen source and its broad degradation spectra of petroleum hydrocarbons. World J Microbiol Biotechnol 38:68. https://doi.org/10.1007/s11274-022-03254-z

Liu Y, Pan J, Wei P, Zhu J, Huang L, Cai J, Xu Z (2012) Efficient expression and purification of recombinant alcohol oxidase in *Pichia pastoris*. Biotechnol Bioprocess Eng 17:693–702. https://doi.org/10.1007/s12257-011-0660-z

Louhasakul Y, Cheirsilp B, Treu L, Kougias PG, Angelidaki I (2020) Metagenomic insights into bioaugmentation and biovalorization of oily industrial wastes by lipolytic oleaginous yeast *Yarrowia lipolytica* during successive batch fermentation. Biotechnol Appl Biochem 67:1020–1029. https://doi.org/10.1002/bab.1878. Epub 2020 Jan 8

Lucena C, Alcalá-Jiménez MT, Romera FJ, Ramos J (2021) Several yeast species induce iron deficiency responses in cucumber plants (*Cucumis sativus* L.). Microorganisms 9:2603. https://doi.org/10.3390/microorganisms9122603. PMID: 34946203; PMCID: PMC8704622

Mannazzu I, Domizio P, Carboni G, Zara S, Zara G, Comitini F, Budroni M, Ciani M (2019) Yeast killer toxins: from ecological significance to application. Crit Rev Biotechnol 39:603–617. https://doi.org/10.1080/07388551.2019.1601679

Marques AR, Resende AA, Gomes FCO, Santos ARO, Rosa CA, Duarte AA, de Lemos-Filho JP, Dos Santos VL (2021) Plant growth-promoting traits of yeasts isolated from the tank bromeliad Vrieseaminarum L.B. Smith and the effectiveness of *Carlosrosaeavrieseae* for promoting bromeliad growth. *Braz J Microbiol* 52:1417–1429. https://doi.org/10.1007/s42770-021-00496-1. *Epub* 2021 May 6. PMID: 33956333; PMCID: PMC8324700

Mendes M, Cassoni AC, Alves S, Pintado ME, Castro PM, Moreira P (2022) Screening for a more sustainable solution for decolorization of dyes and textile effluents using *Candida* and *Yarrowia* spp. J Environ M Manage 307:114421. https://doi.org/10.1016/j.jenvman.2021.114421. *Epub* 2022 Jan 29

Mikolasch A, Berzhanova R, Omirbekova A, Reinhard A, Zühlke D, Meister M, Mukasheva T, Riedel K, Urich T, Schauer F (2021) *Moniliellaspathulata*, an oil-degrading yeast, which promotes growth of barley in oil-polluted soil. Appl Microbiol Biotechnol 105:401–415. https://doi.org/10.1007/s00253-020-11011-1. Epub 2020 Nov 20. PMID: 33219393; PMCID: PMC7778628

Mohd S, Kushwaha AS, Shukla J, Mandrah K, Shankar J, Arjaria N, Saxena PN, Khare P, Narayan R, Dixit S, Siddiqui MH, Tuteja N, Das M, Roy SK, Kumar M (2019) Fungal mediated biotransformation reduces toxicity of arsenic to soil dwelling microorganism and plant. Ecotox Environ Saf 176:108–118. https://doi.org/10.1016/j.ecoenv.2019.03.053

Morath SU, Hung R, Bennett JW (2012) Fungal volatile organic compounds: a review with emphasis on their biotechnological potential. Fungal Biol Rev 26:73–83. https://doi.org/10.1016/j.fbr.2012.07.001

Muneer B, Lali T, Iqbal MJ, Shakoori FR, Shakoori AR (2016) Arsenic processing of yeast isolates IIB-As1 & IIB-As2 and production of glutathione under stress conditions. J Basic Microbiol 56:1124–1131. https://doi.org/10.1002/jobm.201600092

Nadaroglu H, Polat MS (2022) Microbial extremozymes: novel sources and industrial applications. In: Microbial Extremozymes. Academic, pp 67–88

Nakayan P, Hameed A, Singh S (2013) Phosphate-solubilizing soil yeast *Meyerozymaguilliermondii* CC1 improves maize (*Zea mays* L.) productivity and minimizes requisite chemical fertilization. Plant Soil 373:301–315. https://doi.org/10.1007/s11104-013-1792-z

Narsian V, Samaha SMAA, Patel HH (2017) Rock phosphate dissolution by specific yeast. Indian J Microbiol 50:57–62. https://doi.org/10.1007/s12088-009-0019-8

Nassar AH, El-Tarabily KA, Sivasithamparam K (2005) Promotion of plant growth by an auxin-producingisolate of the yeast *Williopsis saturnus* endophytic in maize (*Zea mays* L.) roots. Biol Fertil Soils 42:97–108. https://doi.org/10.1007/s00374-005-0008-y

Nayyar H, Bains TS, Kumar S (2005) Chilling stressed chickpea seedlings: effect of cold acclimation, calcium and abscisic acid on cryoprotective solutes and oxidative damage. Environ Exp Bot 54:275–285. https://doi.org/10.1016/j.envexpbot.2004.09.007

Nguyen Van P, Thi Hong Truong H, Pham TA, Le Cong T, Le T, Thi Nguyen KC (2021) Removal of manganese and copper from aqueous solution by yeast papiliotremahuenov. Mycobiology 49:507–520. https://doi.org/10.1080/12298093.2021.1968624. PMID: 36970636; PMCID: PMC10035953

Nicula NO, Lungulescu EM, Rîmbu GA, Marinescu V, Corbu VM, Csutak O (2023) Bioremediation of wastewater using yeast strains: an assessment of contaminant removal efficiency. Int J Environ Res Public Health 20:4795. https://doi.org/10.3390/ijerph20064795. PMID: 36981703; PMCID: PMC10048942

Nimsi KA, Manjusha K, Kathiresan K, Arya H (2023) Plant growth-promoting yeasts (PGPY), the latest entrant for use in sustainable agriculture: a review. J Appl Microbiol 134:088. https://doi.org/10.1093/jambio/lxac088

Nutaratat P, Amsri W, Srisuk N (2015) Indole-3-acetic acid production by newly isolated red yeast *Rhodosporidium paludigenum*. J Gen Appl Microbiol 61:1–9. https://doi.org/10.2323/jgam.61.1

Oliveira BT, Bizarria R Jr, Silva LG (2019) Rhizosphere yeast *Torulaspora globosa* with plant growth promotion traits and improvement of the development of tomato seedlings under greenhouse conditions. Afr J Agric Res 14:935–942. https://doi.org/10.5897/AJAR2019.13950

Olstorpe M, Jacobsson K, Passoth V, Schnürer J (2011) Controlling fungal growth and mycotoxins in animal feed. In: *Protective cultures, antimicrobial metabolites and bacteriophages for food and beverage biopreservation*. Woodhead Publishing, pp 225–239. https://doi.org/10.1533/9780857090522.2.225

Pang SF, Lui A, Goold G, Chu A, Wong W, Li S, Chan E, Kwok I, Cheung L (2003) NutriSmart®: a fertiliser capable ofre-establishing the sustainability of ecosystems and enhancing the productivity of farmland. In: Proceedings of the 11th Australian agronomy conference, Geelong, VIC, Australia. http://www.regional.org.au/au/asa/2003/p/6/pang.hm

Ramos-Garza J, Aguirre-Noyola JL, Bustamante-Brito R, Zelaya-Molina LX, Maldonado-Hernández J, Morales-Estrada AI, Resendiz-Venado Z, Palacios-Olvera J, Angeles-Gallegos T, Terreros-Moysen P, Cortés-Carvajal M, Martínez-Romero E (2023) Mycobiota of Mexican maize landraces with auxin-producing yeasts that improve plant growth and root development. Plants (Basel) 12:1328. https://doi.org/10.3390/plants12061328. PMID: 36987016; PMCID: PMC10058334

Rehman A, Anjum MS (2011) Multiple metal tolerance and biosorption of cadmium by *Candida tropicalis* isolated from industrial effluents: glutathione as detoxifying agent. Environ Monit Assess 174:585–595. https://doi.org/10.1007/s10661-010-1480-x

Rosa-Magri MM, Tauk-Tornisielo SM, Ceccato-Antonini SR (2011) Bio-prospection of yeasts as biocontrol agents against phytopathogenic molds. Braz Arch Biol Technol 54:1–5. https://doi.org/10.1590/S1516-89132011000100001

Ruiz-Moyano S, Hernández A, Galvan AI, Córdoba MG, Casquete R, Serradilla MJ, Martín A (2020) Selection and application of antifungal VOCs-producing yeasts as biocontrol agents of grey mould in fruits. Food Microbiol 92:103556. https://doi.org/10.1016/j.fm.2020.103556

Sarabia M, Cornejo P, Azcón R, Carreón-Abud Y, Larsen J (2017) Mineral phosphorus fertilization modulates interactions between maize, rhizosphere yeasts and arbuscular mycorrhizal fungi. Rhizosphere 4:89–93. https://doi.org/10.1016/j.rhisph.2017.09.001

Shah D, Shen MW, Chen W, Da Silva NA (2010) Enhanced arsenic accumulation in *Saccharomyces cerevisiae* overexpressing transporters Fps1p or Hxt7p. J Biotechnol 150:101–107. https://doi.org/10.1016/j.jbiotec.2010.07.012

Shi TQ, Peng H, Zeng SY (2017) Microbial production of plant hormones: opportunities and challenges. Bioengineered 8:124–128. https://doi.org/10.1080/21655979.2016.1212138

Šlosarčíková P, Plachá D, Malachová K, Rybková Z, Novotný Č (2020) Biodegradation of reactive Orange 16 azo dye by simultaneous action of *Pleurotusostreatus* and the yeast *Candida zeylanoides*. *Folia Microbiol (Praha)* 65:629–638. https://doi.org/10.1007/s12223-019-00767-3. Epub 2020 Jan 22

Strak MJ, Boyd A, Mileham AJ, Ramonos MA (1990) The plasmid-encoded killer system of Kluyveromyces lactis: a review. Yeast 6:1–29. https://doi.org/10.1002/yea.320060102

Streletskii RA, Kachalkin AV, Glushakova AM (2016) Quantitative determination of indole-3-acetic acid in yeasts using high performance liquid chromatography—tandem mass spectrometry. Microbiology 85:727–736. https://doi.org/10.1134/S0026261716060187

Streletskii RA, Kachalki AV, Glushakova AM, Yurkov AM, Demin VV (2019) Yeasts producing zeatin. Peer J 7:6474. https://doi.org/10.7717/peerj.6474

Sun PF, Fang WT, Shin LY (2014) Indole-3-acetic acid-producing yeasts in the phyllosphere of the carnivorous plant *Drosera indica* L. PLoS One 9:114196. https://doi.org/10.1371/journal.pone.0114196

Todorova TT, Kujumdzieva AV, Vuilleumier S (2010) Non-enzymatic roles for the URE2 glutathione S-transferase in the response of *Saccharomyces cerevisiae* to arsenic. Arch Microbiol 192:909–918. https://doi.org/10.1007/s00203-010-0614-4

Tzelepis G, Karlsson M (2019) Killer toxin-like chitinases in filamentous fungi: structure, regulation and potential roles in fungal biology. Fungal Biol Rev 33:123–132. https://doi.org/10.1016/j.fbr.2018.11.001

Vaksmaa A, Polerecky L, Dombrowski N, Kienhuis MVM, Posthuma I, Gerritse J, Boekhout T, Niemann H (2023) Polyethylene degradation and assimilation by the marine yeast Rhodotorulamucilaginosa. ISME Commun 3:68. https://doi.org/10.1038/s43705-023-00267-z. PMID: 37423910; PMCID: PMC10330194

Verma S, Verma PK, Chakrabarty D (2019) Arsenic bio-volatilization by engineered yeast promotes rice growth and reduces arsenic accumulation in grains. Int J Environ Res 13:475–485. https://doi.org/10.1007/s41742-019-00188-7

Wei M, Dhanasekaran S, Ji Q, Yang Q, Zhang H (2023) Sustainable and efficient method utilizing N-acetyl-L-cysteine for complete and enhanced ochratoxin A clearance by antagonistic yeast. J Hazard Mater 448:130975. https://doi.org/10.1016/j.jhazmat.2023.130975. Epub 2023 Feb 9. PMID: 36860082

Wheiler F, Schmitt MJ (2003) Zygocin, a secreted antifungal toxin of the yeast Zygosaccharomyces bailii, and its effect on sensitive fungal cells. FEMS Yeast Res 3:69–76. https://doi.org/10.1111/j.1567-1364.2003.tb00140.x

Xing D, Magdouli S, Zhang J, Bouafif H, Koubaa A (2023) A comparative study on heavy metal removal from CCA-treated wood waste by *Yarrowialipolytica*: effects of metal stress. J Fungi (Basel) 9:469. https://doi.org/10.3390/jof9040469. PMID: 37108923; PMCID: PMC10145133

Zhang D, Spadaro D, Garibaldi A (2010) Selection and evaluation of new antagonists for their efficacy against postharvest brown rot of peaches. Postharvest Biol Technol 55:174–181. https://doi.org/10.1016/j.postharvbio.2009.09.007

Zhang K, Hu J, Yang S, Xu W, Wang Z, Zhuang P, Grossart HP, Luo Z (2022) Biodegradation of polyester polyurethane by the marine fungus *Cladosporium halotolerans* 6UPA1. J Hazard Mater 437:129406. https://doi.org/10.1016/j.jhazmat.2022.129406. Epub 2022 Jun 18. PMID: 35753302

Zinicovscaia I, Yushin N, Grozdov D, Rodlovskaya E, Khiem LH (2023) Yeast-as bioremediator of silver-containing synthetic effluents. Bioengineering (Basel) 10:398. https://doi.org/10.3390/bioengineering10040398. PMID: 37106585; PMCID: PMC10136145

Mycorrhizal Fungi as a Source of Metabolites for Agricultural Applications

Jorge Poveda, Óscar Santamaría, Julio J. Diez, and Jorge Martín-García

Abstract Agriculture currently faces two major challenges: feeding a continuously growing population and developing in a more sustainable and environmentally friendly manner. In this context, mycorrhizal fungi (MF) can serve as a promising biotechnological strategy to enhance agricultural productivity while reducing the reliance on chemical fertilizers and pesticides. MF can provide beneficial effects to their host plants by supplying nutrients, improving water absorption and inducing tolerance to abiotic stresses, acting as biostimulants, or functioning as biological control agents (BCAs) against pathogens and pests. One of the mechanisms employed by MF to achieve these activities is the production of secondary metabolites, such as plant hormones and biocidal compounds. This chapter aims to compile, analyze and discuss existing studies on the production by MF of metabolites of agricultural significance, including their role as plant growth promoters, antibacterials, antifungals, oomyceticides, or insecticides.

Keywords Plant hormones · Antifungal · Antibacterial · Auxins · Ectomycorrhizal fungi · Arbuscular mycorrhizal fungi

---

J. Poveda (✉)
Department of Plant Production and Forest Resources, Higher Technical School of Agricultural Engineering of Palencia, University Institute for Research in Sustainable Forest Management (iuFOR), University of Valladolid, Madrid, Palencia, Spain

School of Biological Sciences, University of Bristol, Bristol, UK
e-mail: jorge.poveda@uva.es

Ó. Santamaría · J. J. Diez · J. Martín-García
Department of Plant Production and Forest Resources, Higher Technical School of Agricultural Engineering of Palencia, University Institute for Research in Sustainable Forest Management (iuFOR), University of Valladolid, Madrid, Palencia, Spain

J. Poveda et al. (eds.), *Fungal Metabolites for Agricultural Applications*, Fungal Biology, https://doi.org/10.1007/978-3-031-76587-2_3

1 Introduction

In 2022, the global population reached eight billion people. Based on current population growth rates, it is projected that the world population will reach ten billion people by 2050 (WPC 2023). Consequently, to meet the nutritional demands of the growing population, crop production will need to increase by 60–100% by 2050, while concurrently minimizing environmental damage caused by agriculture (Falcon et al. 2022).

For decades, the predominant approach to enhance crop productivity and reduce losses caused by pathogens and pests has been the extensive use of synthetic agrochemicals, such as fertilizers and pesticides. However, the adverse impact of these compounds on the environment and human health has been extensively documented, underscoring the urgency to explore new, safer, and more sustainable strategies (EFSA 2021). In this regard, the European Union (EU) serves as a noteworthy example with its "Farm to Fork Strategy" inside the "Green Deal", aiming to transition the current EU food system towards a sustainable model. The strategy includes objectives to reduce the use of phytosanitary products and synthetic chemical fertilizers by 50% and 20%, respectively, while increasing organic production to 25% by 2030 (Schebesta and Candel 2020). However, these goals should not be achieved at the expense of compromising harvests, which must sustain an ever-growing population. Therefore, it is imperative to explore environmental-friendly alternative approaches. One prominent strategy for developing a more sustainable agriculture revolves around employing microorganisms as biofertilizers and biological control agents (BCAs) against pests and diseases, with the aim of minimizing reliance on agricultural synthetic chemicals (Elnahal et al. 2022).

Microbial biofertilizers encompass bacteria and fungi that enhance soil nutrient availability, nutrient uptake, and nutrient assimilation in plants. They achieve this primarily through biological nitrogen fixation, phosphorus and potassium solubilization, and iron and zinc chelation via siderophore production (Joshi et al. 2021). This group of microorganisms employed as agricultural bioinoculants is also referred to as agricultural probiotics, an explicit term coined to categorize the strategy exclusively based on microorganisms (Sanjuán et al. 2023). This clear distinction arises due to the broader denomination of biostimulants, which encompasses not only microorganisms but also substances that positively influence plant growth, yield, and chemical composition, such as amino acids, protein hydrolysates or chitosan (Shahrajabian et al. 2021). Notably, the EU has designated these microorganisms as biostimulants in Regulation 2019/1009, which encompasses beneficial microorganisms that "enhance plant nutrition processes independently of the product's nutrient content, with the sole objective of improving one or more of the following plant or plant rhizosphere characteristics: a) nutrient use efficiency, b) tolerance to abiotic stress, c) quality traits, and d) availability of bound nutrients in soil or rhizosphere" (EU 2019/1009). Consequently, this definition also includes microorganisms that induce plant tolerance under abiotic stresses (Ali et al. 2022). According to European legislation, these agricultural biostimulants currently allow the registration of microorganisms within the bacterial genera *Azospirillum*, *Azotobacter*, *Rhizobium*, and arbuscular mycorrhizal fungi (Product Function Category PFC 6A) (EU 2019/1009). Although the use of these microorganisms in agriculture still requires substantial research beyond laboratory settings, the development of bioinoculants with tangible effects on crops is progressing (Shahrajabian et al. 2023), ensuring their environmental and health safety (Kumari et al. 2022).

Frequently, microorganisms employed as agricultural bioinoculants do not have a single biological function for plants, and although they are legally classified as biofertilizers or BCAs, they can concurrently serve both roles (Tarigan et al. 2022). Microorganisms classified as BCAs possess the ability to diminish the population of pathogens and pests and/or their detrimental effects on crops. To achieve this, they use various mechanisms, including direct parasitism, production and release of antimicrobial and insecticidal metabolites, competition for space and nutrients, or induction of plant defense mechanisms (Köhl et al. 2019a; Hernández-Rosas et al. 2020). These microorganisms represent a highly promising strategy for the circular economy and performing the Green Deal in Europe. However, European legislation is currently one of the most restrictive in terms of registering these agricultural bioinoculants (Helepciuc and Todor 2022). The commercial development of BCA products needs a rigorous analysis of environmental and human risks, making it challenging in terms of economic and temporal feasibility (Köhl et al. 2019b). Nevertheless, the utilization of BCA products in the field continues to increase annually worldwide, underscoring their benefits and effectiveness (Bejarano and Puopolo 2020).

Both as biofertilizers and BCAs, microorganisms produce a wide array of secondary metabolites with several effects on plants, such as growth promotion, induction of defense responses against biotic stresses, and/or tolerance to biotic stresses, as well as antibacterial, antifungal, oomyceticidal, nematicidal, insecticidal, and herbicidal activities (Singh and Yadav 2020). Notably, microbial production of plant hormones (or analogs) such as auxins, cytokinins, gibberellins, abscisic acid, salicylic acid, or jasmonic acid plays a significant role as plant growth promoters and inducers of plant responses (Vinale and Sivasithamparam 2020; Poveda and González-Andrés 2021). Regarding their biocidal properties, those beneficial microorganisms produce a diverse range of secondary metabolites, including peptaibols, polyketides, alkaloids, terpenes, small non-ribosomal peptides, and pyrones (Keswani et al. 2019; Segaran and Sathiavelu 2019). These secondary metabolites can exist in diffusible form within the liquid medium or as volatile organic compounds (VOCs). Microbial VOCs (MVOCs) have gained significant scientific and technical interest in recent years for their agricultural applications, particularly the groups of benzoids, femilpropanoids, and terpenes (Kaddes et al. 2019).

Consequently, there exists a critical need to develop novel strategies for enhancing agricultural productivity to sustainably feed the future population. Biofertilizers and BCAs, including mycorrhizal fungi and their metabolites, represent one of the most promising alternatives. The direct utilization of the living organisms for this purpose has been extensively studied and reviewed, but the focus on the metabolites themselves is lacking, mainly in mycorrhizal fungi. Thus, the objective of this chapter is to collect and collate the current state-of-art knowledge on the production of metabolites by mycorrhizal fungi with agricultural applications.

2 Mycorrhizal Fungi: Biology, Classification, and Agricultural Applications

Mycorrhizal fungi (MF) encompass over fifty-thousand species of soil fungi that establish mutualistic symbiotic relationships with more than 340 thousand plant species (Genre et al. 2020). This symbiosis is considered to have played a crucial role in the terrestrial colonization by plants (Genre et al. 2020). However, recent studies have identified potential pathogenic behavior of this fungal group in evolutionarily early terrestrial plants like bryophytes (Poveda 2020).

The MF-plant symbiosis is based on the colonization of the soil by fungal hyphae outside the plant roots and the provision of water and nutrients, particularly nitrogen and phosphorus, to the plant, while the host plant supplies the fungus with photo-assimilated carbon compounds (Chaudhary et al. 2022). Despite the evident benefits for both partners, there are certain plants, such as crucifers (Brassicaceae family), that have lost their ability to interact with MF, possibly due to the development of a potent antifungal defense strategy involving secondary metabolites known as glucosinolates (Poveda et al. 2022).

MF are classified into two distinct groups: endomycorrhizal fungi and ectomycorrhizal fungi (EMF), comprising approximately 98% and 2% of the total MF species, respectively (Genre et al. 2020). The primary distinction between these two groups lies in the fact that endomycorrhizal fungi establish structures within the plant cells of the roots they colonize, whereas EMF only colonize the cell apoplast, forming a complex network of hyphae called the Hartig network. Endomycorrhizal fungi can be further categorized into several groups, with the most significant ones, based on their distribution and number of described species, being arbuscular mycorrhizal fungi (AMF), orchid mycorrhizal fungi (OMF), and ericoid mycorrhizal fungi (ErMF), representing approximately 72%, 10%, and 1% of all endomycorrhizal fungi species, respectively (Genre et al. 2020; Chaudhary et al. 2022). AMF can form symbiotic associations with over 80% of terrestrial plants, including more than 90% of seed plants, highlighting their importance for crops (Genre et al. 2020). This fungal group is characterized by forming the most complex symbiotic structure within plant cells known as arbuscules, which are branched hyphae specialized in nutrient exchange. In contrast, OMF and ErMF form simpler intracellular structures in the form of coils and exhibit limited symbiotic relationships, restricted to plants of the Orchidaceae and Ericaceae families, respectively (Genre et al. 2020; Chaudhary et al. 2022).

The MF symbiosis provides significant benefits to plants. A comprehensive database (called MycoDB) has been developed to document the effects of this symbiosis, which can be explored by fungal species (both endomycorrhizae and ectomycorrhizae) and plant species (Chaudhary et al. 2016). In agriculture, numerous commercial bio-inoculants based on are being utilized as biofertilizers/biostimulants and BCAs, either alone or in combination with other microorganisms such as bacteria or endophytic fungi (Díaz-Urbano et al. 2023).

Regarding their use as biofertilizers, the most extensively studied and widely used MF in the agricultural sector are AMF, due to their efficient capacity to supply essential nutrients like P, N, and various trace elements, in addition to water, to their host plants. Regarding P, AMF are able to produce alkaline phosphatases that release phosphates from soil organic matter, as well as organic acids that after their secretion can solubilize mineral phosphates. Concerning N, AMF enhance the absorption by plant host of ammonium, nitrate, and amino acids present in the soil. Furthermore, AMF play a crucial role in the uptake of trace elements such as potassium (K), calcium (Ca), magnesium (Mg), copper (Cu), zinc (Zn), iron (Fe), manganese (Mn), and cobalt (Co) through specific cellular transporters (Fall et al. 2022). Given their biological activity, the majority of patents registered thus far for the use of AMF are focused on their utilization as biofertilizers (Srivastava et al. 2021).

Similarly, EMF are employed as effective biofertilizers due to their ability to provide also their host plants with nutrients, particularly P and N, alongside water. EMF exhibit greater efficiency in P uptake owing to their capacity to absorb inorganic P through specific cellular transporters and solubilize phosphates via the production and release of organic acids into the soil. In terms of N, EMF supply ammonium and nitrate to the plant by decomposing soil organic matter using proteases, chitinases, and other enzymes. These mechanisms make EMF highly suitable for inoculating seedlings in forest nurseries, thereby enhancing the plants' adaptability to their future habitat (Domínguez-Núñez and Albanesi 2019).

Regarding their use as BCAs, both AMF and EMF exhibit various mechanisms of action. Directly, these fungal groups can mitigate the detrimental effects of biotic stresses in plants through competition for root infection sites and space in the rhizosphere. Indirectly, they can modify root architecture, alter the rhizosphere microbiota, and, most significantly, induce plant defense responses at both local and systemic levels (Poveda et al. 2020; Weng et al. 2022). The success of this biological control strategy has been observed not only in controlled greenhouse conditions but also in field settings (Poveda and Baptista 2021), leading to an increasing number of patents each year related to the utilization of AMF as BCAs (Srivastava et al. 2021).

3 Secondary Metabolites Produced by MF

In this section, all the research conducted thus far on the discovery and characterization of metabolites produced by MF that are of interest to plants have compiled and categorized. All the studies analyzed can be observed in Table 1. Furthermore, a summary infographic (Fig. 1) has been created to illustrate the diverse beneficial effects of these fungal metabolites on plants. These effects include their diffusible

Table 1 Metabolites produced by mycorrhizal fungi with plant production benefits: origin, description and applications

Mycorrhizal fungi group	Species	Origin of host plant	Experiment type	Metabolite	Biological activity	References
AMF	*Glomus fasciculatus*	Not indicated	*In vitro*: fungal metabolites production against *Streptomyces cinnamomeous*	Not identified	Antibacterial	Krishna et al. (1982)
	G. mosseae	Not indicated	*In planta*: metabolites application on lettuce, radish and oat seedlings, and barley seeds	Gibberellins Citokinins	Plant growth promotion	Barea and Azcón-Aguilar (1982)
	Rhizophagus irregularis	Not indicated	*In vitro*: fungal metabolites production	Cytokinins (IA) Auxins (IAA) Gibberellins (GA4) Ethylene	Plant growth promotion	Pons et al. (2020)
EMF	*Amanita caesaria*	Not indicated	*In vitro*: fungal metabolites production	Auxins (IAA)	Plant growth promotion	Ulrich (1960)
	A. frostiana	Not indicated	*In vitro*: fungal metabolites production	Auxins (IAA)	Plant growth promotion	Ulrich (1960)
	A. muscaria	Not indicated	*In vitro*: fungal metabolites production	Auxins (IAA)	Plant growth promotion	Ulrich (1960)
		Not indicated	*In vitro*: fungal metabolites production	Auxins Cytokinins	Plant growth promotion	Rudawska (1983)
	A. rubescens	Not indicated	*In vitro*: fungal metabolites production	Auxins (IAA)	Plant growth promotion	Ulrich (1960)
	Boletus badius	Not indicated	*In vitro*: fungal metabolites production	Auxins (IAA)	Plant growth promotion	Ulrich (1960)
	B. bovinus	Not indicated	*In vitro*: fungal metabolites production	Auxins (IAA)	Plant growth promotion	Ulrich (1960)

	B. felleus	Not indicated	*In vitro*: fungal metabolites production	Auxins (IAA)	Plant growth promotion	Ulrich (1960)
	B. granulatus	Not indicated	*In vitro*: fungal metabolites production	Auxins (IAA)	Plant growth promotion	Ulrich (1960)
	B. luteus	Not indicated	*In vitro*: fungal metabolites production	Auxins (IAA)	Plant growth promotion	Ulrich (1960)
		Not indicated	*In vitro*: fungal metabolites production	Not identified	Antibacterial	Santoro and Casida (1962)
	B. variegatus	Not indicated	*In vitro*: fungal metabolites production	Auxins (IAA)	Plant growth promotion	Ulrich (1960)
		Scots pine	*In vitro*: fungal metabolites production	VOCs: ethanol, isobutanol, isoamyl alcohol, acetoin and isobutyric acid	Antifungal	Krupa and Fries (1971)
	Cenococcum graniforme	Not indicated	*In vitro*: fungal metabolites production against *Bacillus cereus* and *B. mycoides*	Not identified	Antibacterial	Krywolap and Casida (1963)
		Trembling Aspen	*In vitro*: fungal metabolites production *In planta*: pines mycorrhization and metabolite extraction from the roots	Not identified	Antibacterial	Krywolap et al. (1964)
		Eastern White pine Red pine Norway spruce	*In vitro*: fungal metabolites production against *Bacillus cereus* *In planta*: pines mycorrhization and metabolite extraction from the aerial part after bacterial infection	Not identified	Antibacterial	Grand and Ward (1969)

(continued)

Table 1 (continued)

Mycorrhizal fungi group	Species	Origin of host plant	Experiment type	Metabolite	Biological activity	References
	Laccaria sp.	Norway spruce Douglas-fir	*In vitro*: fungal cell free filtrates against *Fusarium oxysporum*	Not identified	Antifungal	Sampangiramaiah and Perrin (1990)
	L. bicolor	Not indicated	*In planta*: metabolites application on grey poplar and arabidopsis	VOCs: sesquiterpenes	Plant growth promotion	Ditengou et al. (2015)
	Lactarius sp.	Basswood	*In vitro*: fungal cell free filtrates against *Penicillium implicatum*	Not identified	Antifungal	Park (1970)
	L. deliciosus	Not indicated	*In vitro*: fungal metabolites production against *B. subtilis*	Not identified	Antibacterial	Garrido et al. (1982)
		Not indicated	*In vitro*: fungal metabolites production against *Fusarium oxysporum* *In planta*: pine mycorrhization and damping-off suppression	Not identified	Antifungal	Martín-Pinto et al. (2006)
	L. fuliginosus	Not indicated	*In vivo*: ether: acetone extracts on insects corn earworm and large milkweed bug	Chromenes	Insecticidal	Dowd and Miller (1990)
	L. fumosus	Not indicated	*In vivo*: ether: acetone extracts on insects corn earworm and large milkweed bug	Chromenes	Insecticidal	Dowd and Miller (1990)
	Paxillus involutus	Scots pine	*In vitro*: fungal metabolites production application on soybean callus	Cytokinins (riboside 6, purine, riboside zeatin and zeatin)	Plant growth promotion	Strzelczyk et al. (1985)
		Not indicated	*In vitro*: fungal metabolites production against *Pythium ultimum*	Not identified	Oomiceticidal	Rasanayagam and Jeffries (1992)
		Scots pine	*In vitro*: fungal metabolites production *In planta*: pines mycorrhization	Auxins (IAA)	Plant growth promotion	Rudawska and Kieliszewska-Rokicka1(997)

	P. panuoides	Not indicated	*In vitro*: fungal metabolites production against *B. subtilis*	Not identified	Antibacterial	Garrido et al. (1982)
	Phaeogyroporus beniensis	Not indicated	*In vitro*: fungal metabolites production against *B. subtilis*	Not identified	Antibacterial	Garrido et al. (1982)
	Pisolithus arhizus	Not indicated	*In vitro*: fungal metabolites production against *Truncatella hartigii*	Pisolithin A Pisolithin B	Antifungal	Kope et al. (1991)
	P. tinctorius	Not indicated	*In vitro*: fungal metabolites production against *Rhizoctonia praticola* and *T. hartigii*,	Not identified	Antifungal	Kope and Fortin (1989)
		Not indicated	*In vitro*: fungal metabolites production against *Rhizoctonia praticola*, *T. hartigii*, *Sphaerosporella brunnea*, *Fusarium solani* and *Brunchorstia pinea*	Not identified	Antifungal	Kope and Fortin (1990)
		Fallen	*In vitro*: fungal metabolites production against *F. solani*, *Geotrichum candidum*, *Phanerochaete chrysosporium* and *Verticillium dahliae*	Phenolic metabolite	Antifungal	Suh et al. (1991)
	Rhizopogon luteolus	Not indicated	*In vitro*: fungal metabolites production	Auxins Cytokinins	Plant growth promotion	Rudawska (1983)
		Scots pine	*In vitro*: fungal metabolites production application on soybean callus	Cytokinins (riboside 6, purine and riboside zeatin)	Plant growth promotion	Strzelczyk et al. (1985)
	Rhizopogon roseolus	Not indicated	*In vitro*: fungal metabolites production against *Fusarium oxysporum* and *F. verticillioides*	Not identified	Antifungal	Olaizola et al. (2018)

(continued)

Table 1 (continued)

Mycorrhizal fungi group	Species	Origin of host plant	Experiment type	Metabolite	Biological activity	References
	Russula sardonia	Not indicated	*In vitro*: fungal metabolites production against *B. subtilis*	Not identified	Antibacterial	Garrido et al. (1982)
	Suillus bovinus	Not indicated	*In vitro*: fungal metabolites production	Auxins Cytokinins	Plant growth promotion	Rudawska (1983)
	S. luteus	Not indicated	*In vitro*: fungal metabolites production against *B. subtilis*	Not identified	Antibacterial	Garrido et al. (1982)
		Not indicated	*In vitro*: fungal metabolites production	Auxins Cytokinins	Plant growth promotion	Rudawska (1983)
		Not indicated	*In vitro*: fungal metabolites production	Auxins Gibberellins	Plant growth promotion	Strzelczyk and Pokojska-Burdziej (1984)
		Not indicated	*In vitro*: fungal metabolites production against *Fusarium oxysporum* and *F. verticillioides*	Not identified	Antifungal	Olaizola et al. (2018)
	Tricholoma pessundatum	Scotch pine	*In vitro*: fungal metabolites production against *Bacillus subtilis*	Not identified	Antibacterial	Šašek and Musilek (1967)
	T. saponaeeum	Scotch pine	*In vitro*: fungal metabolites production against *Bacillus subtilis*	Not identified	Antibacterial	Šašek and Musilek (1967)
	T. vaecinum	Scotch pine	*In vitro*: fungal metabolites production against *Bacillus subtilis*	Not identified	Antibacterial	Šašek and Musilek (1967)
	Tuber borchii	Not indicated	*In planta*: fungal metabolites application to hoary rock-rose and arabidopsis	Auxins (IAA) Ethylene	Plant growth promotion	Splivallo et al. (2009)

	T. melanopsorum	Not indicated	*In planta*: fungal metabolites application to hoary rock-rose and arabidopsis	Auxins (IAA) Ethylene	Plant growth promotion	Splivallo et al. (2009)
	Xerocomus chrysenteron	Not indicated	*In vitro*: fungal metabolites production against *B. subtilis*	Not identified	Antibacterial	Garrido et al. (1982)
	Zygorrhynchus moelleri	Not indicated	*In vitro*: fungal metabolites production against *Phytophthora cinnamomi*, *Pythium* spp., *Fusarium oxysporum* f. sp. *lini*, *R. solani* and *Sclerotinia sclerotiorum*	Auxins (IAA) Indole-3-ethanol	Plant growth promotion Oomiceticidal Antifungal	Brown and Hamilton (1992)
ErMF	*Myxotrichum deflexum*	Not indicated	*In vitro*: fungal metabolites production against *Bacillus subtilis*	Alkaloids Flavonoids Phenols Saponins Cardiac glycosides Terpenoids	Antibacterial	Adeoyo et al. (2019)
OMF	Not identified	*Ophrys lutea*	*In vitro*: fungal metabolites production	Auxins (IAA)	Plant growth promotion	Barroso et al. (1986)
Not identified	Not identified	Pinesap	*In planta*: fungal metabolites production	Chloromycorrhizinol A	Not identified	Trofast (1978)
	Not identified	Pinesap	*In vitro*: fungal metabolites production	Mycorrhizin A Chloromycorrhizin A	Antibacterial	Trofast and Wickberg (1977)

IAA indolacetic acid, *IA* isopentenyl adenosine, *GA* gibberellin

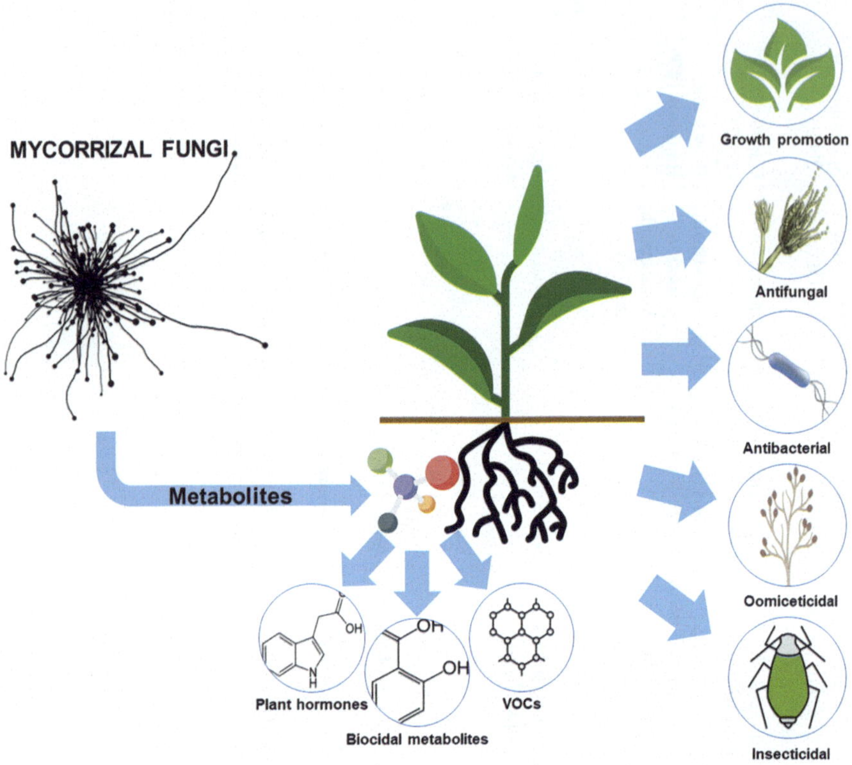

Fig. 1 Infographic on the diversity and agricultural benefit of metabolites produced by mycorrhizal fungi

form in the medium as well as their role as VOCs, in both cases that can promote plant growth and serve as biocides (antifungal, antibacterial, oomyceticides, and insecticides).

3.1 Plant Growth Promoting Activity

The production of phytohormone-like substances is a primary mechanism utilized by rhizosphere microorganisms to promote plant growth in their favor (Ravelo-Ortega et al. 2023). A study conducted in 1960 characterized the *in vitro* production of auxin (indole-acetic acid: IAA) by MF, specifically by 11 different species of EMF (Ulrich 1960). This mechanism has subsequently been investigated in other groups of MF, such as AMF (Pons et al. 2020) and OMF (Barroso et al. 1986), highlighting the role of plant growth promotion in symbiosis. In addition to auxins, MF can produce other growth-promoting hormones *in vitro*, including cytokinins

(Rudawska 1983), gibberellins (Strzelczyk and Pokojska-Burdziej 1984), and ethylene (Pons et al. 2020). So far, the ability to produce all these hormone groups has only been described in the AMF Rhizophagus irregularis (Pons et al. 2020).

The production of these fungal metabolites by MF exerts a direct effect on plant organs and tissues, both *in vitro* and during mycorrhization in the greenhouse or field. Studies using soybean cell callus identified that fungal metabolites produced and released into the culture medium by the EMFs *Paxillus involutus* and *Rhizopogon luteolus* were capable of promoting cell growth and multiplication. This effect was attributed to the production of cytokinins, including riboside 6, purine, riboside zeatin, and zeatin by these ECMs (Strzelczyk et al. 1985). Taking a step further, subsequent experiments demonstrated the ability of *P. involutus* to produce IAA in an *in vitro* culture system, and auxins in nurseries after a controlled mycorrhization of pine seedlings, providing growth promotion on its host plant (Rudawska and Kieliszewska-Rokicka 1997).

This promotion of plant growth in the host plant can occur both before and after the establishment of symbiosis. Research using the edible truffle-forming EMF *Tuber borchii* and *T. melanopsorum*, along with the hoary rock-rose (host) and arabidopsis (non-host) plants, demonstrated that the production of diffusible IAA and ethylene (a volatile hormone) induced primary root shortening, lateral root formation, and root hair stimulation in both plants without direct fungus-plant contact (Splivallo et al. 2009). This mechanism has also been observed in the EMF *Laccaria bicolor*, which significantly induced lateral root production in gray poplar (host) and arabidopsis (non-host) without direct fungus-plant contact. However, in this case, plant hormones were not involved, but rather other fungal metabolites from the sesquiterpene group (Ditengou et al. 2015). Therefore, the production of plant growth-promoting metabolites by MF may play a crucial role in the success of symbiosis, inducing changes in root architecture even in plants that may not be susceptible to forming a symbiotic relationship, such as arabidopsis.

In the absence of MF, the fungal metabolites produced *in vitro* also exhibited bioactivity when applied to various agriculturally significant plants. For instance, the mycelium of *Glomus mosseae* accumulated cytokinins and gibberellins, which were subsequently extracted and applied to lettuce, radish, oat seedlings, and barley seeds. The effects observed in these plants treated with fungal metabolites included hypocotyl elongation in lettuce, reduction of sugars released from embryoless pieces of barley endosperm (effects induced by gibberellins), promotion of cotyledon growth in radish, and chlorophyll retention in oat leaves (effects induced by cytokinins) (Barea and Azcón-Aguilar 1982).

3.2 Biocide Activity

The production of metabolites, whether diffusible in the medium or as VOCs, with antibacterial, antifungal, oomyceticidal, and/or nematicidal properties is a crucial biological control mechanism employed by BCAs, known as antibiosis (Ortiz and

Sansinenea 2021). In the case of bacterial plant pathogens, several studies have demonstrated the ability of MF to produce and release antibacterial metabolites. The first studies in the 1960s demonstrated how different species of EMF can release antibacterial metabolites effective against *Bacillus subtilis*, *B. cereus*, or *B. mycoides* in *in vitro* dual culture essays (Santoro and Casida Jr 1962; Krywolap and Casida Jr 1963; Šašek and Musilek 1967). New EMF species with such activity were described in the 1980s (Garrido et al. 1982). Similarly, antibacterial antibiosis has been observed in AMF, as demonstrated by *Glomus fasciculatus*, which releases biocidal metabolites against *Streptomyces cinnamomeous* in combined *in vitro* cultures (Krishna et al. 1982). However, the specific metabolites involved in the antibacterial activity have not yet been identified in either EMF or AMF. It was not until 2019 that a study identified potential metabolites involved in the antibacterial activity exerted by an ErMF, *Myxotrichum deflexum*, against *B. subtilis in vitro*, which included alkaloids, flavonoids, phenols, saponins, cardiac glycosides, and terpenoids (Adeoyo et al. 2019).

The EMF species that has been most extensively studied for its production of antibacterial metabolites is *Cenococcum graniforme*, which has demonstrated its biocidal activity both *in vitro* and *in planta*. It has been shown that *C. graniforme* can produce antibacterial metabolites that are locally and systemically translocated by the host plant, reducing infection by pathogenic bacteria. These fungal antibacterial metabolites have been isolated from mycorrhized pine roots (Krywolap et al. 1964) but also from needles (Grand and Ward 1969).

Regarding antifungal activity of MF metabolites, only EMF have been studied thus far. The first study in 1970 described the antifungal capacity of cell-free filtrates of liquid cultures of *Lactarius* sp. against *Penicillium implicatum in vitro* (Park 1970). Among the most studied species with antifungal capacity *in vitro* is *Pisolithus tinctorius*, whose metabolites have been shown to be effective biocides against a wide range of plant pathogenic fungi, including *Rhizoctonia praticola*, *Truncatella hartigii*, *Sphaerosporella brunnea*, *Fusarium solani*, *Brunchorstia pinea*, *Geotrichum candidum*, *Phanerochaete chrysosporium*, and *Verticillium dahliae* (Kope and Fortin 1989, 1990; Suh et al. 1991). Phenolic metabolites have been identified as the fungal chemicals responsible for this antifungal effect (Suh et al. 1991). In another species of the same genus, such as *Pisolithus arhizus*, those metabolites have been named as pisolithin A and pisolithin B (Kope et al. 1991). Other antifungal metabolites identified in EMF include indole-3-ethanol, produced by *Zygorrhynchus moelleri*, which inhibited the conidial germination of *Fusarium oxysporum* f. sp. *lini*, *R. solani*, and *Sclerotinia sclerotiorum* (Brown and Hamilton 1992), as well as various VOCs produced by *Boletus variegatus*, such as ethanol, isobutanol, isoarnyl alcohol, acetoin, and isobutyric acid (Krupa and Fries 1971). Importantly, the antifungal activity of the metabolites produced by MF is not limited to *in vitro* action. A study conducted with the EMF *Lactarius deliciosus* described how cell-free filtrates from liquid cultures of the fungus inhibited the germination of *F. oxysporum* conidia, and subsequent mycorrhization of pine seedlings with *L. deliciosus* suppressed damping-off disease caused by *F. oxysporum*, possibly due to the antibiosis observed *in vitro* (Martín-Pinto et al. 2006).

To our knowledge, only two studies to date have investigated the oomyceticidal activity of MF metabolites, both involving EMF. Both studies were published in 1992. It was observed that the metabolites produced *in vitro* by *Paxillus involutus* caused swelling and bursting of the tips of leading hyphae in *Pythium ultimum*, although the specific metabolite involved was not identified (Rasanayagam and Jeffries 1992). Similarly, the metabolite with antifungal activity (indole-3-ethanol) produced by *Z. moelleri* was found to exhibit oomyceticidal activity, inhibiting the germination of *Phytophthora cinnamomi* zoospores and *Pythium* spp. oogonia (Brown and Hamilton 1992).

Regarding pest control, only one study to date has described the insecticidal action of metabolites produced by MF. Ether:acetone extracts of the EMF species *Lactarius fuliginosus* and *L. fumosus* were found to be toxic against the insect pests corn earworm (*Helicoverpa zea*) and large milkweed bug (*Oncopeltus fasciatus*). This effect was attributed to the presence of fungal metabolites known as chromenes, which exhibit antihormonal effects in insects (Dowd and Miller 1990).

All the studies discussed in this section have identified the production of metabolites of interest to plants by MF, highlighting their potential as biotechnological tools. However, after the identification in recent years by Choudhary et al. (2017) of the helper bacteria, involved in the establishment and functioning of mycorrhizae, there are doubts about the actual producer of these metabolites, whether it is the fungus itself or the helper bacteria. These doubts arise from the fact that this group of microorganisms may produce growth factors that stimulate fungal spore germination, mycelial growth, increased root branching, enhanced root colonization, and reduction of soil-mediated stress through the detoxification of antagonistic substances and the inhibition of competitors and antagonists (Choudhary et al. 2017).

4 Conclusions and Future Prospects

The general conclusion of this chapter is that in the search for new biotechnological alternatives to develop a sustainable agriculture, MF represent a tool with great potential, and their worldwide use is increasing every year. While the use of MF as agricultural bio-inoculants is the main focus, it is important to recognize that these fungi produce a wide variety of agriculturally relevant metabolites that are not fully exploited. Although host plants benefit from these metabolites in their symbiosis with MF, there is potential to develop industrial bioreactors for mass production and direct application of these metabolites in the agrosystem.

MF encompasses numerous species with the ability to produce metabolites of agricultural interest, including plant growth-promoting hormones, antibacterials, antifungals, oomyceticides, and insecticides. However, the majority of studies are performed *in vitro* and do not investigate the actual effects of these metabolites on plants. Future research should adopt a more applied approach to address the agricultural sector's need for new, safe and efficient products. Additionally, there is a notable lack of studies focusing on AMF, despite they are the primary group of fungi used as agricultural bio-inoculants. This raises questions about the lack of interest or exploration in this area and whether AMF produce metabolites of interest.

Most of the identified metabolites produced by MF exhibited plant growth-promoting, antibacterial, or antifungal activities. This information provides new perspectives for studying the MF-plant symbiosis, as both groups of metabolites serve as potential strategies employed by the fungus to promote root colonization and prevent mycorrhizospheric competition from bacteria and/or fungi.

The role of these metabolites as oomyceticides or insecticides has received limited discussion thus far, and there is currently no research on their potential as nematicides or herbicides. This opens up new avenues for research with significant potential. For instance, it is known that MF can have detrimental effects on certain plant groups, such as crucifers, leading to a reduction in their growth (Cosme et al. 2018). This could potentially be attributed to the production of specific herbicide metabolites targeting this particular group of plants.

Finally, an important aspect for future research is determining the exact role of helper bacteria in the production of these metabolites by MF. Are these bacteria involved in metabolite synthesis, and do they influence the symbiotic relationship with the fungus? These questions remain to be explored in further studies.

Acknowledgements J.P. thanks Mobility Grants for Researchers UVa-Santander Bank 2023 (MOVILIDAD INVESTIGADORES E INVESTIGADORAS UVa-BANCO SANTANDER 2023) for their help which allowed to carry out a research stay at the University of Bristol (United Kingdom).

References

Adeoyo OR, Pletschke BI, Dames JF (2019) Molecular identification and antibacterial properties of an ericoid associated mycorrhizal fungus. BMC Microbiol 19:1–8. https://doi.org/10.1186/s12866-019-1555-y

Ali S, Moon YS, Hamayun M, Khan MA, Bibi K, Lee IJ (2022) Pragmatic role of microbial plant biostimulants in abiotic stress relief in crop plants. J Plant Interact 17:705–718. https://doi.org/10.1080/17429145.2022.2091801

Barea JM, Azcón-Aguilar C (1982) Production of plant growth-regulating substances by the vesicular-arbuscular mycorrhizal fungus *Glomus mosseae*. Appl Environ Microbiol 43:810–813. https://doi.org/10.1128/aem.43.4.810-813.1982

Barroso J, Neves HC, Pais MS (1986) Production of indole-3-e thanol and indole-3-a cetic acid by the mycorrhizal fungus of *Ophrys lutea* (Orchidaceae). New Phytol 103:745–749. https://doi.org/10.1111/j.1469-8137.1986.tb00849.x

Bejarano A, Puopolo G (2020) Bioformulation of microbial biocontrol agents for a sustainable agriculture. In: How research can stimulate the development of commercial biological control against plant diseases. Springer, pp 275–293

Brown AE, Hamilton JTG (1992) Indole-3-ethanol produced by *Zygorrhynchus moelleri*, an indole-3-acetic acid analogue with antifungal activity. Mycol Res 96:71–74. https://doi.org/10.1016/S0953-7562(09)80999-4

Chaudhary VB, Rúa MA, Antoninka A, Bever JD, Cannon J, Craig A et al (2016) MycoDB, a global database of plant response to mycorrhizal fungi. Sci Data 3:1–10. https://doi.org/10.1038/sdata.2016.28

Chaudhary VB, Holland EP, Charman-Anderson S, Guzman A, Bell-Dereske L, Cheeke TE et al (2022) What are mycorrhizal traits? Trends Ecol Evol 37:573–581. https://doi.org/10.1016/j.tree.2022.04.003

Choudhary DK, Varma A, Tuteja N (2017) Mycorrhizal helper bacteria: sustainable approach. In: Mycorrhiza – function, diversity, state of the Art. Springer, pp 61–74

Cosme M, Fernández I, Van der Heijden MG, Pieterse CM (2018) Non-mycorrhizal plants: the exceptions that prove the rule. Trends Plant Sci 23:577–587. https://doi.org/10.1016/j.tplants.2018.04.004

Díaz-Urbano M, Goicoechea N, Velasco P, Poveda J (2023) Development of agricultural bio-inoculants based on mycorrhizal fungi and endophytic filamentous fungi: co-inoculants for improve plant-physiological responses in sustainable agriculture. Biol Control 105223. https://doi.org/10.1016/j.biocontrol.2023.105223

Ditengou FA, Müller A, Rosenkranz M, Felten J, Lasok H, Van Doorn MM et al (2015) Volatile signalling by sesquiterpenes from ectomycorrhizal fungi reprogrammes root architecture. Nat Commun 6:6279. https://doi.org/10.1038/ncomms7279

Domínguez-Núñez JA, Albanesi AS (2019) Ectomycorrhizal fungi as biofertilizers in forestry. In: Biostimulants in plant science. IntechOpen, pp 117–146

Dowd PF, Miller OK (1990) Insecticidal properties of Lactarius fuliginosus and Lactarius fumosus. Entomol Exp Appl 57(1):23–28

Elnahal AS, El-Saadony MT, Saad AM, Desoky ESM, El-Tahan AM, Rady MM et al (2022) The use of microbial inoculants for biological control, plant growth promotion, and sustainable agriculture: a review. Eur J Plant Pathol 162:759–792. https://doi.org/10.1007/s10658-021-02393-7

European Council (2019) Regulation (EU) 2019/1009 of the European Parliament and of the Council of 5 June 2019 Laying Down Rules on the Making Available on the Market of EU Fertilising Products and Amending Regulations (EC) no 1069/2009 and (EC) no 1107/2009 and Repealing Regulation (EC) no 2003/2003

European Food Safety Authority (EFSA) (2021) The 2019 European Union report on pesticide residues in food. EFSA J 19:e06491. https://doi.org/10.2903/j.efsa.2021.6491

Falcon WP, Naylor RL, Shankar ND (2022) Rethinking global food demand for 2050. Popul Dev Rev 48:921–957. https://doi.org/10.1111/padr.12508

Fall AF, Nakabonge G, Ssekandi J, Founoune-Mboup H, Apori SO, Ndiaye A et al (2022) Roles of arbuscular mycorrhizal fungi on soil fertility: contribution in the improvement of physical, chemical, and biological properties of the soil. Front Fungal Biol 3:723892. https://doi.org/10.3389/ffunb.2022.723892

Garrido N, Becerra J, Marticorena C, Oehrens E, Silva M, Horak E (1982) Antibiotic properties of ectomycorrhizae and saprophytic fungi growing on *Pinus radiata* D Don I. Mycopathologia 77:93–98. https://doi.org/10.1007/BF00437390

Genre A, Lanfranco L, Perotto S, Bonfante P (2020) Unique and common traits in mycorrhizal symbioses. Nat Rev Microbiol 18:649–660. https://doi.org/10.1038/s41579-020-0402-3

Grand LF, Ward WW (1969) An antibiotic detected in conifer foliage and its relation to *Cenococcum graniforme* mycorrhizae. For Sci 15:286–288. https://doi.org/10.1093/forestscience/15.3.286

Helepciuc FE, Todor A (2022) Improving the authorization of microbial biological control products (MBCP) in the European Union within the EU Green Deal framework. Agronomy 12:1218. https://doi.org/10.3390/agronomy12051218

Hernández-Rosas F, Figueroa-Rodríguez KA, García-Pacheco LA, Velasco-Velasco J, Sangerman-Jarquín DM (2020) Microorganisms and biological pest control: an analysis based on a bibliometric review. Agronomy 10:1808. https://doi.org/10.3390/agronomy10111808

Joshi N, Parewa HP, Joshi S, Sharma JK, Shukla UN, Paliwal A, Gupta V (2021) Use of microbial biostimulants in organic farming. In: Advances in organic farming. Woodhead Publishing, pp 59–73

Kaddes A, Fauconnier ML, Sassi K, Nasraoui B, Jijakli MH (2019) Endophytic fungal volatile compounds as solution for sustainable agriculture. Molecules 24:1065. https://doi.org/10.3390/molecules24061065

Keswani C, Singh HB, Hermosa R, García-Estrada C, Caradus J, He YW et al (2019) Antimicrobial secondary metabolites from agriculturally important fungi as next biocontrol agents. Appl Microbiol Biotechnol 103:9287–9303. https://doi.org/10.1007/s00253-019-10209-2

Köhl J, Kolnaar R, Ravensberg WJ (2019a) Mode of action of microbial biological control agents against plant diseases: relevance beyond efficacy. Front Plant Sci 10:845. https://doi.org/10.3389/fpls.2019.00845

Köhl J, Booij K, Kolnaar R, Ravensberg WJ (2019b) Ecological arguments to reconsider data requirements regarding the environmental fate of microbial biocontrol agents in the registration procedure in the European Union. BioControl 64:469–487. https://doi.org/10.1007/s10526-019-09964-y

Kope HH, Fortin JA (1989) Inhibition of phytopathogenic fungi *in vitro* by cell free culture media of ectomycorrhizal fungi. New Phytol 113:57–63. https://doi.org/10.1111/j.1469-8137.1989.tb02395.x

Kope HH, Fortin JA (1990) Antifungal activity in culture filtrates of the ectomycorrhizal fungus *Pisolithus tinctorius*. Can J Plant Sci 68:1254–1259. https://doi.org/10.1139/b90-159

Kope HH, Tsantrizos YS, Fortin JA, Ogilvie KK (1991) p-Hydroxybenzoylformic acid and (R)-(−)-p-hydroxymandelic acid, two antifungal compounds isolated from the liquid culture of the ectomycorrhizal fungus *Pisolithus arhizus*. Can J Microbiol 37:258–264. https://doi.org/10.1139/m91-040

Krishna KR, Balakrishna AN, Bagyarajf DJ (1982) Interaction between a vesicular–arbuscular mycorrhizal fungus and *Streptomyces cinnamomeous* and their effects on finger millet. New Phytol 92:401–405. https://doi.org/10.1111/j.1469-8137.1982.tb03397.x

Krupa S, Fries N (1971) Studies on ectomycorrhizae of pine. I. Production of volatile organic compounds. Canad J Bot 49:1425–1431. https://doi.org/10.1139/b71-200

Krywolap GN, Casida LE Jr (1963) An antibiotic produced by the mycorrhizal fungus *Cenococcum graniforme*. Can J Microbiol 10:365–370. https://doi.org/10.1139/m64-049

Krywolap GN, Grand LF, JrLE C (1964) The natural occurrence of an antibiotic in the mycorrhizal fungus *Cenococcum graniforme*. Can J Microbiol 10:323–328. https://doi.org/10.1139/m64-044

Kumari M, Swarupa P, Kesari KK, Kumar A (2022) Microbial inoculants as plant biostimulants: a review on risk status. Life 13:12. https://doi.org/10.3390/life13010012

Martín-Pinto P, Pajares J, Díez J (2006) *In vitro* effects of four ectomycorrhizal fungi, *Boletus edulis*, *Rhizopogon roseolus*, *Laccaria laccata* and *Lactarius deliciosus* on Fusarium damping off in *Pinus nigra* seedlings. New For 32:323–334. https://doi.org/10.1007/s11056-006-9006-7

Olaizola J, Pajares JA, Díez-Casero JJ (2018) *In vitro* antagonism of edible ectomycorrhizal fungi against *Fusarium oxysporum* and *Fusarium verticillioides*. For Chron 94:117–125. https://doi.org/10.5558/tfc2018-018

Ortiz A, Sansinenea E (2021) Recent advancements for microorganisms and their natural compounds useful in agriculture. Appl Microbiol Biotechnol 105:891–897. https://doi.org/10.1007/s00253-020-11030-y

Park JY (1970) Antifungal effect of an ectotrophic mycorrhizal fungus, *Lactarius* sp., associated with basswood seedlings. Can J Microbiol 16:798–800. https://doi.org/10.1139/m70-134

Pons S, Fournier S, Chervin C, Bécard G, Rochange S, Dit-Frey NF, Puech-Pagès V (2020) Phytohormone production by the arbuscular mycorrhizal fungus *Rhizophagus irregularis*. PLoS One 15:e0240886. https://doi.org/10.1371/journal.pone.0240886

Poveda J (2020) *Marchantia polymorpha* subsp. *ruderalis* (Bischl. & Boissel.-Dub.)-arbuscular mycorrhizal fungi interaction: beneficial or harmful? Symbiosis 82:165–174. https://doi.org/10.1007/s13199-020-00708-6

Poveda J, Baptista P (2021) Filamentous fungi as biocontrol agents in olive (*Olea europaea* L.) diseases: Mycorrhizal and endophytic fungi. Crop Prot 146:105672. https://doi.org/10.1016/j.cropro.2021.105672

Poveda J, González-Andrés F (2021) *Bacillus* as a source of phytohormones for use in agriculture. Appl Microbiol Biotechnol 105:8629–8645. https://doi.org/10.1007/s00253-021-11492-8

Poveda J, Abril-Urias P, Escobar C (2020) Biological control of plant-parasitic nematodes by filamentous fungi inducers of resistance: *Trichoderma*, mycorrhizal and endophytic fungi. Front Microbiol 11:992. https://doi.org/10.3389/fmicb.2020.00992

Poveda J, Díaz-González S, Díaz-Urbano M, Velasco P, Sacristán S (2022) Fungal endophytes of Brassicaceae: molecular interactions and crop benefits. Front Plant Sci 13:932288. https://doi.org/10.3389/fpls.2022.932288

Rasanayagam S, Jeffries P (1992) Production of acid is responsible for antibiosis by some ectomycorrhizal fungi. Mycol Res 96:971–976. https://doi.org/10.1016/S0953-7562(09)80600-X

Ravelo-Ortega G, Raya-González J, López-Bucio J (2023) Compounds from rhizosphere microbes that promote plant growth. Curr Opin Plant Biol 73:102336. https://doi.org/10.1016/j.pbi.2023.102336

Rudawska M (1983) The effect of nitrogen and phosphorus on auxin and cytokinin production by mycorrhizal fungi. Arboretum Kornickie 28:219–236

Rudawska ML, Kieliszewska-Rokicka B (1997) Mycorrhizal formation by *Paxillus involutus* strains in relation to their IAA-synthesizing activity. New Phytol 137:509–517. https://doi.org/10.1046/j.1469-8137.1997.00838.x

Sampangiramaiah K, Perrin I (1990) Interactions between isolates of ectomyrrhizal *Laccarici* spp. and root rot fungi of conifers. In: Current trends in mycorrhizal research. Haryana Agricultural University, pp 124–125

Sanjuán J, Nápoles MC, Pérez-Mendoza D, Lorite MJ, Rodríguez-Navarro DN (2023) Microbials for agriculture: why do they call them biostimulants when they mean probiotics? Microorganisms 11:153. https://doi.org/10.3390/microorganisms11010153

Santoro T, Casida LE Jr (1962) Elaboration of antibiotics by *Boletus luteus* and certain other mycorrhizal fungi. Can J Microbiol 8:43–48. https://doi.org/10.1139/m62-006

Šašek V, Musilek V (1967) Cultivation and antibiotic activity of mycorrhizal basidiomycetes. Folia Microbiol 12:515–523. https://doi.org/10.1007/BF02875713

Schebesta H, Candel JJ (2020) Game-changing potential of the EU's farm to fork strategy. Nat Food 1:586–588. https://doi.org/10.1038/s43016-020-00166-9

Segaran G, Sathiavelu M (2019) Fungal endophytes: a potent biocontrol agent and a bioactive metabolites reservoir. Biocatal Agric Biotechnol 21:101284. https://doi.org/10.1016/j.bcab.2019.101284

Shahrajabian MH, Chaski C, Polyzos N, Petropoulos SA (2021) Biostimulants application: a low input cropping management tool for sustainable farming of vegetables. Biomol Ther 11:698. https://doi.org/10.3390/biom11050698

Shahrajabian MH, Petropoulos SA, Sun W (2023) Survey of the influences of microbial biostimulants on horticultural crops: case studies and successful paradigms. Horticulturae 9:193. https://doi.org/10.3390/horticulturae9020193

Singh J, Yadav AN (eds) (2020) Natural bioactive products in sustainable agriculture. Springer

Splivallo R, Fischer U, Göbel C, Feussner I, Karlovsky P (2009) Truffles regulate plant root morphogenesis via the production of auxin and ethylene. Plant Physiol 150:2018–2029. https://doi.org/10.1104/pp.109.141325

Srivastava S, Johny L, Adholeya A (2021) Review of patents for agricultural use of arbuscular mycorrhizal fungi. Mycorrhiza 31:127–136. https://doi.org/10.1007/s00572-021-01020-x

Strzelczyk E, Pokojska-Burdziej A (1984) Production of auxins and gibberellin-like substances by mycorrhizal fungi, bacteria and actinomycetes isolated from soil and the mycorrhizosphere of pine (*Pinus silvestris* L.). Plant Soil 2:185–194. https://www.jstor.org/stable/42935912

Strzelczyk E, Kampert M, Michalski L (1985) Production of cytokinin-like substances by mycorrhizal fungi of pine (*Pinus sylvestris* L.) in cultures with and without metabolites of actinomycetes. Acta Microbiol Pol 34:177–185

Suh HW, Crawford DL, Korus RA, Shetty K (1991) Production of antifungal metabolites by the ectomycorrhizal fungus *Pisolithus tinctorius* strain SMF. J Ind Microbiol Biotechnol 8:29–35. https://doi.org/10.1007/BF01575588

Tarigan SI, Toth S, Szalai M, Kiss J, Turoczi G, Toepfer S (2022) Biological control properties of microbial plant biostimulants. A review. Biocont Sci Technol 32:1351–1371. https://doi.org/10.1080/09583157.2022.2129589

Trofast J (1978) Chloromycorrhizinol A, a furochroman from an isolate of the roots of *Monotropa hypopitys*. Phytochemistry 17:1359–1361. https://doi.org/10.1016/S0031-9422(00)94589-8

Trofast J, Wickberg B (1977) Mycorrhizin a and chloromycorrhizin a, two antibiotics from a mycorrhizal fungus of monotropa hypopitys L. Tetrahedron 33:875–879. https://doi.org/10.1016/0040-4020(77)80038-0

Ulrich JM (1960) Auxin production by mycorrhizal fungi. Physiol Plant 13:429–443. https://doi.org/10.1111/j.1399-3054.1960.tb08065.x

Vinale F, Sivasithamparam K (2020) Beneficial effects of *Trichoderma* secondary metabolites on crops. Phytother Res 34:2835–2842. https://doi.org/10.1002/ptr.6728

Weng W, Yan J, Zhou M, Yao X, Gao A, Ma C et al (2022) Roles of arbuscular mycorrhizal fungi as a biocontrol agent in the control of plant diseases. Microorganisms 10:1266. https://doi.org/10.3390/microorganisms10071266

WPC (2023) World population clock. Accesible online: https://www.worldometers.info/world-population/

Fungal Metabolites as Plant Growth Promoters in Crops

Tamara Sánchez-Gómez, Óscar Santamaría, Jorge Martín-García, and Jorge Poveda

Abstract World population is projected to reach almost ten billion by 2050, approximately 24% more than today. This situation leads to a pressing need to increase crop production and yields, but not at any cost: it must be done minimizing environmental damage and, consequently, also harms to human health. Trying to achieve this goal, the European Union (EU)'s Green Deal has as one of its objectives for the present decade to reduce the application of chemical inputs in agriculture. This fact creates the need to search for organic alternatives, gaining importance at this point plant growth promoting microorganisms (PGPM) and their metabolites. Within this group, plant growth promoting fungi (PGPF) such as endophytes or arbuscular mycorrhizae, are particularly important. Among endophytes, the genera *Trichoderma*, *Penicillium*, or *Aspergillus* have been intensively studied with this purpose, whereas within arbuscular mycorrhizae, the genera *Glomus* and *Rhizophagus* are the most prominent. Both groups have the ability to generate specific secondary metabolites with plant-growth promoting activity in many crops, such as phytohormones, non-ribosomal peptides, poliketides, terpenes and alkaloids. The production, commercialization, and application of PGPF and their metabolites may be relatively simple and cost-effective. Consequently, they may represent a feasible organic alternative to replace many of the current synthetic fertilizers.

Keywords Plant growth-promoting fungi · Organic fertilizers · Secondary metabolites · Endophytes · Phytohormones

T. Sánchez-Gómez (✉) · Ó. Santamaría · J. Martín-García · J. Poveda
Recognized Research Group AGROBIOTECH, UIC-370 (JCyL), Department of Plant Production and Forestry Resources, University Institute for Research in Sustainable Forest Management (iuFOR), Higher Technical School of Agricultural Engineering (ETSIIAA), University of Valladolid, Palencia, Spain
e-mail: tamara.sanchez@uva.es

J. Poveda et al. (eds.), *Fungal Metabolites for Agricultural Applications*, Fungal Biology, https://doi.org/10.1007/978-3-031-76587-2_4

1 Introduction

Global population is expected to reach around 9.7 billion by 2050, ~24% higher than today (United Nations 2022). This situation leads to a pressing need to increase crop production. A similar increase was already achieved in the 1960s, with the *wrongly* named Green Revolution, which began in the USA and brought a more intense use of mechanization, large-scale monocultures, and a massive use of chemical fertilizers and pesticides. While this new industrial agriculture managed to exponentially increase productivity, which of course allowed to reduce the global hunger, the practices used have led to significant soil degradation, massive loss of biodiversity, and increased water pollution, among other environmental impacts (McNeill 2009; Glaeser 2010). The current framework differs from the previous one: crop yields and efficiency of farming systems must be increased, but not at any cost. It must be done without damaging neither the environment nor human health (or at least minimizing this damage as much as possible).

At a regulatory level, one of the priorities of the 2030 Agenda is to reduce synthetic inputs in agriculture. According to data from the Food and Agriculture Organization of the United Nations, the use of synthetic pesticides and fertilizers has increased by 36% and 40% respectively in the last two decades (FAO 2021). Thus, the European Union (EU)'s Green Deal and, particularly, the Farm to Fork Strategy, has as objectives for the present decade to reduce the application of chemical pesticides and fertilizers by 50% and 20%, respectively (European Commission 2019, 2020).

Although organic farming still accounts for a small percentage of global agriculture, specifically 1.6% (Willer et al. 2023), it is a sector that has been continuously booming over the last decade, from 36.7 million hectares recorded in 2011, to 76.4 in 2021 (Willer et al. 2013, 2023). This upward trend in organic practices is favoured by governmental policies and by the scarcity and rising prices of chemical fertilizers in recent years (FAO 2022). In terms of policies, the most remarkable in this respect is the aforementioned Farm to Fork Strategy, which aims to achieve an increase of approximately 15% in the agricultural area cultivated under organic farming system in the EU (from the 9.9% registered in 2021 to the 25%) (EUROSTAT 2023). Within those organic fertilization practices, the use of plant growth promoting microorganisms and their natural metabolites could become an important non-contaminating alternative to achieve higher crop yields.

2 Plant Growth Promoting Microorganisms: Definition and Classification

Plant growth promoting microorganisms (PGPM) are bacteria or fungi that benefit the growth of crops. They can do it direct- or indirectly, through different mechanisms such as improving nutrient uptake and assimilation, secreting phytohormones that promote growth and flowering or producing other secondary metabolites that help in plant defence mechanisms and accessory processes. The PGPM group can be subdivided into plant growth-promoting rhizobacteria (PGPR) and plant growth-promoting fungi (PGPF), the latter mainly consisting of endophytes and mycorrhizal-arbuscular fungi (Mishra et al. 2017; El-Saadony et al. 2022).

PGPR are those rhizobacteria that can enhance plant growth by a wide variety of mechanisms. They belong mainly to the phyla Proteobacteria and Actinobacteria. Within the Proteobacteria group, some of the genera of PGPR with the highest abundance are *Pseudomonas* and *Rhizobium*; within the Actinobacteria phylum, the more abundant genera are *Streptomyces* and *Nocardia*, among others (Debasis et al. 2019; Chakraborty and Ramteke 2023). Quite a few species of the genus *Bacillus* (phylum Firmicutes) also have growth-promoting functions in plants, especially *B. subtilis* (Sansinenea 2019; Hashem et al. 2019). Although there is a wide spectrum of PGPR with very interesting promoter activities, this chapter is focused on fungal metabolites and the reason why this group has not been researched in depth.

PGPF could be defined as fungi able to colonize the rhizosphere and roots of a plant in a reasonable time, to compete effectively with native soil microbiota, to promote crop growth, and to control pathogens and pests, while being harmless to humans (Hossain et al. 2017). The best PGPF genera known so far belong to the groups of arbuscular mycorrhizal fungi (AMF), such as *Glomus* spp. or *Rhizophagus* spp. (Emmanuel and Babalola 2020; Noceto et al. 2021), or endophytes, such as *Trichoderma* spp., *Penicillium* spp., or *Aspergillus* spp., among many others (Poveda et al. 2021).

These organisms also provide a variety of benefits to their host plants, such as an increase in crop yields by promoting seed germination and vigour and root morphogenesis, and also an improvement in photosynthesis and flowering processes (Hossain et al. 2017). These benefits can be explained by multiple mechanisms of action, some of them directly improving the nutrition and growth in plants, while others acting indirectly by helping to enhance non-essential but clearly beneficial processes for the plant, such as the activation of systemic resistance, or the production of antibiotics and other secondary metabolites (Baron and Rigobelo 2021).

According to the European legal framework, the recent REGULATION (EU) 2019/1009 concerning the European fertilizers market, which came into effect in July 2022, microorganisms have been included as components of fertilizer products for the first time. Initially, only 4 groups of microorganisms (mycorrhizal fungi, *Azotobacter* spp., *Rhizobium* spp., and *Azospirillum* spp.) are included in the regulation. However, it is expected that many other genera and species will be included in the regulation as research in this field progresses.

3 Secondary Fungal Metabolites as Plant Growth Promoters

Metabolites can be described as the products obtained from the processes involved in the primary or secondary metabolism of an organism, in this case a fungus. Primary fungal metabolites are molecules directly involved in the basic physiological processes of the fungus. They are typically formed as a result of energy metabolism and are essential for survival and proper development. This group includes crucial compounds, such as proteins, carbohydrates, nucleic acids, and lipids. Secondary fungal metabolites are the products of secondary metabolic pathways of fungi. They do not play a strictly decisive role in growth, development and reproduction as primary metabolites do, but have important subordinate functions, such as defence, competition, and interspecies interactions (Daley et al. 2017). These secondary metabolites usually present low molecular weight and high specificity at species level, and sometimes even at strain level (Keller et al. 2005).

There are numerous secondary fungal metabolites with very diverse chemical nature and functions, so there are multiple possibilities for their classification. The one followed in this chapter is based on their chemical structure, similarly to that described in other publications (Mousa and Raizada 2013; Daley et al. 2017). According to this classification, the main groups of metabolites with a role in plant growth promotion are: non-ribosomal peptides (siderophores), poliketides, terpenes, and alkaloids. In addition to these general groups of secondary metabolites, it has been considered appropriate to add the group of phytohormones produced by fungi, as some of them (specifically indole-3-acetic acid, gibberellins and cytokinins) are important plant growth enhancers (Fig. 1).

Many secondary metabolites derived from fungi have been extensively studied due to their potential applications in human pharmaceuticals. An illustrative example is the well-known cyclosporine, a non-ribosomal peptide synthesized by the fungus *Tolypocladium inflatum* that serves as immunosuppressive medication widely employed in organ transplantations (Tedesco and Haragsim 2012). In the realm of biological alternatives to synthetic chemical inputs in agriculture, a substantial part of these metabolites has garnered attention due to their potential in combating pests and pathogens (Liu and Li 2004). Nevertheless, a subset of these

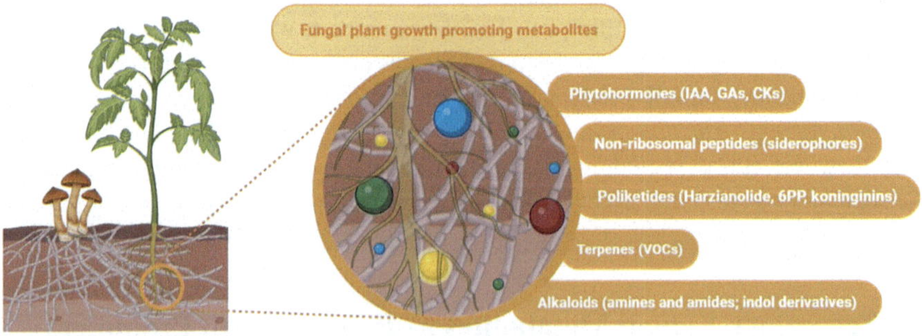

Fig. 1 Main fungal secondary metabolites with growth-promoting activity in plants. *IAA* indole-3-acetic acid, *GAs* gibberellins, *CKs* cytokinins, *6PP* 6-pentyl-α-pyrone, *VOCs* volatile organic compounds

compounds has also exhibited a direct capability to promote plant growth, a topic that will be thoroughly explored in this section. An overview of the secondary metabolite described hereafter can be observed in Table 1.

3.1 Phytohormones

Phytohormones are typically produced by plants and are essential for their metabolic processes. They are subdivided into different groups: auxins, cytokinins (CKs), gibberellins (GAs), abscisic acid (ABA), ethylene (ET), salicylic acid (SA), jasmonic acid (JA), brasinoesteroids (BRs), and strigolactones (SLs). Some of them can also be produced by other organisms (such as fungi) as products of their secondary metabolism and can act as efficient plant growth promoters. Among this group, CKs, GAs and the auxin indole-3-acetic acid (IAA) can be highlighted (Khan et al. 2023).

IAA is an important phytohormone with capacity to control plant development in both beneficial and deleterious ways. It plays a central role in cell division, tropism control, elongation, fruit development and senescence. IAA is also in charge of initiating roots, leaves and flowers (Jordán and Casaretto 2006; Duca et al. 2014). This phytohormone is produced not only by plants themselves, but also by numerous bacteria and fungi. For example, some ectomycorrhizal fungi, such as *Boletus* spp., *Coprinus* spp. or *Amanita* spp., have demonstrated their ability to produce IAA (Ulrich 1960). There are also studies showing that some AMF belonging to Glomeraceae, in association with bacteria, can produce this phytohormone (Turrini et al. 2018; Giovannini et al. 2020; Nacoon et al. 2020). Therefore, it can be considered in many cases as a fungal plant growth promoting metabolite (Fu et al. 2015).

The majority of research conducted thus far on the IAA production by fungi has been focused on fungal endophytes. Within this group, many species belonging to the genera *Trichoderma*, *Penicillium*, *Alternaria and Aspergillus* have shown their capacity for IAA production. Different strains of *T. harzianum* or *T. viride* have demonstrated their competence in producing IAA directly (Saber et al. 2017; Bader et al. 2020) or in enhancing IAA production in the host plant (Zhang et al. 2013; Kumar et al. 2017). Waqas et al. (2012) confirmed the production of IAA by *Penicillium* sp. Years later, the production of the phytohormone by *P. sclerotiorum* species (Nath et al. 2015), *P. chrysogenum* and *P. crustosum* (Hassan 2017) and *P. ruqueforti* (Ikram et al. 2018) was also demonstrated. In the same way, different strains of *Alternaria alternata* have been shown to be able to naturally produce IAA (Khan et al. 2015b; Qiang et al. 2019). Within the *Aspergillus* group, numerous species such as *A. awamori* (Mehmood et al. 2019) or *A. niger* (Nath et al. 2015) have also been verified as IAA producers. Some *Fusarium* species, such as *F. oxysporum* (Poveda et al. 2022) and *F. avenaceum* (García-Latorre et al. 2023), have shown the

Table 1 Secondary fungal metabolites with growth-promoting activity in plants

Group of secondary fungal metabolites		Fungal group	Fungal species	Host and/or experimental plant	Biological activity	References
Phytohormones	Indole-3-acetic acid (IAA)	Ectomycorrhizal	*Boletus* spp.	Not specified	Direct IAA production	Ulrich (1960)
			Coprinus spp.	Not specified	Direct IAA production	Ulrich (1960)
			Amanita spp.	Not specified	Direct IAA production	Ulrich (1960)
		AMF	Glomeraceae family	Not specified	Direct IAA production	Turrini et al. (2018); Giovannini et al. (2020)
			Glomus multisubtensum strain KKU-UD-JA-DBr	Isolated from *Helianthus tuberosus*	Direct IAA production	Nacoon et al. (2020)
			Rhizophagus intraradices strain KKU-Wh	Isolated from *Helianthus tuberosus*	Direct IAA production	
		Endophytes	*Trichoderma harzianum* strain WKY1	Not specified	Direct IAA production Anthracnose disease control in sorghum	Saber et al. (2017)
			T. harzianum strains FCCT16, FCCT58, FCCT199–2 and FCCT-363-2	Not specified	Direct IAA production P solubilization	Bader et al. (2020)
			T. harzianum strain T-E5	Tested on cucumber plants	Enhance IAA production in plant host	Zhang et al. (2013)
			T. viride strain VKF3	Isolated from mangrove species	Direct IAA production	Kumar et al. (2017)
			Penicillium spp. strain LWL3	Tested on rice plants	Direct IAA production Direct GAs production	Waqas et al. (2012)
			P. sclerotiorum	Isolated from *Camellia sinensis*	Direct IAA production Mineral solubilization	Nath et al. (2015)

Phytohormones	Indole-3-acetic acid (IAA)	Endophytes	*P. chrysogenum*	Isolated from *Teucrium polium*	Direct IAA production Antimicrobial P solubilization	Hassan (2017)
			P. crustosum	Isolated from *Teucrium polium*	Direct IAA production Antimicrobial P solubilization	
			P. ruqueforti	Not specified	Direct IAA production Improvement of nutrient uptake Block heavy metals	Ikram et al. (2018)
			Phoma glomerata strain LWL2	Tested on rice plants	Direct IAA production	Waqas et al. (2012)
			Aspergillus awamori	Isolated from *Withenia somnifera* Tested on maize plants	Direct IAA production	Mehmood et al. (2019)
			A. niger	Isolated from *Camellia sinensis*	Direct IAA production K solubilization	Nath et al. (2015)
			Alternaria alternata strain LQ1230	Isolated from *Elymus dahuricus* Tested on wheat plants	Direct IAA production	Qiang et al. (2019)
			A. alternata strain RSF-6 L	Isolated from *Solanum nigrum* Tested on rice plants	Direct IAA production	Khan et al. (2015b)
			Fusarium tricinctum strain RSF-4 L	Isolated from *Solanum nigrum* Tested on rice plants	Direct IAA production	

(continued)

Table 1 (continued)

Group of secondary fungal metabolites		Fungal group	Fungal species	Host and/or experimental plant	Biological activity	References
			F. oxysporum strain H890	Isolated from kale roots. Tested on *Brassica napus, B. rapa, B. oleracea* var. *acephala, B. juncea, B. nigra,* and *B. carinata.*	Direct IAA production	Poveda et al. (2022)
			Acrocalymma vagum strain H22		Direct IAA production	
			Setophoma terrestris strain H64		Direct IAA production P solubilization	
			Pyrenophora gallaeciana strain H441		Direct IAA production	
			Sarocladium terricola Strain E025	Isolated from herbaceous plants in Spanish dehesas	Direct IAA production	García-Latorre et al. (2023)
	Cytokinins (CKs)	Ectomycorrhizal	*Rhizopogon ochraceorubens*	Not specified	Direct CKs production	Crafts and Miller (1974)
		Endophyte	*Claviceps purpurea*	Isolated and tested on rye plants	Direct CKs production	Hinsch et al. (2015)
	Gibberellins (GAs)	Endophytes	*F. moniliforme*	Not specified	Direct GAs production	Mitter et al. (2002)
			F. proliferatum strain strain BRL1	Isolated from *Oxalis corniculata*	Direct GAs production	Bilal et al. (2018)
			F. solani	Isolated from soil	Direct GAs production	Bhalla et al. (2010)
			Penicillium citrinum strain KACC43900	Isolated from *Ixeris repenes* roots Tested on rice and *Atriplex gmelinii* seedlings	Direct GAs production	Khan et al. (2008)

			P. janthinellum strain LK5	Isolated from *Solanum lycopersicum* roots Tested on rice and *Solanum sitiens* plants	Direct GAs production	Khan et al. (2013)
			P. chrysogerum strain F1	Isolated from *Camellia sinensis*	Direct GAs production	Nath et al. (2015)
			Phoma glomerata strain LWL2	Tested on rice plants	Direct GAs production Improvement of nutrient uptake Stress mitigation	Waqas et al. (2012)
			P. herbarum strain TK-2-4	Isolated from soybean plants Tested on rice plants	Direct GAs production	Hamayun et al. (2009a, b)
			Aspergillus clavatus strain Y2H0002	Isolated from *Nymphoides peltate* Tested on rice plants	Direct GAs production	You et al. (2015)
			A. niger strain CSR3	Isolated from *Cannabis sativa* Tested on maize plants	Direct GAs production	Lubna Asaf et al. (2018)
			Cladosporium sphaerospermum strain DK-1-1	Isolated and tested on soybean plants	Direct GAs production	Hamayun et al. (2009a, b)
			Paecilomyces formosus strain LHL10	Isolated and tested on cucumber plants	Direct GAs production	Khan et al. (2012)

(continued)

Table 1 (continued)

Group of secondary fungal metabolites		Fungal group	Fungal species	Host and/or experimental plant	Biological activity	References
Non-ribosomal peptides (NRP)	Siderophores	Endophytes	*Aspergillus fumigatus*	Still untested in plants	Siderophore biosynthesis	Schrettl et al. (2007)
			A. fumigatus	Still untested in plants	Siderophore biosynthesis	Blatzer et al. (2011)
			A. fumigatus	Still untested in plants	Siderophore biosynthesis	Haas (2014)
			A. niger strain 581PDA1	Isolated from wheat plants	Siderophore biosynthesis	Ripa et al. (2019)
			Beauveria brongniartii strain SP9	Isolated from soil tested on pepper plants	Siderophore biosynthesis	Toscano-Verduzco et al. (2020)
			B. bassiana strains RGM-557 y RGM-731	Tested on tomato plants	Siderophore biosynthesis P solubilization	Barra-Bucarei et al. (2020)
			B. bassiana strain ARSEF 2860	Not specified	Siderophore biosynthesis	Krasnoff et al. (2020)
			Fusarium oxysporum strain H890	Isolated from kale roots. Tested on six *Brassica spp.*	Siderophore biosynthesis	Poveda et al. (2022)
			Pyrenophora gallaeciana strain H441			
			Fusarium avenaceum strain E060	Isolated from herbaceous plants in Spanish dehesas	Siderophore biosynthesis	García-Latorre et al. (2023)
			Sarocladium terricola strain E025			

Poliketides	Harzianolide	Endophytes	*Trichoderma harzianum* strains T22, T39 and A6; and T. atroviride strain P1	Tested on pea seedlings	Plant growth promotion Induction of systemic defence responses	Vinale et al. (2008)
			T. harzianum strain SQR-T037	Tested on tomato seedlings	Plant growth promotion Induction of systemic defence responses	Cai et al. (2013)
	6-pentyl-α-pyrone (6PP)	Endophytes	*T. harzianum* strains T22, T39 and A6; and *T. atroviride* strain P1	Tested on pea, tomato, and canola seedlings	Plant growth promotion Induction of systemic defence responses	Vinale et al. (2008)
			T. harzianum strain M10 and *T. atroviride* strain P1. *T. harzianum* strain T22 for field tests.	Tested on grapevine plants	Plant growth promotion Increase in crop yield	Pascale et al. (2017)
			T. harzianum strain T22	Tested on tomato plants	Increase in crop yield Higher content of antioxidant compounds	Carillo et al. (2020)
Poliketides	Koninginins	Endophytes	*T. koningii*	Tested on tomato seeds and seedlings	Improvement of germination and initial growth Antibiotic activity	Vinale et al. (2007)
Terpenes	VOCs	Endophytes	*Cladosporium halotolerans* strains NGPF1	Tested on *Nicotiana benthamiana* plants	Increase in growth, root development and production of phytohormones	Jiang et al. (2021)
		AMF	*Funneliformis mosseae* (together with rhizobacterium *Ensifer meliloti*)	Tested on grapevine plants	Induction of the production of beneficial VOCs Improvement of plant defence	Velásquez et al. (2020)

(continued)

Table 1 (continued)

Group of secondary fungal metabolites			Fungal group	Fungal species	Host and/or experimental plant	Biological activity	References
Alkaloids	Amines and amides	Peramine	Endophytes	*Acremonium lolii*	Tested on ryegrass plants	Insect deterrent activity (indirect plant growth promotion)	Rowan 1993
				Neotyphodium spp.	Isolated and tested on Arizona fescue plants	Plant fitness enhancements Antiherbivore activity	Faeth et al. (2002); Bush et al. (1997)
		Ergot alkaloids		*Claviceps* spp. and *Neotyphodium* spp.	Not specified	Toxic activity against insects, nematodes, and mammalian herbivores (indirect plant growth promotion)	Mousa and Raizadal. (2013)
	Indole derivatives	Lolines	Endophytes	*Neotyphodium* spp. (and its sexual form *Epichloë* spp.)	Isolated from Poaceae family	Enhancement of drought tolerance and nutrients uptake Insect deterrent activity	Schardl et al. (2004)

ability to produce this phytohormone as well. Outside these well-known genera there are other fungal species with the same capability, for example *Acrocalymma vagum, Setophoma terrestris,* the recently discovered *Pyrenophora gallaeciana* (Poveda et al. 2022) or the species *Sarocladium terricola* (García-Latorre et al. 2023) (Table 1).

Cytokinins (CKs) are another type of phytohormones (chemically defined as derivatives of the adenine base) that promote cell division, shoot initiation and delayed senescence, among other effects (Jordán and Casaretto 2006). More than 200 different natural and synthetic CKs are known today, most of them produced in the meristem of the roots and transported to other parts of the plant through the vascular system (Greene 1980). The production of this type of hormones by fungi has not been studied as deeply as IAA or GAs. However, there is some research showing that several ectomycorrhizal fungivarious fungal species (such as *Rhizopogon* sp.) in fact have the ability to produce CKs. Examples include the ectomycorrhizal fungus *Rhizopogon ochraceorubens* (Crafts and Miller 1974). Another interesting study corroborated that the parasitic fungus *Claviceps purpurea*, a species amply known for, in addition to its well-known capacity to produce numerous alkaloids with psychoactive effects, production of phytohormones (Hinsch et al. 2015).

GAs are tetracyclic diterpenic acids with a key role in plant growth and in multiple developmental processes, including seed germination, longitudinal stem growth, floral transition, and fruit patterning (Martínez et al. 2016). Although more than 100 types of gibberellins are currently known, not all of them have the capacity to regulate plant growth. The most active and most widely used are GA_1, GA_3, GA_4 and GA_7 (Gao et al. 2017).

GAs were discovered in the 1920s in rice fields in Japan, which were affected by the fungal pathogen *Gibberella fujikuroi* (=*Fusarium fujikuroi*). One of the most visible symptoms of the disease was an exaggerated growth of the stems, which caused the plants to eventually buckle. The research team in charge discovered the pathogenic fungus to produce the so-called GAs, substances responsible for the abnormal growth of rice plants. Subsequently, these compounds were isolated from plant tissue (Arrondo-Odriozola 1987) and became more important in the 1960s for their role as plant-growth promoters during the aforementioned Green Revolution in the United States (Dash and Rai 2022). Like IAA, GAs can also be classified as fungal plant growth promoting metabolites, since there are numerous fungi that produce them.

In addition to *F. fujikuroi*, there are other species within this genus that have demonstrated their ability to produce GAs, such as *F. moniliforme* (Mitter et al. 2002), *F. proliferatum* (Bilal et al. 2018) or *F. solani* (Bhalla et al. 2010). Numerous species within the genus *Penillium* possess this capability too. Khan et al. (2008) observed the production of GAs by *P. citrinum*. Years later, the production was also found in *P. janthinellum* (Khan et al. 2013) and in *P. chrysogenum* (Nath et al. 2015). The genus *Phoma* is another GA-producing group, especially *P. glomerata* (Waqas et al. 2012) and *P. herbarum* (Hamayun et al. 2009a, b). Many other genera of endophytic fungi, such as *Aspergillus, Cladosporium* or *Paecilomyces*, have also been found to naturally generate this type of phytohormones (You et al. 2015; Khan et al. 2015; Poveda et al. 2021) (Table 1).

3.2 Non-ribosomal Peptides (NRP)

NRP are an important group of fungal secondary metabolites, showing notable structural diversity and constituting a broad category that encompasses some of the most potent antibiotics and other significant pharmaceutical compounds, spanning from penicillin to the abovementioned cyclosporine. These peptides are synthesized independently of the ribosome, in a nucleic acids-independent way by a group of multimodular megaenzymes called non-ribosomal peptide synthetases (Marahiel 2009). Although there are numerous non-ribosomal peptides produced by fungi, siderophores stand out for their role in promoting plant growth (Bills et al. 2014; Řezanka et al. 2019).

According to the generally accepted definition, siderophores are NRP produced by microorganisms for scavenging iron from environment (iron-quelators). They have been defined as low molecular weight compounds possessing a high affinity for iron (III), whose biosynthesis is regulated by iron levels, allowing an increased iron supply to cells (Hider 1984).

Iron is an essential micronutrient for plants. Despite its prevalence in the Earth's crust, iron in soils is found in many cases as iron (III), almost insoluble at neutral and alkaline pH levels. Consequently, it remains inaccessible to plants in such conditions. This is why some organisms produce siderophores, which can form complexes with iron, enhancing its solubility. While plants have the capability to produce siderophores on their own, the presence of siderophore-producing fungi in the soil increases the availability of this micronutrient, thereby fostering improved crop growth (Řezanka et al. 2019). Siderophores are also necessary for the solubilization of other micronutrients such as copper [often found in its Cu (II) form] or molybdenum (Hider 1984).

Depending on the oxygen ligands for Fe (III) coordination, microbial siderophores are typically classified as catecholates, hydroxamates, and a´-carboxylates (Winkelmann 2002; Saha et al. 2016). The majority of fungal siderophores belong to the hydroxamate-class. Within this class, they can be grouped into four structural families: fusarinines, coprogens, ferrichromes and rhodotorulic acid (Haas 2014). One of the most notorious fungal genera for its ability to naturally synthesize siderophores is *Aspergillus*. In fact, numerous studies have corroborated the production of different fumarins and ferrichromes by *A. fumigatus* (Schrettl et al. 2007; Blatzer et al. 2011; Haas 2014). There is also evidence of siderophore production by the strain *A. niger* 581PDA1 isolated from wheat plants (Ripa et al. 2019). Within the genus *Beauveria*, siderophore production has also been demonstrated at least by *B. brongniartii* (Toscano-Verduzco et al. 2020) and several strains of *B. bassiana* (Barra-Bucarei et al. 2020; Krasnoff et al. 2020). Beyond *Aspergillus* and *Beauveria* groups, there are other species capable of producing siderophores. Examples include *Fusarium oxysporum* and *Pyrenophora gallaeciana* isolated from kale roots, which have been confirmed to produce these molecules and to function as growth promoters in kale and various other *Brassica* species (Poveda et al. 2022). Some strains of *Fusarium avenaceum* and *Sarocladium terricola* isolated from herbaceous plants in Spanish dehesas posess the same capability (García-Latorre et al. 2023) (Table 1).

3.3 Poliketides

Polyketides make up a highly diverse group of natural products with structural carbon skeletons which are assembled from simple acyl building blocks, primarily sourced from one of the simplest building blocks available in nature: acetic acid. Many of these compounds or derivatives thereof have become important therapeutics for clinical use, such as statins, widely used in patients with hypercholesterolemia (Manzoni and Rollini 2002; Barrios-González and Miranda 2010). In contrast, certain polyketides have gained notoriety as food-spoiling toxins or virulence factors, such as aflatoxins (Hertweck 2009).

Regarding their influence on plant growth, there are numerous polyketides produced by *Trichoderma* spp. that have demonstrated beneficial properties in this sense. Multiple species of *Trichoderma* have the ability to produce different types of PGP polyketides, such as harzianolide, pyrones like 6-pentyl-α-pyrone (6PP) and koninginins (A, B, C, D, E and G) by *T. harzianum*, *T. viride* and *T. koningii, respectively* (Cutler et al. 1989; Cutler et al. 1991; Parker et al. 1995a, b; Cardoza et al. 2005). These three species of *Trichoderma* are precisely the most often reported for producing these metabolites (Cardoza et al. 2005; Keswani et al. 2014; Vinale and Sivasithamparam 2020)

Harzianolide possess auxin-like activities in plants, inducing positive effects on stem growth and activation of defence systemic responses (Vinale et al. 2008; Cai et al. 2013). The polyketide 6PP, in addition to these effects (Vinale et al. 2008), improves yields and fruit quality in some crops (Pascale et al. 2017; Carillo et al. 2020). Concerning koninginins, while their most studied aspect is the antifungal activity, there are also studies showcasing their significance in promoting seed germination and early seedling growth in certain crops (Vinale et al. 2007) (Table 1).

3.4 Non-phytohormone Terpenes

Chemically, some of the most important phytohormones (such as the aforementioned ABA and GAs) belong to the terpenes group. However, in this section it is intended to collect those other terpenes which are not phytohormones, but which are also relevant in plant growth promotion.

Terpenes or terpenoids are the most abundant and most widely occurring secondary metabolites in nature. Most of them are produced by plants, but also filamentous fungi produce a wide variety of these compounds, which can be used for multiple applications (Cardoza et al. 2005). Terpenes are basically subdivided into six main groups, listed here in increasing order of number of isoprene units: monoterpenes, sesquiterpenes, diterpenes, sesterterpenes, triterpenes, and carotenoids (Hanson 2003). Monoterpenes are mainly aromatic compounds produced by plants, such as geraniol, menthol or camphor, among many others. Sesquiterpenes are also aromatic substances found in essential oils derived from plants. Within this group caryophyllene (from cloves), humulene (from hops) or cedrene (from cedar) are found. There are also some sesquiterpenes produced by fungi, such as botrylanes (*Botrytis cinerea*) and trichothecenes (*Fusarium* spp.). These are in most cases toxic, either to plants or animals. Regarding diterpenes, most of them are wood resin products, such us abietic acid (from *Abies* spp.) or primaric acid (from *Pinus* spp.). However, there are other diterpenes which can be produced not only by plants, but also by fungi (Hanson 2003).

Sesterterpenes are considered a rare and relatively small group within terpenes, but their sources are widespread, having been isolated from terrestrial fungi, lichens, higher plants, insects, and various marine organisms, especially sponges. They exhibit diverse biological properties such as anti-inflammatory, cytotoxic, antifeedant, platelet aggregation, and antimicrobial effects (Liu et al. 2007). Therefore, some of them can indirectly affect plant growth promotion, as in the case of sesterterpene aleurodiscal (produced by the fungus *Aleurodiscus mirabilis*), with high potential against several plant pathogenic fungi (Lauer et al. 1989). Triterpenes were first isolated from fish, and later found in mammals and plant oils. More recently they have also been found in the fruiting bodies of *Polyporus* spp. or *Ganoderma* spp. The well-known ginseng fungus (*Panax ginseng*) also produces triterpene glycosides. Derivatives of the triterpenes are steroids. One noteworthy compound among them is ergosterol, which is synthesized by a wide range of fungal species and has also indirect plant growth promoting effects, mainly on the expression of defence-related genes in different crops (Lochman and Mikes 2006; Rossard et al. 2010). Carotenoids, pigments found in many plants, are also precursors of vitamin A, and have usually important antioxidant properties (Hanson 2003). Moreover, several studies indicate that carotenoids play an important role in plant responses to different types of biotic and abiotic stresses, such as water stress, salt stress, or stress caused by competition; these facts would ultimately have an indirect effect on plant growth promotion (Uarrota et al. 2018).

Most terpenes are also considered volatile organic compounds (VOCs), either by themselves or in combination with other molecules derived from fatty acids or aromatic compounds of different chemical nature. Little attention has been paid to the direct effects of distinct fungal VOCs on plant growth promotion. However, there are some interesting studies in this regard, such as the work of Jiang et al. (2021), which tested the impact of VOCs produced by the fungus *Cladosporium halotolerans* NGPF1 on *Nicotiana benthamiana* plants growth, resulting in increased growth, root development and production of phytohormones such as auxins and gibberellins; or the study of Velásquez et al. (2020), which showed that the combined inoculation of the arbuscular mycorrhizal fungus *Funneliformis mosseae* and the rhizobacterium *Ensifer meliloti* in *Vitis vinifera* plants induced the production of numerous beneficial VOCs and, interestingly, also produced an increase in monoterpene alcohols, related to plant defence (Table 1).

3.5 *Alkaloids*

Alkaloids are defined as basic-pH (alkaline) compounds synthesized by living organisms containing one or more heterocyclic nitrogen atoms, derived from amino acids (with some exceptions). Alkaloids constitute a very large and diverse group of secondary metabolites, with more than 12,000 substances isolated. Alkaloid function in plants is mostly linked to defence mechanisms, being their toxicity an effective mechanism to inhibit the action of predators, like herbivores or insects. This fact obviously implies an indirect improvement in plants growth (Agostini-Costa et al. 2012).

Alkaloids are subdivided into two major groups: amines and amides, and indole – derivatives. There are also other alkaloids, such as pyridines or quinazolines, which can eventually have important applications. Despite the fact that most of alkaloids are studied and used for their pharmacological properties in humans, some of them have special relevance in agriculture, for their insecticidal, plant-growth promoting, or the aforementioned anti-herbivory properties (Zhang et al. 2012; Mousa and Raizada 2013).

Included in amines and amides group, peramine and ergot alkaloids have an important insect deterrent activity (Bush et al. 1997) or are toxic for insects, nematodes, and mammalian herbivores (Mousa and Raizada 2013), respectively. While the peramine production has been demonstrated by the fungi *Acremonium lolii* (Rowan 1993) and *Neotyphodium* spp. (Faeth et al. 2002), ergot alkaloids, such as the well-know ergovaline, are produced by fungi of Clavicipitaceae family, commonly by *Claviceps* spp. and *Neotyphodium* spp. (Bush et al. 1997; Zhang et al. 2012; Mousa and Raizada 2013).

Lolines, indole-derivatives also produced by *Neotyphodium* spp. and its sexual form *Epichloë* spp. (Mousa and Raizada 2013), have also been demonstrated to repel insects and herbivorous mammals, although they generally have a milder effect than ergot alkaloids (Bush et al. 1997). Lolines seem also to be involved directly in plants growth-regulatory activities, fostering enhancements of drought tolerance and nutrient status, and improved growth particularly of the root in several plant species (Schardl et al. 2004) (Table 1).

4 Biotechnological Production of Some PGP Metabolites

Despite the large amount of research studies on PGPM, only four groups are currently authorized for marketing and use as biofertilizers in the Eurepean Union. Of these four groups, only one concerns PGPF, specifically mycorrhizal fungi (REGULATION (EU) 2019/1009). However, it is true that other fungal species registered and authorized as phytosanitary products may also have plant growth-promoting functions, as is the case with several species of *Trichoderma* (IMPLEMENTING REGULATION (EU) 540/2011). This lack of transferability from the scientific to the applied field may be due to the scarcity of experiments conducted under field conditions, among other reasons. This scarcity of information on the functions of microorganisms under real conditions leads to a reticence of companies to start producing them (Rodrigo et al. 2021). It is also worth noting that, as can be seen from both Regulations, none of the few current EU-authorized fungus-based products are metabolites, but rather different formulations of the living organism.

The ideal commercial formulations for PGPF (commercially known as fungal biofertilizers or biostimulants) should meet the following criteria: (i) stabilize the microorganism and extend its shelf life; (ii) protect the active ingredient from external environmental factors at the target site of action, and (iii) facilitate the application of the biofertilizer under field conditions (Burges 2012). Biofertilizers based on living organisms can be marketed under different formulations, which can be generally grouped into solid, liquid or gel formulations. Solid formulations are the most commonly used, as they are much less susceptible to contamination by other organisms than liquid formulations. Among the different techniques applied to obtain solid formulations are included the use of solid carriers (such as vermiculite or biochar) surrounding the fungal propagules or drying techniques (such as freeze-drying among others) (Chaudhary and Shukla 2020; Martinez et al. 2023).

Regarding the fungal-produced metabolites, there are numerous of them that have been produced for decades, mainly for therapeutic or alimentary purposes. A good example is the well-known antibiotics, produced by fermentation processes in bioreactors with different fungal species (Devi et al. 2020). However, mass production of fungal metabolites for agricultural purposes is not so widespread, and still in the research and development stages in most of cases. An exception to this pattern are phytohormones. When used at low concentrations, they are probably the most effective plant-growth promoting metabolites. On the contrary, it is important to clarify that, if they are used at high doses, a herbicidal effect is produced (Auxin Mimics herbicides) (HRAC 2023). They can be obtained by a variety of methods, including extraction from higher plants or algae, chemical synthesis, and microbial fermentation. In plants, hormone concentrations are exceptionally low, which implies the need for a huge amount of plant material to produce a minimum quantity of hormone. Conversely, chemical synthesis poses substantial challenges and drawbacks due to the intricate chemical structures of phytohormones, which complicate operations, resulting in elevated expenses and reduced purity. As a result, there is a growing interest in exploring microbial production (using PGPM species), a novel alternative for phytohormone production. As an example of a recent study with an industrial focus in this regard, the work of Pandya et al. (2023) can be highlighted. In this study, the mass production of gibberellins (specifically GA_3) by different fermentation processes from *Fusarium* spp. is studied and optimized. Methods based on microbial production effectively overcome some of the aforementioned limitations, producing final products with good bioactivity and purity, while significantly reducing production costs (Shi et al. 2017; Tsavkelova et al. 2006).

Concerning other PGP metabolites, in recent years there have been some attempts by different research groups to develop and patent formulations of these types of products, either as biofertilizers or as organic pesticides. A good example is the study of Lidoy et al. (2023), where some flavonoids (phenolic compounds produced by fungi) were used to promote spore germination and symbiosis establishment of the AMF *Rhizophagus irregularis*, considered a PGPF. The inclusion of theses flavonoids in the formulation of AMF-based products improved the efficiency of the commercial inoculants. Another example is the patented of several extracts obtained from two strains of endophytic fungi (*Alternaria alternata* HTF58 and *Fusarium acuminatum* HRO8) as possible components for antifungal phytosanitary products for agricultural use (Cosoveanu and Cabrera-Pérez 2019).

5 Conclusion

Today's global agriculture urgently needs a solution that provides food for an ever-growing human population, while safeguarding environmental sustainability and human health. Research in the field of plant growth-promoting microorganisms (such as rhizobacteria, and mycorrhizal and endophytic fungi) represents a crucial step in this direction. These organisms offer extraordinary potential as alternatives to traditional chemical fertilisers, which are not only harmful to the nature and people's health, but also increasingly expensive and scarce. In this sense, the use of not only the microorganisms *per se*, but also the metabolites they produce can become a powerful biological tool to increase crop yields. Some of the most important fungal plant growth promoting metabolites are the phytohormones IAA, CKs and GAs; the siderophores (non-ribosomal peptides for scavenging the micronutrient iron from the environment); some poliketides such as harzianolide, 6PP-pyrone or koninginins; various VOCs and some alkaloids which can be considered as indirect plant growth promoters, although they mainly support the plant's defence system. By better understanding the metabolic pathways of production of these metabolites by fungi, the implementation of an industrial production system of such compounds will be possible and could finally lead to a realistic commercialization and application, thus offering farmers feasible substitutes for conventional chemical inputs.

Acknowledgments T. S-G. thanks the University of Valladolid and Banco Santander for financing her Predoctoral Contract (CONTPR-2022-426, Call 2022), which made it possible to carry out the exhaustive bibliographic review and the composition of this publication.

References

Agostini-Costa TDS, Vieira RF, Bizzo HR, Silveira D, Gimenes MA (2012) Secondary metabolites. J Chromatogr A 1:149–156

Arrondo-Odriozola E (1987) Giberelina. Rev Catalana Micol 11:23–27

Bader AN, Salerno GL, Covacevich F, Consolo VF (2020) Native *Trichoderma harzianum* strains from Argentina produce indole-3 acetic acid and phosphorus solubilization, promote growth and control wilt disease on tomato (*Solanum lycopersicum* L.). J King Saud Univ Sci 32(1):867–873. https://doi.org/10.1016/j.jksus.2019.04.002

Baron NC, Rigobelo EC (2021) Endophytic fungi: a tool for plant growth promotion and sustainable agriculture. Mycology 13(1):39–55. https://doi.org/10.1080/21501203.2021.1945699

Barra-Bucarei L, González MG, Iglesias AF, Aguayo GS, Peñalosa MG, Vera PV (2020) *Beauveria bassiana* multifunction as an endophyte: growth promotion and biologic control of *Trialeurodes vaporariorum*, (Westwood) (Hemiptera: Aleyrodidae) in tomato. Insects 11(9):591. https://doi.org/10.3390/insects11090591

Barrios-González J, Miranda RU (2010) Biotechnological production and applications of statins. Appl Microbiol Biotechnol 85:869–883. https://doi.org/10.1007/s00253-009-2239-6

Bhalla K, Singh SB, Agarwal R (2010) Quantitative determination of gibberellins by high performance liquid chromatography from various gibberellins producing *Fusarium* strains. Environ Monit Assess 167:515–520. https://doi.org/10.1007/s10661-009-1068-5

Bilal L, Asaf S, Hamayun M, Gul H, Iqbal A, Ullah I, Lee IJ, Hussain A (2018) Plant growth promoting endophytic fungi *Aspergillus fumigatus* TS1 and *Fusarium proliferatum* BRL1 produce gibberellins and regulates plant endogenous hormones. Symbiosis 76:117–127. https://doi.org/10.1007/s13199-018-0545-4

Bills G, Li Y, Chen L, Yue Q, Niu XM, An Z (2014) New insights into the echinocandins and other fungal non-ribosomal peptides and peptaibiotics. Nat Prod Rep 31(10):1348–1375. https://doi.org/10.1039/C4NP00046C

Blatzer M, Schrettl M, Sarg B, Lindner HH, Pfaller K, Haas H (2011) SidL, an *Aspergillus fumigatus* transacetylase involved in biosynthesis of the siderophores ferricrocin and hydroxyferricrocin. Appl Environ Microbiol 77(14):4959–4966. https://doi.org/10.1128/AEM.00182-11

Burges HD (2012) Formulation of microbial biopesticides: beneficial microorganisms, nematodes and seed treatments. Springer. https://doi.org/10.1007/978-94-011-4926-6

Bush LP, Wilkinson HH, Schardl CL (1997) Bioprotective alkaloids of grass-fungal endophyte symbioses. Plant Physiol 114(1):1. https://doi.org/10.1104/pp.114.1.1

Cai F, Yu G, Wang P, Wei Z, Fu L, Shen Q, Chen W (2013) Harzianolide, a novel plant growth regulator and systemic resistance elicitor from *Trichoderma harzianum*. Plant Physiol Biochem 73:106–113. https://doi.org/10.1016/j.plaphy.2013.08.011

Cardoza RE, Hermosa MR, Vizcaíno JA, Sanz L, Monte E, Gutiérrez S (2005) Secondary metabolites produced by *Trichoderma* and their importance in the biocontrol process. Microorganism for industrial enzymes and biocontrol, pp 1–22. ISBN: 81-308-0040-3

Carillo P, Woo SL, Comite E, El-Nakhel C, Rouphael Y, Fusco GM, Borzacchiello A, Lanzuise S, Vinale F (2020) Application of *Trichoderma harzianum*, 6-pentyl-α-pyrone and plant biopolymer formulations modulate plant metabolism and fruit quality of plum tomatoes. Plan Theory 9(6):771. https://doi.org/10.3390/plants9060771

Chakraborty U, Ramteke P (2023) Plant probiotics: recent and future prospects. Front Plant Sci 14. https://doi.org/10.3389/fpls.2023.1254184

Chaudhary T, Shukla P (2020) Commercial bioinoculant development: techniques and challenges. In: Shukla P (ed) Microbial enzymes and biotechniques. Springer, Singapore. https://doi.org/10.1007/978-981-15-6895-4_4

Cosoveanu A, Cabrera-Pérez R (2019) Hongos endófitos HTF58 *Alternaria alternata* y HRO8 *Fusarium acuminatum* de *Artemisa thuscula* y *Artemisa austriaca* como antifúngicos de uso agrícola. Spanish Patent No. ES2696982B2. Spanish Patent and Trademark Office. https://patents.google.com/patent/ES2696982B2/es

Crafts CB, Miller CO (1974) Detection and identification of cytokinins produced by mycorrhizal fungi. Plant Physiol 54(4):586–588. https://doi.org/10.1104/pp.54.4.586

Cutler HG, Himmelsbach DS, Arrendale RF, Cole PD, Cox RH (1989) Koninginin a: a novel plant growth regulator from *Trichoderma koningii*. Agric Biol Chem 53(10):2605–2611. https://doi.org/10.1080/00021369.1989.10869746

Cutler HG, Himmelsbach DS, Yagen B, Arrendale RF, Jacyno JM, Cole PD, Cox RH (1991) Koninginin B: a biologically active congener of koninginin A from *Trichoderma koningii*. J Agric Food Chem 39(5):977–980. https://doi.org/10.1021/jf00005a035

Daley DK, Brown KJ, Badal S (2017) Fungal metabolites. Pharmacogn J 20:413–421. https://doi.org/10.1016/B978-0-12-802104-0.00020-2

Dash PK, Rai R (2022) Green revolution to grain revolution: florigen in the frontiers. J Biotechnol 343:38–46. https://doi.org/10.1016/j.jbiotec.2021.10.002

Debasis M, Snežana A, Panneerselvam P, Manisha SA, Tanja V, Ganeshamurthy AN, Devvret V, Radha TK, Divya J (2019) Plant growth promoting microorganisms (PGPMs) helping in sustainable agriculture: current perspective. Int J Agric Vet Sci 7(2):50–74

Devi R, Kaur T, Guleria G, Rana KL, Kour D, Yadav N, Yadav AN, Saxena AK (2020) Chapter 9 – Fungal secondary metabolites and their biotechnological applications for human health. In: Rastegari AA, Yadav AN, Yadav N (eds) New and future developments in microbial biotechnology and bioengineering. https://doi.org/10.1016/B978-0-12-820528-0.00010-7

Duca D, Lorv J, Patten CL, Rosa D, Glick BR (2014) Indole-3-acetic acid in plant–microbe interactions. Antonie Van Leeuwenhoek 106:85–125. https://doi.org/10.1007/s10482-013-0095-y

El-Saadony MT, Saad AM, Soliman SM, Salem HM, Ahmed AI, Mahmood M, El-Tahan AM, Ebrahim AAM, El-Mageed TAA, Negm SH, Selim S, Babalghith AO, Elrys AS, El-Tarabily KA, Abu Qamar SF (2022) Plant growth-promoting microorganisms as biocontrol agents of plant diseases: mechanisms, challenges and future perspectives. Front Plant Sci 13:923880. https://doi.org/10.3389/fpls.2022.923880

Emmanuel OC, Babalola OO (2020) Productivity and quality of horticultural crops through co-inoculation of arbuscular mycorrhizal fungi and plant growth promoting bacteria. Microbiol Res 239:126569. https://doi.org/10.1016/j.micres.2020.126569

European Commission (2019) The European Green Deal. COM/2019/640 final. https://eur-lex.europa.eu/legal-content/EN/TXT/HTML/?uri=CELEX:52019DC0640

European Commission (2020) Farm to Fork strategy. COM/2020/381 final. https://eur-lex.europa.eu/legal-content/EN/TXT/HTML/?uri=CELEX:52020DC0381

European Union (2011) COMMISSION IMPLEMENTING REGULATION 540/2011 of 25 May 2011 implementing Regulation (EC) No 1107/2009 of the European Parliament and of the Council as regards the list of authorized active substances as regards the list of authorized active substances. https://eur-lex.europa.eu/eli/reg_impl/2011/540/oj

European Union (2019) REGULATION 2019/1009 of the European Parliament and of the Council of 5 June 2019 laying down rules on the making available on the market of EU fertilising products and amending Regulations (EC) No 1069/2009 and (EC) No 1107/2009 and repealing Regulation (EC) No 2003/2003. https://eur-lex.europa.eu/eli/reg/2019/1009/oj

EUROSTAT (2023) Developments in organic farming. Publications Office of the European Union, Luxembourg. Retrieved from https://ec.europa.eu/eurostat/statistics-explained/index.php?title=Developments_in_organic_farming

Faeth SH, Bush LP, Sullivan TJ (2002) Peramine alkaloid variation in *Neotyphodium*-infected Arizona fescue: effects of endophyte and host genotype and environment. J Chem Ecol 28:1511–1526. https://doi.org/10.1023/A:1019916227153

FAO (2021) World food and agriculture—statistical yearbook 2021. Rome. https://doi.org/10.4060/cb4477en

FAO (2022) El mercado mundial de fertilizantes: balance de la situación de un mercado en dificultades. Rome

Fu SF, Wei JY, Chen HW, Liu YY, Lu HY, Chou JY (2015) Indole-3-acetic acid: a widespread physiological code in interactions of fungi with other organisms. Plant Signal Behav 10(8). https://doi.org/10.1080/15592324.2015.1048052

Gao X, Zhang Y, He Z, Fu X (2017) Gibberellins. In: Li J, Li C, Smith SM (eds) Hormone metabolism and signaling in plants. Academic, pp 107–160. https://doi.org/10.1016/B978-0-12-811562-6.00004-9

García-Latorre C, Rodrigo S, Marin-Felix Y, Stadler M, Santamaria O (2023) Plant-growth promoting activity of three fungal endophytes isolated from plants living in dehesas and their effect on *Lolium multiflorum*. Sci Rep 13(1):7354. https://doi.org/10.1038/s41598-023-34036-8

Giovannini L, Palla M, Agnolucci M, Avio L, Sbrana C, Turrini A, Giovannetti M (2020) Arbuscular mycorrhizal fungi and associated microbiota as plant biostimulants: research strategies for the selection of the best performing inocula. Agronomy 10(1):106. https://doi.org/10.3390/agronomy10010106

Glaeser B (2010) Agriculture between the green revolution and ecodevelopment- which way to go? In: Glaeser B (ed) The green revolution revisited: critique and alternatives (2), Taylor & Francis. ISBN: 0-203-84044-5

Greene EM (1980) Cytokinin production by microorganisms. Bot Rev 46(1):25–74. https://doi.org/10.1007/BF02860866

Haas H (2014) Fungal siderophore metabolism with a focus on *Aspergillus fumigatus*. Nat Prod Rep 31(10):1266–1276. https://doi.org/10.1039/C4NP00071D

Hamayun M, Khan SA, Ahmad N, Tang DS, Kang SM, Na CI, Sohn EY, Hwang YH, Shin DH, Lee BH, Kim JG, Lee IJ (2009a) *Cladosporium sphaerospermum* as a new plant growth promoting endophyte from the roots of *Glycine max* (L.) Merr. World J Microbiol Biotechnol 25:627–632. https://doi.org/10.1007/s11274-009-9982-9

Hamayun M, Khan SA, Khan AL, Rehman G, Sohn EY, Shah AA, Kim SK, Joo GJ, Lee IJ (2009b) *Phoma herbarum* as a new gibberellin-producing and plant growth-promoting fungus. J Microbiol Biotechnol 19(10):1244–1249

Hanson JR (2003) Natural products: the secondary metabolites. R Soc Chem 17:1–28. ISBN: 0-85404-490-6

Hashem A, Tabassum B, Abd-Allah EF (2019) *Bacillus subtilis*: a plant-growth promoting rhizobacterium that also impacts biotic stress. Saudi J Biol Sci 26(6):1291–1297. https://doi.org/10.1016/j.sjbs.2019.05.004

Hassan SED (2017) Plant growth-promoting activities for bacterial and fungal endophytes isolated from medicinal plant of *Teucrium polium* L. J Adv Res 8(6):687–695. https://doi.org/10.1016/j.jare.2017.09.001

Hertweck C (2009) The biosynthetic logic of polyketide diversity. Angew Chem Int Ed 48(26):4688–4716. https://doi.org/10.1002/anie.200806121

Hider RC (1984) Siderophore mediated absorption of iron. Siderophores from microorganisms and plants. Struct Bond 58:25–87. https://doi.org/10.1007/BFb0111310

Hinsch J, Vrabka J, Oeser B, Novák O, Galuszka P, Tudzynski P (2015) De novo biosynthesis of cytokinins in the biotrophic fungus *Claviceps purpurea*. Environ Microbiol 17(8):2935–2951. https://doi.org/10.1111/1462-2920.12838

Hossain MM, Sultana F, Islam S (2017) Plant growth-promoting fungi (PGPF): phytostimulation and induced systemic resistance. Plant Microb Interact Agro Ecol Perspect 2:135–191. https://doi.org/10.1007/978-981-10-6593-4_6

HRAC (2023) Global herbicides classification. Retrieved from: https://hracglobal.com/tools/classification-lookup/?s=&mode=10909&letter=&number=#classificationLookup

Ikram M, Ali N, Jan G, Jan FG, Rahman IU, Iqbal A, Hamayun M (2018) IAA producing fungal endophyte *Penicillium ruqueforti* Thom., enhances stress tolerance and nutrients uptake in wheat plants grown on heavy metal contaminated soils. PLoS One 13(11). https://doi.org/10.1371/journal.pone.0208150

Jiang L, Lee MH, Kim CY, Kim SW, Kim PI, Min SR, Lee J (2021) Plant growth promotion by two volatile organic compounds emitted from the fungus *Cladosporium halotolerans* NGPF1. Front Plant Sci 12:794349. https://doi.org/10.3389/fpls.2021.794349

Jordán M, Casaretto J (2006) Hormonas y reguladores del crecimiento: auxinas, giberelinas y citocininas. In: Squeo FA, and Cardemil L (eds) Fisiología Vegetal, vol 15, pp 1–28

Keller NP, Turner G, Bennett JW (2005) Fungal secondary metabolism—from biochemistry to genomics. Nat Rev Microbiol 3(12):937–947. https://doi.org/10.1038/nrmicro1286

Keswani C, Mishra S, Sarma BK, Singh SP, Singh HB (2014) Unraveling the efficient applications of secondary metabolites of various *Trichoderma* spp. Appl Microbiol Biotechnol 98:533–544. https://doi.org/10.1007/s00253-013-5344-5

Khan SA, Hamayun M, Yoon H, Kim HY, Suh SJ, Hwang SK, Kim JM, Lee IJ, Choo YS, Yoon UH, Kong WS, Lee BM, Kim JG (2008) Plant growth promotion and *Penicillium citrinum*. BMC Microbiol 8:231. https://doi.org/10.1186/1471-2180-8-231

Khan AL, Hamayun M, Kang SM, Kim YH, Jung HY, Lee JH, Lee IJ (2012) Endophytic fungal association via gibberellins and indole acetic acid can improve plant growth under abiotic stress: an example of *Paecilomyces formosus* LHL10. BMC Microbiol 12:3. https://doi.org/10.1186/1471-2180-12-3

Khan AL, Waqas M, Khan AR, Hussain J, Kang SM, Gilani SA, Hamayun M, Shin JH, Kamran M, Al-Harrasi A, Yun BW, Adnan M, Lee IJ (2013) Fungal endophyte *Penicillium janthinellum* LK5 improves growth of ABA-deficient tomato under salinity. World J Microbiol Biotechnol 29:2133–2144. https://doi.org/10.1007/s11274-013-1378-1

Khan AL, Hussain J, Al-Harrasi A, Al-Rawahi A, Lee IJ (2015) Endophytic fungi: resource for gibberellins and crop abiotic stress resistance. Crit Rev Biotechnol 35(1):62–74. https://doi.org/10.3109/07388551.2013.800018

Khan AR, Ullah I, Waqas M, Shahzad R, Hong SJ, Park GS, Jung BK, Lee IJ, Shin JH (2015b) Plant growth-promoting potential of endophytic fungi isolated from *Solanum nigrum* leaves. World J Microbiol Biotechnol 31:1461–1466. https://doi.org/10.1007/s11274-015-1888-0

Khan MIR, Singh A, Poor P (eds) (2023) Plant hormones in crop improvement. Elsevier. ISBN: 9780323919555

Krasnoff SB, Howe KJ, Heck ML, Donzelli BG (2020) Siderophores from the entomopathogenic fungus *Beauveria bassiana*. J Nat Prod 83(2):296–304. https://doi.org/10.1021/acs.jnatprod.9b00698

Kumar NV, Rajam KS, Rani ME (2017) Plant growth promotion efficacy of indole acetic acid (IAA) produced by a mangrove associated fungi *Trichoderma viride* VKF3. Int J Curr Microbiol Appl Sci 6(11):2692–2701. https://doi.org/10.20546/ijcmas.2017.611.317

Lauer U, Anke T, Sheldrick WS, Scherer A, Steglich W (1989) Aleurodiscal: an antifungal sesterterpenoid from *Aleurodiscus mirabilis* (Berk. & Curt.) Höhn. J Antibiot 42(6):875–882. https://doi.org/10.7164/antibiotics.42.875

Lidoy J, Berrio E, García M, España-Luque L, Pozo MJ, Lopez-Raez JA (2023) Flavonoids promote *Rhizophagus irregularis* spore germination and tomato root colonization: a target for sustainable agriculture. Front Plant Sci 13:1094194. https://doi.org/10.3389/fpls.2022.1094194

Liu X, Li S (2004) Fungal secondary metabolites in biological control of crop pests. In: An Z (ed) Handbook of industrial mycology. CRC Press, pp 742–767. https://doi.org/10.1201/9780203970553

Liu Y, Wang L, Jung JH, Zhang S (2007) Sesterterpenoids. Nat Prod Rep 24(6):1401–1429. https://doi.org/10.1039/B617259H

Lochman J, Mikes V (2006) Ergosterol treatment leads to the expression of a specific set of defence-related genes in tobacco. Plant Mol Biol 62:43–51. https://doi.org/10.1007/s11103-006-9002-5

Lubna Asaf S, Hamayun M, Gul H, Lee IJ, Hussain A (2018) *Aspergillus niger* CSR3 regulates plant endogenous hormones and secondary metabolites by producing gibberellins and indole-acetic acid. J Plant Interact 13:100–111. https://doi.org/10.1080/17429145.2018.1436199

Manzoni M, Rollini M (2002) Biosynthesis and biotechnological production of statins by filamentous fungi and application of these cholesterol-lowering drugs. Appl Microbiol Biotechnol 58:555–564. https://doi.org/10.1007/s00253-002-0932-9

Marahiel MA (2009) Working outside the protein-synthesis rules: insights into non-ribosomal peptide synthesis. J Pept Sci 15(12):799–807. https://doi.org/10.1002/psc.1183

Martínez C, Espinosa-Ruiz A, Prat S (2016) Gibberellins and plant vegetative growth. Annu Plant Revs 49:285–322. https://doi.org/10.1002/9781119210436.ch10

Martinez Y, Ribera J, Schwarze FWMR, De France K (2023) Biotechnological development of *Trichoderma*-based formulations for biological control. Appl Microbiol Biotechnol 107(18):5595–5612. https://doi.org/10.1007/s00253-023-12687-x

McNeill JR (2009) The green revolution. In: Rangarajan M (ed) Environmental issues in India: a reader. Pearson Education, India. ISBN: 978–81–317-0810-1

Mehmood A, Hussain A, Irshad M, Hamayun M, Iqbal A, Khan N (2019) In vitro production of IAA by endophytic fungus *Aspergillus awamori* and its growth promoting activities in *Zea mays*. Symbiosis 77:225–235. https://doi.org/10.1007/s13199-018-0583-y

Mishra J, Singh R, Arora NK (2017) Plant growth-promoting microbes: diverse roles in agriculture and environmental sustainability. In: Kumar V, Kumar M, Sharma S, Prasad R (eds) Probiotics and plant health. Springer, Singapore. https://doi.org/10.1007/978-981-10-3473-2_4

Mitter N, Srivastava AC, Renu-Ahamad S, Sarbhoy AK, Agarwal DK (2002) Characterization of gibberellin producing strains of *Fusarium moniliforme* based on DNA polymorphism. Mycopathology 153:187–193. https://doi.org/10.1023/A:1014946217539

Mousa WK, Raizada MN (2013) The diversity of anti-microbial secondary metabolites produced by fungal endophytes: an interdisciplinary perspective. Front Microbiol 4:65. https://doi.org/10.3389/fmicb.2013.00065

Nacoon S, Jogloy S, Riddech N, Mongkolthanaruk W, Kuyper TW, Boonlue S (2020) Interaction between phosphate solubilizing bacteria and arbuscular mycorrhizal fungi on growth promotion and tuber inulin content of *Helianthus tuberosus* L. Sci Rep 10:4916. https://doi.org/10.1038/s41598-020-61846-x

Nath R, Sharma GD, Barooah M (2015) Plant growth promoting endophytic fungi isolated from tea (*Camellia sinensis*) shrubs of Assam. India Appl Ecol Environ Res 13(3):877–891. https://doi.org/10.15666/aeer/1303_877891

Noceto PA, Bettenfeld P, Boussageon R, Hériché M, Sportes A, van Tuinen D, Courty PM, Wipf D (2021) Arbuscular mycorrhizal fungi, a key symbiosis in the development of quality traits in crop production, alone or combined with plant growth-promoting bacteria. Mycorrhiza 31(6):655–669. https://doi.org/10.1007/s00572-021-01054-1

Pandya JB, Patani AN, Raval VH, Rajput KN, Panchal RR (2023) Understanding the fermentation potentiality for gibberellic acid (GA3) production using fungi. Curr Microbiol 80:385. https://doi.org/10.1007/s00284-023-03454-2

Parker SR, Cutler HG, Schreiner PR (1995a) Koninginin C: a biologically active natural product from *Trichoderma koningii*. Biosci Biotechnol Biochem 59(6):1126–1127. https://doi.org/10.1271/bbb.59.1126

Parker SR, Cutler HG, Schreiner PR (1995b) Koninginin E: isolation of a biologically active natural product from *Trichoderma koningii*. Biosci Biotechnol Biochem 59(9):1747–1749. https://doi.org/10.1271/bbb.59.1747

Pascale A, Vinale F, Manganiello G, Nigro M, Lanzuise S, Ruocco M, Marra R, Lombardi N, Woo SL, Lorito M (2017) *Trichoderma* and its secondary metabolites improve yield and quality of grapes. Crop Prot 92:176–181. https://doi.org/10.1016/j.cropro.2016.11.010

Poveda J, Eugui D, Abril-Urías P, Velasco P (2021) Endophytic fungi as direct plant growth promoters for sustainable agricultural production. Symbiosis 85(1):1–19. https://doi.org/10.1007/s13199-021-00789-x

Poveda J, Rodríguez VM, Díaz-Urbano M, Sklenář F, Saati-Santamaría Z, Menéndez E, Velasco P (2022) Endophytic fungi from kale (*Brassica oleracea* var. *acephala*) modify roots-glucosinolate profile and promote plant growth in cultivated *Brassica* species. First description of *Pyrenophora gallaeciana*. Front Microbiol 13:981507. https://doi.org/10.3389/fmicb.2022.981507

Qiang X, Ding J, Lin W, Li Q, Xu C, Zheng Q, Li Y (2019) Alleviation of the detrimental effect of water deficit on wheat (*Triticum aestivum* L.) growth by an indole acetic acid-producing endophytic fungus. Plant Soil 439:373–391. https://doi.org/10.1007/s11104-019-04028-7

Reglamento (UE) (2019/1009 del Parlamento Europeo y del Consejo, de 5 de junio de 2019, por el que se establecen disposiciones relativas a la puesta a disposición en el mercado de los productos fertilizantes UE y se modifican los Reglamentos (CE) n. o 1069/2009 y (CE) n. o 1107/2009 y se deroga el Reglamento (CE) n. o 2003/2003. Diario oficial de la Unión Europea, L170–1, de 25 de junio de 2019. http://data.europa.eu/eli/reg/2019/1009/oj

Řezanka T, Palyzová A, Faltýsková H, Sigler K (2019) Siderophores: amazing metabolites of microorganisms. Stud Nat Prod Chem 60:157–188. https://doi.org/10.1016/B978-0-444-64181-6.00005-X

Ripa FA, Cao WD, Tong S, Sun JG (2019) Assessment of plant growth promoting and abiotic stress tolerance properties of wheat endophytic fungi. Biomed Res Int 2019:6105865–6105812. https://doi.org/10.1155/2019/6105865

Rodrigo S, García-Latorre C, Santamaria O (2021) Metabolites produced by fungi against fungal phytopathogens: review, implementation, and perspectives. Plan Theory 11(1):81. https://doi.org/10.3390/plants11010081

Rossard S, Roblin G, Atanassova R (2010) Ergosterol triggers characteristic elicitation steps in *Beta vulgaris* leaf tissues. J Exp Bot 61(6):1807–1816. https://doi.org/10.1093/jxb/erq047

Rowan DD (1993) Lolitrems, peramine and paxilline–mycotoxins of the ryegrass endophyte interaction. Agric Ecosyst Environ 44:103–122. https://doi.org/10.1016/0167-8809(93)90041-M

Saber WI, Ghoneem KM, Rashad YM, Al-Askar AA (2017) *Trichoderma harzianum* WKY1: an indole acetic acid producer for growth improvement and anthracnose disease control in sorghum. Biocontrol Sci Tech 27(5):654–676. https://doi.org/10.1080/09583157.2017.1321733

Saha M, Sarkar S, Sarkar B, Sharma BK, Bhattacharjee S, Tribedi P (2016) Microbial siderophores and their potential applications: a review. Environ Sci Pollut Res 23:3984–3999. https://doi.org/10.1007/s11356-015-4294-0

Sansinenea E (2019) Bacillus spp.: as plant growth-promoting bacteria. In: Singh H, Keswani C, Reddy M, Sansinenea E, García-Estrada C (eds) Secondary metabolites of plant growth promoting rhizomicroorganisms. Springer, Singapore. https://doi.org/10.1007/978-981-13-5862-3_11

Schardl CL, Leuchtmann A, Spiering MJ (2004) Symbioses of grasses with seedborne fungal endophytes. Annu Rev Plant Biol 55:315–340. https://doi.org/10.1146/annurev.arplant.55.031903.141735

Schrettl M, Bignell E, Kragl C, Sabiha Y, Loss O, Eisendle M, Wallner A, Herbert NA Jr, Haynes K, Haas H (2007) Distinct roles for intra-and extracellular siderophores during *aspergillus fumigatus* infection. PLoS Pathog 3(9):e128. https://doi.org/10.1371/journal.ppat.0030128

Shi TQ, Peng H, Zeng SY, Ji RY, Shi K, Huang H, Ji XJ (2017) Microbial production of plant hormones: opportunities and challenges. Bioengineered 8(2):124–128. https://doi.org/10.1080/21655979.2016.1212138

Tedesco D, Haragsim L (2012) Cyclosporine: a review. J Transp Secur 2012:230386. https://doi.org/10.1155/2012/230386

Toscano-Verduzco FA, Cedeño-Valdivia PA, Chan-Cupul W, Hernández-Ortega HA, Ruiz-Sánchez E, Galindo-Velasco E, Cruz-Crespo E (2020) Phosphates solubilization, indol-3-acetic acid and siderophores production by *Beauveria brongniartii* and its effect on growth and fruit quality of *Capsicum chinense*. J Hortic Sci Biotechnol 95(2):235–246. https://doi.org/10.1080/14620316.2019.1662737

Tsavkelova EA, Klimova SY, Cherdyntseva TA, Netrusov AI (2006) Hormones and hormone-like substances of microorganisms: a review. Appl Biochem Microbiol 42:229–235. https://doi.org/10.1134/S000368380603001X

Turrini A, Avio L, Giovannetti M, Agnolucci M (2018) Functional complementarity of arbuscular mycorrhizal fungi and associated microbiota: the challenge of translational research. Front Plant Sci 9:1407. https://doi.org/10.3389/fpls.2018.01407

Uarrota VG, Stefen DLV, Leolato LS, Gindri DM, Nerling D (2018) Revisiting carotenoids and their role in plant stress responses: from biosynthesis to plant signalling mechanisms during stress. In: Gupta D, Palma J, Corpas F (eds) Antioxidants and antioxidant enzymes in higher plants. Springer, Cham. https://doi.org/10.1007/978-3-319-75088-0_10

Ulrich JM (1960) Auxin production by mycorrhizal fungi. Physiol Plant 13(3):429–443. ISSN: 0031-9317

United Nations (2022) World population prospects 2022: summary of results. UN DESA/POP/2022/TR/NO. 3. eISBN: 9789210014380

Velásquez A, Vega-Celedón P, Fiaschi G, Agnolucci M, Avio L, Giovannetti M, D'Onofrio C, Seeger M (2020) Responses of *Vitis vinifera* cv. Cabernet Sauvignon roots to the arbuscular mycorrhizal fungus *Funneliformis mosseae* and the plant growth-promoting rhizobacterium *Ensifer meliloti* include changes in volatile organic compounds. Mycorrhiza 30:161–170. https://doi.org/10.1007/s00572-020-00933-3

Vinale F, Sivasithamparam K (2020) Beneficial effects of *Trichoderma* secondary metabolites on crops. Phytother Res 34(11):2835–2842. https://doi.org/10.1002/ptr.6728

Vinale F, Chiessa G, Sivasithamparam K, Ghisalberti EL, Marra R, Conte P, Piccolo A, Aloj V, Turra D, Lanzuise S, Ruocco M, Woo SL, Lorito M (2007) Antibiotic and plant growth promotion activity of *Trichoderma* koninginins. J Plant Pathol 89(3)

Vinale F, Sivasithamparam K, Ghisalberti EL, Marra R, Barbetti MJ, Li H, Woo SL, Lorito M (2008) A novel role for *Trichoderma* secondary metabolites in the interactions with plants. Physiol Mol Plant Pathol 72(1–3):80–86. https://doi.org/10.1016/j.pmpp.2008.05.005

Waqas M, Khan AL, Kamran M, Hamayun M, Kang SM, Kim YH, Lee IJ (2012) Endophytic fungi produce gibberellins and indoleacetic acid and promotes host-plant growth during stress. Molecules 17:10754–10773. https://doi.org/10.3390/molecules170910754

Willer H, Lernoud J, Kilcher L (eds) (2013) The world of organic agriculture. Statistics and emerging trends 2013. FiBL-IFOAM Report. Research Institute of Organic Agriculture FiBL, Frick, and IFOAM – Organics International, Bonn

Willer H, Schlatter B, Trávníček J (eds) (2023) The world of organic agriculture. Statistics and emerging trends 2023. Research Institute of Organic Agriculture FiBL, Frick, and IFOAM – Organics International, Bonn. https://doi.org/10.5281/zenodo.7572890

Winkelmann G (2002) Microbial siderophores-mediated transport. Biochem Soc Trans 30(4):691–695

You YH, Kwak TW, Kang SM, Lee MC, Kim JG (2015) *Aspergillus clavatus* Y2H0002 as a new endophytic fungal strain producing gibberellins isolated from *Nymphoides peltata* in fresh water. Mycobiology 43:87–91. https://doi.org/10.5941/MYCO.2015.43.1.87

Zhang Y, Han T, Ming Q, Wu L, Rahman K, Qin L (2012) Alkaloids produced by endophytic fungi: a review. Nat Prod Commun 7(7). https://doi.org/10.1177/1934578X1200700742

Zhang F, Yuan J, Yang X, Cui Y, Chen L, Ran W, Shen Q (2013) Putative *Trichoderma harzianum* mutant promotes cucumber growth by enhanced production of indole acetic acid and plant colonization. Plant Soil 368:433–444. https://doi.org/10.1007/s11104-012-1519-6

Fungal Metabolites as Inductors of Plant Abiotic Stresses Tolerance in Crops

María Diaz-Urbano, Pablo Velasco, and Víctor M. Rodríguez

Abstract Abiotic stresses impact negatively plant growth and productivity worldwide. One promising method to improve plant tolerance to these stresses is through inoculation with endophytic fungi. Endophytic fungi form symbiotic relationships with host plants and can produce bioactive metabolites that induce physiological changes conferring stress resistance. This review summarizes key studies on bioactive compounds produced by symbiotic fungi that enable plants to maintain growth, photosynthesis, and antioxidant activity under stressful conditions. Specific metabolites produced by endophytes such as phytohormones, antioxidants and osmolytes are highlighted for their roles in signalling cascades that counteract damaged caused by stresses. Directions for future research include characterising metabolites of endophyte strains that promote resistance, profiling the mechanisms that make plants more resistant to abiotic stresses, and field experiments to translate controlled results to field conditions. With further elucidation of the metabolic crosstalk between plants and endophytes, microbial inoculation can become an effective and sustainable tool to enhance agricultural drought tolerance.

Keywords Fungal endophytes · Metabolomics · Stress tolerance · Osmolytes

1 Introduction

In the next century the primary sector of the global economy will have to face important threats that will require a revolution of the productivity methods to make the sector more sustainable without losing productivity. In the mid-twentieth, thanks to improved technologies, global agriculture experienced a burst by introducing new and more productive varieties, especially in the third world, in the so-called

M. Diaz-Urbano · P. Velasco · V. M. Rodríguez (✉)
Group of Genetics, Breeding and Biochemistry of Brassicas,
Mision Biologica de Galicia (MBG-CSIC), Pontevedra, Spain
e-mail: vmrodriguez@mbg.csic.es

J. Poveda et al. (eds.), *Fungal Metabolites for Agricultural Applications*, Fungal Biology, https://doi.org/10.1007/978-3-031-76587-2_5

green revolution period. This burst allowed to provide enough food for a world population in continuous growth. Predictions estimate that the fast growth of world population will come to an end by 2050; however, by that time the world population will reach the amount of 10,000 million people from the current 8000 million in 2023. In this context, the agricultural sector will have to face a great challenge to ensure the availability of food. A major threat to reach this goal is the advance of anthropogenic induced climate change. As early as 2007, the IPCC Fourth Assessment Report: Climate Change indicated: "An increase of droughts over low latitudes and mid-latitude continental interiors in summer is likely, but sensitive to model land-surface formulation" (IPPC 2007). Projections for the 2090s made by Burke et al. (2006), using different scenarios, show regions of strong wetting and drying with a net overall global drying trend. Other authors consider that by 2050, over 50% of arable land could be impacted by desertification (Dai 2013). To compensate for the losses due to climate events farmers alter their inputs and practices. A field survey from China, one of the largest world-wide agriculture producers, reveals that, when faced with drought periods, in order to mitigate the potential negative impacts, 35% of farmers increase crop production inputs (e.g., fertilizers, pesticides, and labour) (Chen et al. 2013), increasing the impact of agriculture on global climate change.

Abiotic stresses are defined as changes in the environmental conditions that negatively impact plant growth and survival. Within certain limits, plants can adapt to variations on temperature, soil salinity or accumulation of metals or water availability, but severe or prolonged exposure to these environmental changes can disrupt plants internal homeostasis causing physiological damage. To withstand these environmental stresses, plants have evolved intricate regulatory pathways that allow them to resist environments in a timely manner (Zhang et al. 2022). Plant resistance to abiotic stresses can be classified as plant evolutionary adjustments that are heritable, or plant acclimation, that consists on transitory changes on plant physiology and metabolism that improve plant performance and survival in stressful environments (Bandurska 2022). Likewise, stress adaptation mechanisms can be divided into two categories, stress avoidance, which consists on changes that prevent or retard the physiological damage caused by the stressor (e. g., stomatal closure under water deficient conditions that prevent water-losses through leaves) and stress tolerance, which consist on physiological changes that allow the plant to cope with the ongoing stress in plant cells (e. g., accumulation of compatible metabolites that protect the plant cell against damage caused by the stressors) (Bandurska 2022). At cellular levels abiotic stress cause oxidative stress, damage to cellular components such as membrane lipids, proteins, and nucleic acids and dysfunction of the metabolism (Mathivanan 2021), so plants have evolved response mechanisms to cope with them. Desaturation of membrane lipids, activation of reactive species scavengers, induction of molecular chaperones, and accumulation of compatible solutes are more generalized and conserved cellular defence responses (He et al. 2018).

The rhizosphere constitutes a unique ecological environment that allows the plant to establish an association with different microorganisms. This microbiota plays an important role in supporting plant development and enhancing the plant resilience to different stresses that are either biotic or abiotic. Root associations with microorganisms include both endo- and ectosymbionts that are mostly bacteria or fungi. By definition, fungal symbionts colonize plant tissues without causing apparent diseases. Plants' fungal symbionts can be classified into two major classes: mycorrhizal fungi, obligate symbionts that only establish associations with plant roots and endophytic fungi, facultative symbionts that may be associated with roots, stems and/or leaves. In the last decades, endophytic fungi are attracting increasing attention due to their potential to be used as alternatives to chemicals in agriculture. These fungi form mutualistic relationships with plants that can provide a variety of benefits, such as nutrient acquisition, defense against pathogens and improved drought tolerance (Hardoim et al. 2015). The ability of different endophytic filamentous fungi to increase the availability of water in plants has been reported in several crops. In grasses, such as *Festuca arundinacea* and *Lolium arundinaceum*, fungus *Epichloë coenophiala* is able to increase water availability by affecting the soil water release path, *Neotyphodium* spp. in *Lolium perenne* was able to increase water availability by root growth promotion, and inoculation of *Brachiaria* spp. with *Sarocladium* promote the maintenance of plant water status and the increase of dry matter content, total non-structural carbohydrate content and biomass (Dastogeer and Wylie 2017). In addition, by colonizing plant roots, endophytic fungi maintain a molecular dialogue with their hosts, inducing the expression of genes related to tolerance to this abiotic stress. Endophytic fungi are known to produce a wide range of secondary metabolites and bioactive compounds, including alkaloids, polyphenols, terpenoids, and peptides (Sharma and Singh 2021). These metabolites can play pivotal roles in mediating plant-fungal interactions and enhancing plant stress tolerance. They can act through direct or indirect mechanisms. Direct mechanisms involve fungal hyphae assisting the uptake and transport of water and nutrients, especially phosphorus and nitrogen (Pratap et al. 2016). Indirect mechanisms include endophytes producing plant growth hormones and regulating osmolytes that minimize the effects of drought on the host (Dastogeer and Wylie 2017). A metabolomic study in *Piriformospora indica*-colonized chinese cabbage (*Brassica rapa*) plants showed the effect of fungi colonization in plant metabolism and how these changes can benefit the plant survival under stress (Hua et al. 2017). The present review will focus specifically on bioactive metabolites that induce physiological changes in plants colonized by associated endophytes that allow to maintain growth and productivity under abiotic stresses (Fig. 1).

2 Endophytes Can Produce Phytohormones That Modulate Plant Response to Abiotic Stresses

Phytohormones are defined as signalling molecules that are produced in extremely low concentration and regulate different processes in the plant (Fahad et al. 2015). These compounds are essential for various physiological functions and are involved

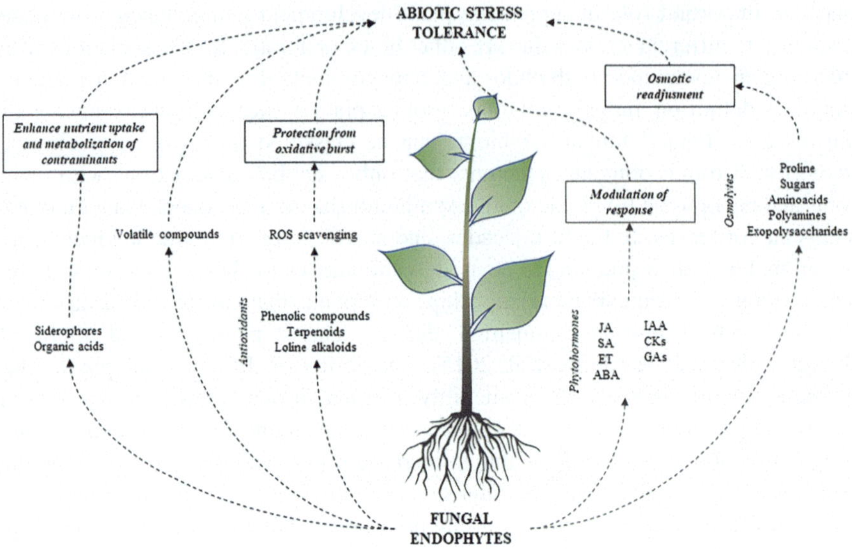

Fig. 1 A schematic representation of metabolite families produced by symbiotic fungi that enhance plant tolerance to abiotic stresses

in a wide range of processes, from seed germination to fruit ripening. Broadly speaking, phytohormones can be classified as growth or defensive-related compounds (Waqas et al. 2015; Jogawat et al. 2021). Belonging to the former group, auxins (IAA), gibberelins (GAs) and citokinins (CKs) function by driving cell division, elongation and differentiation which stimulates the generation of new tissues. These hormones also promote shoot and root elongation and stimulate the differentiation of vascular tissues (Mukherjee et al. 2022). The latter group includes salicylic acid (SA), jasmonic acid (JA), abscisic acid (ABA) or ethylene (ET). SA and JA have been associated to the plant response to biotic stresses. Although it is generally established that the SA pathway acts against biotrophic pathogens and the JA pathway against necrotrophs, there is increasing evidence of a continuous antagonistic and synergistic interaction of both hormonal defense pathways against each pathogen (Li et al. 2019). ET is a volatile hormone that cooperates with JA in the activation of defenses against necrotrophic pathogens and antagonizes SA-dependent resistance against biotrophic pathogens. ABA is involved in the response of the plant to environmental stresses. This phytohormone regulates seed germination and dormancy, opening and closing of stomata and act as an endogenous messenger in the response to abiotic stresses (Tuteja 2007; Mukherjee et al. 2022). However, in the last decades has become more evident that phytohormones do not act through isolate pathways responding to specific stress factors but the proper adaptation of plants to changing environments relies on an intricate balance and complex interactions of all these phytohormones. The role of phytohormones in response to drought is an example of this complexity.

ABA has been considered the central hormone driven the plant response to osmotic stress. When plant suffer drought stress the intracellular level of ABA rise up, binding the protein complex ABA-receptor PYR/PYL/RCARs. The formation of this complex produces a signalling cascade that activates a drought tolerance response (Ullah et al. 2018). However, since the early 2000s the implication of other phytohormones in this tolerant response became apparent. Exogenous application of JA increased drought resistance in different plant species (Bandurska et al. 2003; Ai et al. 2008). This effect has been associated with the accumulation of antioxidant compounds in response to the increased levels of JA and also to the crosstalk between oxylipins and ABA, regulating stomata closure (Savchenko et al. 2014). It also has been reported that foliar application of SA increases drought tolerance in different plant species (Khalvandi et al. 2021; Feng et al. 2023). The signal transduction of drought tolerance in *Arabidopsis thaliana* requires SA (Chini et al. 2004).

Some fungi have demonstrated the ability to synthesize phytohormones that promote plant growth (Hasan 2002; Lubna et al. 2018; Rehman et al. 2022). The capacity of some endophytic fungus to produce phytohormones under abiotic stresses has been associated to the alleviation of stress symptoms in the host plant. Qiang et al. (2019) isolated an IAA-producing strain of the fungus *Alternaria alternata* which promote growth on wheat (*Triticum aestivum* L.) at both control and drought-stressed plants. Interestingly, this fungus has a significant effect promoting root growth exclusively when plants are under water-limiting conditions, effect that can be associated to the increase production of IAA. Zhou et al. (2021) reported that seedlings of *Pinus tabulaeformis* treated with the fermentation broth supernatant obtained from a fungal isolate of the genus *Phoma* showed higher chlorophyll content and turgor than control seedlings after 15 days of osmotic-treatment. This fermentation broth was rich in ABA secreted by the fungus and its application to stressed plants yielded similar alleviation effects that treatment with exogenous ABA. Javed et al. (2022) studied three hormone-producing fungus isolated from the xerophytic plant *Carthamus oxyacantha* L. Secretion of phytohormones to the medium is stimulated when these fungi are grown in a medium with lower osmotic potential (8% PEG-8000) mimicking the effect of drought or salinity. The effect of stress on the secretion of the different hormones was not homogeneous since *Meyerozyma guilliermondi* produced significantly higher levels of ABA and IAA under osmotic stress whereas *Aspergillus aculeatus* increases the production of GA. The combination of the three fungi has the highest impact promoting plant growth under osmotic stress. The endophytic fungus *Paecilomyces formosus* also produces bioactive GAs and IAA. These fungi promote growth of the dwarf Waito-C (GAs mutant) and Dongjin (GAs normal) rice cultivars under heat stress (Waqas et al. 2015) and also mitigated the negative impact of salinity stress on cucumber plants (Khan et al. 2012).

Endophytes are also able to promote plant growth under stress by modifying phytohormone content indirectly, such as, by altering ET production and changing the balance of CKs and ABA or IAA signalling (Mastouri et al. 2012; Waqas et al. 2012; Khan et al. 2018; Lata et al. 2018; Mehmood et al. 2019). The synthesis of these signalling molecules can alter the hormonal balance of plants. In fact, numerous studies suggest that exogenous synthesis of IAA and GA by endophytes leads to a considerable reduction of endogenous JA and ABA levels in plants. This has been linked to the promotion of growth and resistance to abiotic stress conditions, such as heat, drought or salinity stress (Khan et al. 2015; Waqas et al. 2015; Lubna et al. 2018).

Some microorganisms can alter the biosynthesis of ET by producing the enzyme 1- aminocyclopropane-1-carboxylate (ACC) deaminase. This enzyme hydrolyzes the ET precursor ACC to ammonia and ketobutyrate, leading to the reduction of endogenous ET in the plant (Orozco-Mosqueda et al. 2020). Limitation, but not prevention, of the synthesis of ET has been associated to alleviation of stress symptoms and growth promotion under water-limiting environments. This has been extensively studied considering the effect of tolerance promotion of rhizobacteria against abiotic stresses (Heydarian et al. 2016; Tiwari et al. 2018; Duan et al. 2021), but little is known about the potential of ACC deaminase-producing fungal endophytes promoting abiotic stress tolerance in plants. Recently, Rehman et al. (2022) reported the isolation of two fungal endophytes from wild safflower (*Carthamus oxyacanthus*) that show ACC deaminase activity. These fungi, classified as *Trichoderma gamsii* and *Fusarium proliferatum*, promote drought-tolerance in the tree *Moringa oleifera*. Interestingly enough, these fungi also produce significant levels of IAA, that increase in more than a 20% when fungi are grown in osmotic media. ET and IAA have an inhibitory crosstalk pathway in such a way that high plant endogenous ET levels produced under stress limit the IAA mediated growth promotion (Orozco-Mosqueda et al. 2020). Limiting ET production through the activity of the fungal ACC-deaminase may decrease this feedback inhibition under drought stress.

3 Oxidative Stress Protection Mediated by Endophytic Fungi

Oxidative stress in biological systems occurs when cells are exposed to abiotic stresses that lead to the excessive formation of reactive oxygen species (ROS). ROS are by-products of the aerobic metabolism produced mainly in cellular organelles (i.e. chloroplasts, mitochondria and peroxisomes). Under favorable conditions living organisms have scavenging mechanisms that can maintain ROS production at basal levels. However, under abiotic stress this delicate balance is disrupted leading to overproduction of ROS, which causes membrane degradation, lipid peroxidation and DNA modifications (Marcińska et al. 2013). Several works have investigated the effect of different endophytic fungi on the redox system to increase the resistance of plants to abiotic stresses. For instance, Dastogeer et al. (2018) evaluated the interactions of two endophytic fungi extracted from wild *Nicotiana* species found in drought-prone areas of northern Australia, within *Nicotinana benthamiana* plants. Inoculation with both of the two fungal strains significantly increased plants tolerance to water stress as manifested by their significant delay in wilting of shoot tips. Under water stress, the interaction between endophytes and plants correlates with an increasing activity of the antioxidant enzymes catalase, peroxidase and polyphenol oxidase being more prominent as water stress progress. In the same way, Sadeghi et al. (2020) investigated the effects of inoculating mandarin endophytic fungi, *Penicillium citrinum*, *Aurobassium pullunts* and *Dothideomycetes* sp., individually and in combination with each other on the reactive oxygen species (ROS) scavenging and antioxidant functions in plants under drought stress. H_2O_2, O_2^- and lipid peroxidation level were significantly reduced in the plants inoculated with fungal endophytes. There is, however, a lack of information about the role of specific fungal metabolites in the antioxidant response in these papers. Endophytic fungi are able to produce secondary antioxidant metabolites that block the damage caused by the excessive production of ROS, including phenolic compounds, tannins, hydroxyanthraquinones and terpenoids (Kumar and Prasher 2023). Based on *in vitro* analysis several authors have reported the antioxidant capacity of the crude extracts obtained from a significant number of endophytic fungi. The antioxidants pestacin and isopestacin were extracted from endophytic *Pestalotiopsis microspore*, which are scavengers of superoxide and hydroxyl free radicals (Strobel and Daisy 2003). Also, the crude extracts of *Nigrospora* sp. and *Chaetomium nigricolor* show high antioxidant activity via DPPH scavenging assay (Atiphasaworn et al. 2017; Dhayanithy et al. 2019). Accumulation of phenolic compounds, specially hydroxinamic acids and derivatives, has been associated with a higher antioxidant capacity of endophytes crude extracts (Huang et al. 2007; Sayed et al. 2022). Moreover, fungal endophytes are well known for producing a wide variety of alkaloids. Loline alkaloids are a family of specific metabolites produced by endophytic fungi belonging to the family Clavicipitaceae (Schardl et al. 2007). These alkaloids were initially identified on *Lolium temulentum* and subsequently in other grass species. Originally considered plant metabolites, the fact that loline alkaloids are absent in grass plants that do not establish symbiotic associations with fungal endophytes and the discovery that *Neotyphodium*, anamorphs of *Epichloe*, can produce loline alkaloids in defined culture media, indicates that lolines are the product of clavicipitaceous endophytes metabolism (Blankenship 2004). Accumulation of lolines in grass seeds infected with *Epichloë* endophytes significantly correlates with enhanced levels of tocochromanols, suggesting that antioxidant capacity in grass seeds is associated with the concentration of these alkaloids (Gundel et al. 2018). Accumulation of loline alkaloids has also been associated to drought tolerance in different plant species infected with clavicipitaceous endophytes. The onset of water stress produced a significant increase in the accumulation of loline alkaloids in plant tissues of different grass species that is maintained during stress treatment (Nagabhyru et al. 2013; Lin et al. 2022).

Natural antioxidant compounds were also identified in other species of endophytes. Qin et al. (2019) identified a coumarin analogue, named (Z)-N-(4-hydroxystyryl) formamide (NFA), from the endophytic fungus *Aspergillus fumigatus* that promotes drought tolerance in rice. Priming seeds with the crude extract obtained from this fungus maintain membrane integrity, and regulate the contents of proline, NADPH oxidase, antioxidases and heat-shock-proteins under drought stress in rice plants. Authors postulate that the drought resistance of rice conferred by NFA might be attributed to NFA's dual effects: it induces NADPH oxidase after a short period of drought, while activates antioxidant enzyme system to eliminate ROS after a long period of drought. This compound also alleviates oxidative symptoms of rice and Arabidopsis under flooding stress when sprayed directly onto plant seedlings (Xue et al. 2020, 2021). In this last paper, Xue et al. (2021) also reported the isolation of a macrocyclic alkaloid, named chaetoglobosin A (CheA), from *Chaetomium globosum* which shows similar antioxidant capacity than that observed for NFA and also reduces oxidative symptoms on rice plants under flooding stress.

4 Endophytic Fungi Produce Osmotically-Active Metabolites

Fungi require continuous water influx for hyphae growth and expansion through soil or plant tissues (Davis et al. 2000). In water scarce environments fungi perform an internal osmotic readjustment, accumulating small hydrophilic compounds, called compatible osmolytes or osmoprotectants, in the cytosol reducing the internal water potential (Ψ_w) (Jennings 1993). The principal compatible osmolytes produced by fungi are polyols, small low molecular weight sugars and proline (Jennings 1993). These compounds also play an important role maintaining the structure of macromolecules, preventing protein denaturation and disruption of the cell wall structure due to dehydration. Likewise, water uptake in the plant roots require a negative water potential (Ψ_{roots}) compared to that of the soil (Ψ_{soil}). Under drought stress or salinity stresses, the Ψ_{soil} declines limiting the capacity of the roots to uptake water (Oliver et al. 2010). This causes dehydration of the plant tissues and loss of turgor. Most terrestrial plants, especially angiosperms, cannot survive on environments with a water potential (Ψ_w) lower than -10 MPa (Proctor and Pence 2022). Plants with beneficial endophytes can survive with less water by making more efficient use of it, as they can access water that an uncolonized plant could not. The literature reporting fungi mediated accumulation of osmolytes in stressed plants is extensive in both mono and dicots plants (Bae et al. 2009; Bayat et al. 2009; Nagabhyru et al. 2013; Molina-Montenegro et al. 2016), whereas other studies have reported a decrease of osmolytes accumulation in plants colonized with endophytes (Naveed et al. 2013; Ferus et al. 2019). These different behaviours have been interpreted as responses of stress tolerance or avoidance, respectively (Ruiz-Sánchez et al. 2010). However, little information is available in the literature about the role of fungal metabolites alleviating the symptoms of stressed plants.

In media with low osmotic potential some fungi are able to produce high amounts of osmolytes that can protect fungi cells from dehydration. Plants inoculated with these fungi showed higher growth rates under drought stress or salinity and higher accumulation of osmolytes in the roots than control plants, opening the possibility that an active transfer of these osmolytes from fungi to the plant tissue could activate stress tolerance mechanisms in the plant (Javed et al. 2022). One of the osmolytes that are produced by symbiotic fungi under stress conditions are small sugars and polyols. Ocón et al. (2007) reported that trehalose turnover is increased in arbuscular mycorrhizal fungi (AMF) under osmotic, thermal or mineral stress. This non-reducing dissacharide was associated to stress response in different organism, including plants. Liu et al. (2014) hypothesized that some AMF can improve plant resistance to low temperature through AMF-enhanced trehalose accumulation, which may induce the expression of aquaporin genes that enhances the water uptake capacity of mycorrhizal plants at low temperatures. Richardson et al. (1992) reported the accumulation of two sugar alcohols (i.e., mannitol and arabitol) in tall fescue (*Festuca arundinacea*) in plants exclusively infected with the endophyte *Acremonium coenophialum*, suggesting that these polyols are produced by the fungus. Interestingly, accumulation of arabitol was only detected in infected plants exposed to water stress. Since the biochemical pathways to produce these compounds exist on both, plants and fungi, it is difficult to establish whether an active transfer of sugars and polyols exist in the fungi-plant interaction. Complex interactions between biosynthetic pathways in the symbiont association may take place in a way that endophyte metabolism could also modify the levels of plant osmolytes indirectly. For instance, specific fungal alkaloids are biosynthesized from proline and aspartate and share common biosynthetic intermediates with polyamines, so this leads to the question whether these alkaloids could also interfere with the metabolism of these osmolytes in the plant and what would be the impact on stress tolerance in plants. The work of Nagabhyru et al. (2013) evidences a complex relationship between loline alkaloids biosynthesis and proline accumulation in *Neotyphodium*-infected plants. In most cases the increased levels of these alkaloids under water stress treatment accompanied a higher increase of proline in the plant tissue, which may suggest a positive feed-back loop that activates the accumulation of proline. It was also reported that the fungus *Epichloë bromicola* confers salinity stress tolerance in wild barley trough the conversion of the polyamine putrescine to spermine and spermidine. This fungus also converted soluble forms of polyamines to insoluble bound forms (Chen et al. 2019). These results reflect the complexity of the symbiont interactions in alkaloid biosynthesis under stress.

Endophytes can also promote tolerance to abiotic stresses trough modification of the surrounding environment. Microbial endophytes are known to produce exopolysaccharides (EPS), carbohydrates polymers that play an important role mediating the symbiotic relationship with plant host (Liu et al. 2017). The extracellular EPS matrix also plays a protective role in stressful environments. In water-scarce environments EPS create a protecting matrix that slow down the rate of water-loss due to disequilibrium in Ψ_w (Roberson and Firestone 1992). This matrix can also help to sequester Na^+ ions on saline environments reducing its toxicity (Grover et al. 2011). The reduced disponibility of Na^+ reduces its uptake through plant roots maintaining high K^+/Na^+ ratio which increase plant survival in salt stressed conditions. A clear evidence of the role of EPS promoting tolerance to salinity was reported by Sun et al. (2020). Using a reverse genetic approach these authors demonstrate that the rice tolerance to salinity mediated by the bacteria *Pantoea alhagi* is related directly to the accumulation of EPS. Alleviation of stress symptoms of plants associated to EPS-producing endophytic bacteria has been studied under saline stress (Upadhyay et al. 2011; Sun et al. 2020) or Cd toxicity (Wei et al. 2023). In contrast, little information is available about the stress induced tolerance mediated by EPS-producing endophytic fungi, though a considerable number of these fungi have been reported to produce EPS in different culture media (Smiderle et al. 2012; Mahapatra and Banerjee 2013; Orlandelli et al. 2016; Sadeghi et al. 2020). Recently, a series of papers published by a group of the University of Vidyasagar in India, reported the isolation of EPS from two species of endophytic fungi, *Mucor* sp. HELF2 and *Colletotrichum alatae* (Santra and Banerjee 2021, 2023). Foliar application of the extracted EPS to rice seedlings alleviate drought symptoms increasing root and shoot growth, total chlorophyll content, accumulation of osmolytes and promoting the activity of antioxidant enzymes. These results encourage further investigations about the use of EPS-producing fungi to alleviate abiotic symptoms in crops.

5 Role of Endophytes on Nutrient Uptake Under Abiotic Stress and Bioremediation

Endophytic fungi are able to improve the uptake of macronutrients, such as phosphorus, nitrogen, potassium and magnesium, or micronutrients, such as zinc, iron, and copper, from the soil and organic matter and increase the supply of these nutrients to the plant host (Baron and Rigobelo 2022). Mechanisms of soil transformation by endophytic fungi can be classified into mineralization (transformation of organic to inorganic forms), solubilization (inorganic form to soluble form) and mobilization (uptake) (Devi et al. 2022). Abiotic stresses can have a strong impact on the capacity of plants to uptake nutrients from soil. During drought, reduction of nutrient uptake may occur by the reduction of nutrient supply through mineralization or by reducing nutrient diffusion in the soil (Bista et al. 2018). Flooding and low temperatures can also interfere with mineral uptake indirectly by reducing root size and volume reducing the capacity of absorption of various mineral nutrients per root mass unit (Alam 1999; Ji et al. 2023). Many studies have shown that endophytes can assist plants by transforming and solubilizing nutrients making them conveniently accessible to plants. Two major secondary metabolites produced by endophytic fungi have been associated to a higher mobilization of nutrients under stressful conditions, organic acids and siderophores.

Organic acids harbour carboxyl and hydroxyl functional groups that show high affinity for cations favouring nutrient mobilization via a ligand exchange reaction (Alori et al. 2017). A study performed by Zhao et al. (2014) indicates that the release of organic acids (oxalic, formic and lactic acids) from *Trichoderma asperellum* solubilize iron in saline soils that can be assimilated directly by plants. The endophytic fungus *Penicillium oxalicum* is able to growth in alkaline or saline soils realising organic acids that can drastically change the pH of the media, which opens the possibility to use this fungus in soil remediation (Liao et al. 2018). Fungi have also been reported to be highly effective on releasing P from insoluble organic compounds, which is related to the secretion of organic acids. The fungi *Trichoderma harzianum* and *Aspergillus niger* are able to solubilize P from different insoluble sources. P is solubilized through the production of organic acids, such as citric, succinic, propionic, malic and acetic acid (Gaind 2016). Wheat seedlings inoculated with these fungi show a better root and shoot development under salt stress than control plants, which is correlated to a higher P solubilization.

Siderophores are low-weight, high-affinity iron chelating molecules, produced by fungi and bacteria. The complex iron siderophore can be absorbed by plant roots favouring iron uptake under Fe-limiting conditions. Several species of *Trichoderma* have been reported to produce siderophores under stressful environments. An extensive screening of 70 salt tolerant *Trichoderma* spp. identified three isolates with high capacity to produce siderophores that alleviate salt stress symptoms in maize plants (Kumar et al. 2017). The siderophore-containing crude extract obtained from *T. asperellum* also alleviates stress symptoms in cucumber under salt stress (Qi and Zhao 2013). To confirm that siderophores play an important role mobilizing insoluble iron under stress conditions, authors performed the experiments using insoluble Fe^{+3} as iron source. In these experimental conditions, cucumber plants inoculated with *T. asperellum* showed significantly better biomass parameters than those recorded in control plants. Three isolates from the drought-tolerant *Penicillium citrinum* exhibited the propensity to solubilise phosphate and zinc and produce siderophores which contributes to wheat plants' adaptation to abiotic stresses (Kaur and Saxena 2023).

In addition to iron, siderophores can also form stable complexes with other metals such as Al, Cd, Cu, Ga, In, Pb and Zn. Though some of these metals are essential nutrients for the plants at low concentrations, above certain thresholds these elements are toxic for the plant. Different studies also reveal the great potential of the use of endophytes to remove contaminants, pollutants, and toxins from soil and water (bioremediation). Paradigmatic examples are fungi isolated from plants that are able to growth in highly metal-contaminated mined soils or field irrigated with industrial/wastewater. Most of these fungi have the capacity of producing siderophores that chelate heavy metals and make them less accessible to plant roots. *Phialocephala fortinii* and *Rhizodermea veluwensis*, isolated from roots of *Clethra barbinervis*, a tree species that grows naturally at several mine sites, produce siderophores that can chelate Cu, Zn, and Pb from contaminated soil (Yamaji et al. 2016). Similarly, *Trichoderma* spp. is able to promote the growth of chickpea in contaminated soils by preventing heavy metals from entering into the plant cells (Syed et al. 2023).

Aside from production of siderophores, fungi can enhance plant tolerance to contaminated soils trough biological transformation of contaminants. For example, the arbuscular mycorrhiza *Rhizophagus irregularis* can transform insoluble toxic arsenic (As) to a variety of organic forms, including dimethylarsinic acid, dimethylarsine, and trimethylarsine that are release to the atmosphere avoiding accumulation of As in the plant tissues (Li et al. 2021).

6 Endophytic Fungal Volatiles

Organic volatiles are low molecular compounds with a low boiling point at atmospheric pressure, so they tend to evaporate at room temperature. In the last decades these compounds have been extensively studied due to their importance in plant physiology. Organic volatiles can play different roles attracting pollinators, repelling herbivores, attracting herbivore-predators or in plant-plant communication. There are around 300 organic volatiles isolated from fungi, most of them characterized for their implication in microbial (i.e., fungi-bacteria communication), plant-fungi or fungi-animal interactions (Roy and Banerjee 2019). In contrast, much less attention has been paid to how fungal organic volatiles can modulate plant response to abiotic stresses. Two studies have demonstrated organic volatiles produced by endophytic fungi enhance Arabidopsis tolerance to salinity stress. In both, the authors used bi-partite Petri plates to physically separate the fungi from the plants to confirm that volatile compounds are responsible from the observed tolerance response. Jalali et al. (2017) evaluate the effect of volatiles produced by 13 strains of *Trichoderma* spp. Volatile compounds produced by three species, T. *koningii*, T. *viridescence* and *Hypocrea orientalis* (teleomorph of *Trichoderma*), improved biomass production of Arabidopsis plants under control and salt stress conditions. Li and Kang (2018) reveals that volatiles produced by the endophytic fungi *Fusarium oxysporum* and *Verticillium dahliae* promote growth and chlorophyll content of Arabidopsis plants under salt stress. These authors go one step further and, using Arabidopsis mutants defective in auxin fluxes and response, reported that fungal organic volatiles regulate plant response to salt stresses through modulation of auxin transport. However, further studies are required to identify the volatiles produced by these fungi and the mechanisms of perception in the plants.

7 Future Perspectives

By colonizing plant roots, endophytic fungi maintain a molecular dialogue with their hosts inducing the expression of genes and modulating the metabolic response to abiotic stresses. Thanks to its potential to promote plant growth and stress resilience, the use of endophytic fungi may be a promising tool to increase crop yields in a context of sustainable agriculture. However, the efficient use of these microorganisms requires broadening our knowledge on different aspects. The mechanisms involved in extending stress tolerance to host plants, especially in crops, still remain elusive. The application of technical advances on genetics and 'omics' technologies will help to elucidate the signalling mechanisms of plant hormones, bioactive secondary metabolites and gene expression during endophyte-plant interactions in stressful environments. The use of genome-editing technologies to obtain mutant fungal strains has a huge potential to put some light on the mechanisms of plant stress-tolerance modulated by fungal endophytes. There is also a limitation since not all fungi endophytes are culturable which difficult the identification of bioactive compounds produced by those fungi. Advances in techniques of shotgun metagenomics could be a breakthrough to harness unculturable endophytes. Deep metabolomics analyses, combined with radiochemical labeling, are also necessary to identify specific fungal metabolites that can induce tolerance in the host plant. Also, more field experiments are necessary to translate controlled results to real agricultural environments. The high potential of endophytic fungi as bio-stimulants merits comprehensive assessment and further exploration. Yet, our overall understanding of the potential roles of endophytic fungi in the field is limited and further studies are necessary (Table 1).

Table 1 Mechanisms of endophytic fungi in promoting plant growth under stress

Mechanisms	Mode of action	Endophytic fungi	References
Phytohormones production	Endophytes can produce phytohormones like IAA, GAs, and CKs that promote plant growth	*Paecilomyces Formosus, Alternaria alternata, Phoma, Meyerozyma guilliermondi, Aspergillus aculeatus*	Qiang et al. (2019); Zhou et al. (2021); Javed et al. (2022); Waqas et al. (2015)
	Endophytes produce defensive phytohormones like salicylic acid and jasmonic acid that modulate plant stress response		
	Phytohormone production is influenced by abiotic stresses on the endophytes		
	Endophytes, through ACC deaminase activity, reduce endogenous ET levels, alleviating stress symptoms and promoting growth	*Trichoderma gamsii, Fusarium proliferatum*	Rehman et al. (2022); Orozco-Mosqueda et al. (2020)
Oxidative stress protection	Endophytes increase activity of antioxidant enzymes in plants under stress	*Penicillium citrinum, Aurobassium pullunts and Dothideomycetes*	Dastogeer et al. (2018); Sadeghi et al. (2020)
	Endophytes produce secondary metabolites like alkaloids, phenolics, and terpenoids that scavenge reactive oxygen species	*Nigrospora sp., Epichloë and Chaetomium nigricolor*	Syed et al. (2023); Xue et al. (2020, 2021)
	Compounds like loline alkaloids, pestacin, and NFA confer oxidative protection	*Pestalotiopsis microspore, Nigrospora sp., Aspergillus fumigatus, Chaetomium globosum, Neotyphodium*	Strobel and Daisy (2003); Atiphasaworn et al. (2017); Qin et al. (2019)

(continued)

Table 1 (continued)

Mechanisms	Mode of action	Endophytic fungi	References
Osmolyte production	Endophytes produce compatible osmolytes like sugars, polyols and proline under low water potential	*Acremonium coenophialum, Neotyphodium, Epichloë bromicola*	Ocón et al. (2007); Richardson et al. (1992); Nagabhyru et al. (2013)
	Osmolytes likely transferred to plants, improving water uptake and stress tolerance		
	EPS produced by some endophytes reduce water loss rate and Na^+ toxicity	*Pantoea alhagi*	Sun et al. (2020); Grover et al. (2011); Santra and Banerjee (2021, 2023)
	EPS applied externally promotes drought and salt tolerance in plants	*Mucor sp, Colletotrichum alatae*	
Nutrient uptake	Endophytes solubilize or mobilize nutrients like N, P, K, Fe through organic acids and siderophores	*Trichoderma asperellum, Penicillium oxalicum, Trichoderma harzianum and Aspergillus niger*	Zhao et al. (2014); Gaind (2016); Kumar et al. (2017)
	Chelation of nutrients by siderophores improves uptake under stressful conditions	*Trichoderma spp., Penicillium citrinum*	Qi and Zhao (2013)
	Endophytes can alter metal absorption and support bioremediation	*Phialocephala fortinii, Rhizodermea veluwensis, Trichoderma sp*	Yamaji et al. (2016); Li et al. (2021)
Volatile compounds	Fungal volatiles compounds confer salt tolerance	*T. koningii, T. viridescence, H. orientalis*	Jalali et al. (2017); Li and Kang (2018)
	Volatiles may modulate auxin transport and salt stress response	*Fusarium oxysporum, Verticillium dahliae*	

References

Ai L, Li ZH, Xie ZX, Tian XL, Eneji AE, Duan LS (2008) Coronatine alleviates polyethylene glycol-induced water stress in two rice (*Oryza sativa* L.) cultivars. J Agron Crop Sci 194(5):360–368. https://doi.org/10.1111/j.1439-037X.2008.00325.x

Alam SM (1999) Nutrient uptake by plants under stress conditions. In: Handbook of plant and crop stress. Press, pp 285–313

Alori ET, Glick BR, Babalola OO (2017) Microbial phosphorus solubilization and its potential for use in sustainable agriculture. Front Microbiol 8. https://www.frontiersin.org/articles/10.3389/fmicb.2017.00971

Atiphasaworn P, Monggoot S, Gentekaki E, Brooks S, Pripdeevech P (2017) Antibacterial and antioxidant constituents of extracts of endophytic fungi isolated from *Ocimum basilicum* var. *thyrsiflora* leaves. Curr Microbiol 74(10):1185–1193. https://doi.org/10.1007/s00284-017-1303-1

Bae H, Sicher RC, Kim MS, Kim S-H, Strem MD, Melnick RL, Bailey BA (2009) The beneficial endophyte *Trichoderma hamatum* isolate DIS 219b promotes growth and delays the onset of the drought response in *Theobroma cacao*. J Exp Bot 60(11):3279–3295. https://doi.org/10.1093/jxb/erp165

Bandurska H (2022) Drought stress responses: coping strategy and resistance. In: Plants, MDPI, vol 11, Issue 7. https://doi.org/10.3390/plants11070922

Bandurska H, Stroiński A, Kubiś J (2003) The effect of jasmonic acid on the accumulation of ABA, proline and spermidine and its influence on membrane injury under water deficit in two barley genotypes. Acta Physiol Plant 25(3):279–285. https://doi.org/10.1007/s11738-003-0009-0

Baron NC, Rigobelo EC (2022) Endophytic fungi: a tool for plant growth promotion and sustainable agriculture. J Mycol 13(1):39–55. https://doi.org/10.1080/21501203.2021.1945699

Bayat F, Mirlohi A, Khodambashi M (2009) Effects of endophytic fungi on some drought tolerance mechanisms of tall fescue in a hydroponics culture. Russ J Plant Physiol 56(4):510–516. https://doi.org/10.1134/S1021443709040104

Bista DR, Heckathorn SA, Jayawardena DM, Mishra S, Boldt JK (2018) Effects of drought on nutrient uptake and the levels of nutrient-uptake proteins in roots of drought-sensitive and -tolerant grasses. Plan Theory 7(2). https://doi.org/10.3390/plants7020028

Blankenship JD (2004) Loline alkaloid biosynthesis in *Neotyphodium uncinatum*, a fungal endophyte of *Lolium pratense*. Ph.D. Thesis. University of Kentucky

Burke EJ, Brown SJ, Christidis N (2006) Modeling the recent evolution of global drought and projections for the twenty-first century with the hadley centre climate model. J Hydrometeorol 7(5):1113–1125. https://doi.org/10.1175/JHM544.1

Chen Y, Wu Z, Okamoto K, Han X, Ma G, Chien H, Zhao J (2013) The impacts of climate change on crops in China: a Ricardian analysis. Glob Planet Change 104:61–74. https://doi.org/10.1016/j.gloplacha.2013.01.005

Chen T, Li C, White JF, Nan Z (2019) Effect of the fungal endophyte *Epichloë bromicola* on polyamines in wild barley (*Hordeum brevisubulatum*) under salt stress. Plant Soil 436(1):29–48. https://doi.org/10.1007/s11104-018-03913-x

Chini A, Grant JJ, Seki M, Shinozaki K, Loake GJ (2004) Drought tolerance established by enhanced expression of the CC–NBS–LRR gene, ADR1, requires salicylic acid, EDS1 and ABI1. Plant J 38(5):810–822. https://doi.org/10.1111/j.1365-313X.2004.02086.x

Dai A (2013) Increasing drought under global warming in observations and models. Nat Clim Chang 3(1):52–58. https://doi.org/10.1038/nclimate1633

Dastogeer KMG, Wylie SJ (2017) Plant-fungi association: role of fungal endophytes in improving plant tolerance to water stress. In: Plant-microbe interactions in agro-ecological perspectives, vol 1. Springer, pp 161–176. https://doi.org/10.1007/978-981-10-5813-4_8

Dastogeer KMG, Li H, Sivasithamparam K, Jones MGK, Wylie SJ (2018) Fungal endophytes and a virus confer drought tolerance to *Nicotiana benthamiana* plants through modulating osmolytes, antioxidant enzymes and expression of host drought responsive genes. Environ Exp Bot 149:95–108. https://doi.org/10.1016/j.envexpbot.2018.02.009

Davis DJ, Burlak C, Money NP (2000) Osmotic pressure of fungal compatible osmolytes. Mycol Res 104(7):800–804. https://doi.org/10.1017/S0953756299002087

Devi R, Kaur T, Kour D, Yadav A, Yadav AN, Suman A, Ahluwalia AS, Saxena AK (2022) Minerals solubilizing and mobilizing microbiomes: A sustainable approach for managing minerals' deficiency in agricultural soil. J Appl Microbiol 133(3):1245–1272. https://doi.org/10.1111/jam.15627

Dhayanithy G, Subban K, Chelliah J (2019) Diversity and biological activities of endophytic fungi associated with *Catharanthus roseus*. BMC Microbiol 19(1):22. https://doi.org/10.1186/s12866-019-1386-x

Duan B, Li L, Chen G, Su-Zhou C, Li Y, Merkeryan H, Liu W, Liu X (2021) 1-Aminocyclopropane-1-carboxylate deaminase-producing plant growth-promoting rhizobacteria improve drought stress tolerance in grapevine (*Vitis vinifera* L.). Front Plant Sci 12:706990. https://doi.org/10.3389/fpls.2021.706990

Fahad S, Hussain S, Bano A, Saud S, Hassan S, Shan D, Khan FA, Khan F, Chen Y, Wu C, Tabassum MA, Chun MX, Afzal M, Jan A, Jan MT, Huang J (2015) Potential role of phytohormones and plant growth-promoting rhizobacteria in abiotic stresses: consequences for changing environment. Environ Sci Pollut Res 22(7):4907–4921. https://doi.org/10.1007/s11356-014-3754-2

Feng Y, Zhao Y, Li Y, Zhou J, Shi H (2023) Improving photosynthesis and drought tolerance in *Nicotiana tabacum* L. by foliar application of salicylic acid. Life 16(1):2224936. https://doi.org/10.1080/26895293.2023.2224936

Ferus P, Barta M, Konôpková J (2019) Endophytic fungus *Beauveria bassiana* can enhance drought tolerance in red oak seedlings. Trees 33(4):1179–1186. https://doi.org/10.1007/s00468-019-01854-1

Gaind S (2016) Phosphate dissolving fungi: mechanism and application in alleviation of salt stress in wheat. Microbiol Res 193:94–102. https://doi.org/10.1016/j.micres.2016.09.005

Grover M, Ali SZ, Sandhya V, Rasul A, Venkateswarlu B (2011) Role of microorganisms in adaptation of agriculture crops to abiotic stresses. World J Microbiol Biotechnol 27(5):1231–1240. https://doi.org/10.1007/s11274-010-0572-7

Gundel PE, Seal CE, Biganzoli F, Molina-Montenegro MA, Vázquez-de-Aldana BR, Zabalgogeazcoa I, Bush LP, Martínez-Ghersa MA, Ghersa CM (2018) Occurrence of alkaloids in grass seeds symbiotic with vertically-transmitted Epichloë fungal endophytes and its relationship with antioxidants. Front Ecol Evol 6. https://www.frontiersin.org/articles/10.3389/fevo.2018.00211

Hardoim PR, van Overbeek LS, Berg G, Pirttilä AM, Compant S, Campisano A, Döring M, Sessitsch A (2015) The hidden world within plants: ecological and evolutionary considerations for defining functioning of microbial endophytes. Microbiol Mol Biol Rev 79(3):293–320. https://doi.org/10.1128/mmbr.00050-14

Hasan HAH (2002) Gibberellin and auxin production by plant root-fungi and their biosynthesis under salinity-calcium interaction. Acta Microbiol Immunol Hung 48(3):101–106. https://doi.org/10.17221/4207-pse

He M, He C-Q, Ding N-Z (2018) Abiotic stresses: general defenses of land plants and chances for engineering multistress tolerance. Front Plant Sci 9. https://www.frontiersin.org/articles/10.3389/fpls.2018.01771

Heydarian Z, Yu M, Gruber M, Glick BR, Zhou R, Hegedus DD (2016) Inoculation of soil with plant growth promoting bacteria producing 1-Aminocyclopropane-1-carboxylate deaminase or expression of the corresponding acdS gene in transgenic plants increases salinity tolerance in Camelina sativa. Front Microbiol 7:1966. https://doi.org/10.3389/fmicb.2016.01966

Hua MD-S, Senthil Kumar R, Shyur L-F, Cheng Y-B, Tian Z, Oelmüller R, Yeh K-W (2017) Metabolomic compounds identified in *Piriformospora indica*-colonized chinese cabbage roots delineate symbiotic functions of the interaction. Sci Rep 7(1):9291. https://doi.org/10.1038/s41598-017-08715-2

Huang W-Y, Cai Y-Z, Hyde KD, Corke H, Sun M (2007) Endophytic fungi from *Nerium oleander* L (Apocynaceae): main constituents and antioxidant activity. World J Microbiol Biotechnol 23(9):1253–1263. https://doi.org/10.1007/s11274-007-9357-z

IPPC (2007) IPCC fourth assessment report: climate change. FAO on Behalf of the IPPC Secretariat

Jalali F, Zafari D, Salari H (2017) Volatile organic compounds of some *Trichoderma* spp. increase growth and induce salt tolerance in *Arabidopsis thaliana*. Fungal Ecol 29:67–75. https://doi.org/10.1016/j.funeco.2017.06.007

Javed J, Rauf M, Arif M, Hamayun M, Gul H, Ud-Din A, Ud-Din J, Sohail M, Rahman MM, Lee I-J (2022) Endophytic fungal consortia enhance basal drought-tolerance in moringa oleifera by upregulating the antioxidant enzyme (APX) through heat shock factors. Antioxidants 11(9). https://doi.org/10.3390/antiox11091669

Jennings DH (1993) Stress tolerance of fungi. In: Anaerobic fungi. Inc, pp 257–273

Ji W, Hu X, Kang M, Qiu X, Liu B, Tang L, Zhu Y, Cao W, Liu L (2023) Effects of pre-anthesis low-temperature stress on the mineral components in wheat grains. Front Plant Sci 14. https://www.frontiersin.org/articles/10.3389/fpls.2023.1221466

Jogawat A, Yadav B, Chhaya L, N., Singh, A. K., Narayan, O. P. (2021) Crosstalk between phytohormones and secondary metabolites in the drought stress tolerance of crop plants: A review. Physiol Plant 172(2):1106–1132. https://doi.org/10.1111/ppl.13328

Kaur R, Saxena S (2023) *Penicillium citrinum*, a drought-tolerant endophytic fungus isolated from wheat (*Triticum aestivum* L.) leaves with plant growth-promoting abilities. Cur Microbiol 80(5):184. https://doi.org/10.1007/s00284-023-03283-3

Khalvandi M, Siosemardeh A, Roohi E, Keramati S (2021) Salicylic acid alleviated the effect of drought stress on photosynthetic characteristics and leaf protein pattern in winter wheat. Heliyon 7(1):e05908. https://doi.org/10.1016/j.heliyon.2021.e05908

Khan A, Hamayun M, Kang SM, Kim YH, Jung HY, Lee JH, Lee IJ (2012) Endophytic fungal association via gibberellins and indole acetic acid can improve plant growth under abiotic stress: an example of *Paecilomyces formosus* LHL10. BMC Microbiol 12:1–14. https://doi.org/10.1186/1471-2180-12-3

Khan AL, Waqas M, Lee IJ (2015) Resilience of *Penicillium resedanum* LK6 and exogenous gibberellin in improving *Capsicum annuum* growth under abiotic stresses. J Plant Res 128(2):259–268. https://doi.org/10.1007/s10265-014-0688-1

Khan N, Bano A, Zandi P (2018) Effects of exogenously applied plant growth regulators in combination with PGPR on the physiology and root growth of chickpea (*Cicer arietinum*) and their role in drought tolerance. J Plant Interact 13(1):239–247. https://doi.org/10.1080/17429145.2018.1471527

Kumar V, Prasher IB (2023) Phytochemical analysis and antioxidant activity of endophytic fungi isolated from *Dillenia indica* Linn. Appl Biochem Biotechnol 196:332–349. https://doi.org/10.1007/s12010-023-04498-7

Kumar K, Manigundan K, Amaresan N (2017) Influence of salt tolerant *Trichoderma* spp. on growth of maize (*Zea mays*) under different salinity conditions. J Basic Microbiol 57(2):141–150. https://doi.org/10.1002/jobm.201600369

Lata R, Chowdhury S, Gond SK, White JF (2018) Induction of abiotic stress tolerance in plants by endophytic microbes. Lett Appl Microbiol 66(4):268–276. https://doi.org/10.1111/lam.12855

Li N, Kang S (2018) Do volatile compounds produced by *Fusarium oxysporum* and *Verticillium dahliae* affect stress tolerance in plants? Mycol 9(3):166–175. https://doi.org/10.1080/21501203.2018.1448009

Li N, Han X, Feng D, Yuan D, Huang L-J (2019) Signaling crosstalk between salicylic acid and ethylene/jasmonate in plant defense: do we understand what they are whispering? Int J Mol Sci 20(3). https://doi.org/10.3390/ijms20030671

Li J, Chen B, Zhang X, Hao Z, Zhang X, Zhu Y (2021) Arsenic transformation and volatilization by arbuscular mycorrhizal symbiosis under axenic conditions. J Hazard Mater 413:125390. https://doi.org/10.1016/j.jhazmat.2021.125390

Liao J, Jiang J, Xue S, Qingyu C, Wu H, Manikandan R, Hartley W, Huang L (2018) A novel acid-producing fungus isolated from bauxite residue: the potential to reduce the alkalinity. Geomicrobiol J 35(10):840–847. https://doi.org/10.1080/01490451.2018.1479807

Lin W, Gao C, Wang J, Xu W, Wang M, Li M, Ma B, Tian P (2022) Effects of drought stress on Peramine and Lolitrem B in *Epichloë*-endophyte-infected perennial ryegrass. Life 12(8). https://doi.org/10.3390/life12081207

Liu Z, Ma L, He X, Tian C (2014) Water strategy of mycorrhizal rice at low temperature through the regulation of PIP aquaporins with the involvement of trehalose. Appl Soil Ecol 84:185–191. https://doi.org/10.1016/j.apsoil.2014.07.010

Liu J, Wang X, Pu H, Liu S, Kan J, Jin C (2017) Recent advances in endophytic exopolysaccharides: production, structural characterization, physiological role and biological activity. Carbohydr Polym 157:1113–1124. https://doi.org/10.1016/j.carbpol.2016.10.084

Lubna AS, Hamayun M, Gul H, Lee I-J, Hussain A (2018) *Aspergillus niger* CSR3 regulates plant endogenous hormones and secondary metabolites by producing gibberellins and indoleacetic acid. J Plant Interact 13(1):100–111. https://doi.org/10.1080/17429145.2018.1436199

Mahapatra S, Banerjee D (2013) Fungal exopolysaccharide: production, composition and applications microbiol. Insights 6:1–16. https://doi.org/10.4137/MBI.S10957

Marcińska I, Czyczyło-Mysza I, Skrzypek E, Grzesiak MT, Janowiak F, Filek M, Dziurka M, Dziurka K, Waligórski P, Juzoń K, Cyganek K, Grzesiak S (2013) Alleviation of osmotic stress

effects by exogenous application of salicylic or abscisic acid on wheat seedlings. Int J Mol Sci 14(7):13171–13193. https://doi.org/10.3390/ijms140713171
Mastouri F, Björkman T, Harman GE (2012) *Trichoderma harzianum* enhances antioxidant defense of tomato seedlings and resistance to water deficit. Mol Plant-Microbe Interact 25(9):1264–1271. https://doi.org/10.1094/MPMI-09-11-0240
Mathivanan S (2021) Abiotic stress-induced molecular and physiological changes and adaptive mechanisms in plants. In: Abiotic stress in plants, Ch. 15. IntechOpen. https://doi.org/10.5772/intechopen.93367
Mehmood A, Hussain A, Irshad M, Hamayun M, Iqbal A, Khan N (2019) In vitro production of IAA by endophytic fungus *Aspergillus awamori* and its growth promoting activities in Zea mays. Symbiosis 77(3):225–235. https://doi.org/10.1007/s13199-018-0583-y
Molina-Montenegro MA, Oses R, Torres-Díaz C, Atala C, Zurita-Silva A, Ruiz-Lara S (2016) Root-endophytes improve the ecophysiological performance and production of an agricultural species under drought condition. AoB PLANTS 8:plw062. https://doi.org/10.1093/aobpla/plw062
Mukherjee A, Gaurav AK, Singh S, Yadav S, Bhowmick S, Abeysinghe S, Verma JP (2022) The bioactive potential of phytohormones: A review. Biotechnol Rep (Amst) 35:e00748. https://doi.org/10.1016/j.btre.2022.e00748
Nagabhyru P, Dinkins RD, Wood CL, Bacon CW, Schardl CL (2013) Tall fescue endophyte effects on tolerance to water-deficit stress. BMC Plant Biol 13:127. https://doi.org/10.1186/1471-2229-13-127
Naveed M, Hussain MB, Zahir Z, Mitter B, Sessitsch A (2013) Drought stress amelioration in wheat through inoculation with *Burkholderia phytofirmans* strain PsJN. Plant Growth Regul 73:121–131. https://doi.org/10.1007/s10725-013-9874-8
Ocón A, Hampp R, Requena N (2007) Trehalose turnover during abiotic stress in arbuscular mycorrhizal fungi. New Phytol 174(4):879–891. https://doi.org/10.1111/j.1469-8137.2007.02048.x
Oliver MJ, Cushman JC, Koster KL (2010) Dehydration tolerance in plants BT- plant stress tolerance. In: Methods and protocol. Humana Press, pp 3–24. https://doi.org/10.1007/978-1-60761-702-0_1
Orlandelli RC, Vasconcelos AFD, Azevedo JL, Corradi da Silva M de L, Pamphile JA (2016) Screening of endophytic sources of exopolysaccharides: preliminary characterization of crude exopolysaccharide produced by submerged culture of *Diaporthe* sp. JF766998 under different cultivation time. Biochimie Open 2:33–40. https://doi.org/10.1016/j.biopen.2016.02.003
Orozco-Mosqueda M d C, Glick BR, Santoyo G (2020) ACC deaminase in plant growth-promoting bacteria (PGPB): an efficient mechanism to counter salt stress in crops. Microbiol Res 235:126439. https://doi.org/10.1016/j.micres.2020.126439
Pratap D, Harikesh S, Singh B, Prabha R (2016) Microbial inoculants in sustainable agricultural productivity. In: Research perspectives. Springer, pp 233–274. https://doi.org/10.1007/978-81-322-2644-4
Proctor MCF, Pence VC (2022) Vegetative tissues: bryophytes, vascular resurrection plants and vegetative propagules. In: Desiccation and plant survival. CABI Publishing, Wallingford, pp 207–237. https://doi.org/10.1079/9780851995342.0207
Qi W, Zhao L (2013) Study of the siderophore-producing *Trichoderma asperellum* Q1 on cucumber growth promotion under salt stress. J Basic Microbiol 53(4):355–364. https://doi.org/10.1002/jobm.201200031
Qiang X, Ding J, Lin W, Li Q, Xu C, Zheng Q, Li Y (2019) Alleviation of the detrimental effect of water deficit on wheat (Triticum aestivum L.) growth by an indole acetic acid-producing endophytic fungus. Plant Soil 439(1):373–391. https://doi.org/10.1007/s11104-019-04028-7
Qin W, Liu C, Jiang W, Xue Y, Wang G, Liu S (2019) A coumarin analogue NFA from endophytic *aspergillus fumigatus* improves drought resistance in rice as an antioxidant. BMC Microbiol 19(1):50. https://doi.org/10.1186/s12866-019-1419-5
Rehman B, Javed J, Rauf M, Khan SA, Arif M, Hamayun M, Gul H, Khilji SA, Sajid ZA, Kim WC, Lee IJ (2022) ACC deaminase-producing endophytic fungal consortia promotes drought stress tolerance in *M. oleifera* by mitigating ethylene and H_2O_2. Front Plant Sci 13:1–22. https://doi.org/10.3389/fpls.2022.967672

Richardson MD, Chapman GW Jr, Hoveland CS, Bacon CW (1992) Sugar alcohols in endophyte-infected tall fescue under drought. Crop Sci 32(4):1060–1061. https://doi.org/10.2135/cropsci1992.0011183X003200040045x

Roberson EB, Firestone MK (1992) Relationship between desiccation and exopolysaccharide production in a soil *Pseudomonas* sp. Appl Environ Microbiol 58(4):1284–1291. https://doi.org/10.1128/aem.58.4.1284-1291.1992

Roy S, Banerjee D (2019) Volatile organic compounds from endophytic fungi – recent advancement in white biotechnology through fungi. In: Perspective for value-added products and environments. Springer, pp 149–175. https://doi.org/10.1007/978-3-030-14846-1_5

Ruiz-Sánchez M, Aroca R, Muñoz Y, Polón R, Ruiz-Lozano JM (2010) The arbuscular mycorrhizal symbiosis enhances the photosynthetic efficiency and the antioxidative response of rice plants subjected to drought stress. J Plant Physiol 167(11):862–869. https://doi.org/10.1016/j.jplph.2010.01.018

Sadeghi F, Samsampour D, Askari Seyahooei M, Bagheri A, Soltani J (2020) Fungal endophytes alleviate drought-induced oxidative stress in mandarin (*Citrus reticulata* L.): toward regulating the ascorbate–glutathione cycle. Sci Hortic 261:108991. https://doi.org/10.1016/j.scienta.2019.108991

Santra HK, Banerjee D (2021) Production, optimization, characterization and drought stress resistance by β-Glucan-rich Heteropolysaccharide from an Endophytic fungi Colletotrichum alatae LCS1 isolated from clubmoss (*Lycopodium clavatum*). Front Fungal Biol 2:796010. https://doi.org/10.3389/ffunb.2021.796010

Santra HK, Banerjee D (2023) Broad spectrum bioactivity of a novel β-glucan rich heteropolysaccharide, Pestalopine isolated from endophytic fungi *Pestalotiopsis chamaeropsis* CEL6. Arch Microbiol 205(8):284. https://doi.org/10.1007/s00203-023-03620-x

Savchenko T, Kolla VA, Wang C-Q, Nasafi Z, Hicks DR, Phadungchob B, Chehab WE, Brandizzi F, Froehlich J, Dehesh K (2014) Functional convergence of oxylipin and abscisic acid pathways controls stomatal closure in response to drought. Plant Physiol 164(3):1151–1160. https://doi.org/10.1104/pp.113.234310

Sayed AM, Sherif NH, El-Gendy AO, Shamikh YI, Ali AT, Attia EZ, El-Katatny MH, Khalifa BA, Hassan HM, Abdelmohsen UR (2022) Metabolomic profiling and antioxidant potential of three fungal endophytes derived from *Artemisia annua* and *Medicago sativa*. Nat Prod Res 36(9):2404–2408. https://doi.org/10.1080/14786419.2020.1831495

Schardl CL, Grossman RB, Nagabhyru P, Faulkner JR, Mallik UP (2007) Loline alkaloids: currencies of mutualism. Phytochemistry 68(7):980–996. https://doi.org/10.1016/j.phytochem.2007.01.010

Sharma P, Singh SP (2021) Role of the endogenous fungal metabolites in the plant growth improvement and stress tolerance. In: Fungi bio-prospects in sustainable agriculture, environment and Nano-technology: volume 3: fungal metabolites, functional genomics and nano-technology. INC. https://doi.org/10.1016/B978-0-12-821734-4.00002-2

Smiderle FR, Olsen LM, Ruthes AC, Czelusniak PA, Santana-Filho AP, Sassaki GL, Gorin PAJ, Iacomini M (2012) Exopolysaccharides, proteins and lipids in *Pleurotus pulmonarius* submerged culture using different carbon sources. Carbohydr Polym 87(1):368–376. https://doi.org/10.1016/j.carbpol.2011.07.063

Strobel G, Daisy B (2003) Bioprospecting for microbial endophytes and their natural products. Microbiol Mol Biol Rev 67(4):491–502. https://doi.org/10.1128/MMBR.67.4.491-502.2003

Sun L, Lei P, Wang Q, Ma J, Zhan Y, Jiang K, Xu Z, Xu H (2020) The endophyte Pantoea alhagi NX-11 alleviates salt stress damage to Rice seedlings by secreting exopolysaccharides. Front Microbiol 10:3112. https://doi.org/10.3389/fmicb.2019.03112

Syed A, Elgorban AM, Bahkali AH, Eswaramoorthy R, Iqbal RK, Danish S (2023) Metal-tolerant and siderophore producing *Pseudomonas fluorescence* and *Trichoderma* spp. improved the growth, biochemical features and yield attributes of chickpea by lowering Cd uptake. Sci Rep 13(1):4471. https://doi.org/10.1038/s41598-023-31330-3

Tiwari G, Duraivadivel P, Sharma S, Hariprasad P (2018) 1-Aminocyclopropane-1-carboxylic acid deaminase producing beneficial rhizobacteria ameliorate the biomass characters of Panicum maximum Jacq. by mitigating drought and salt stress. Scientific Reports 8(1):17513. https://doi.org/10.1038/s41598-018-35565-3

Tuteja N (2007) Abscisic acid and abiotic stress signaling. Plant Signal Behav 2(3):135–138. https://doi.org/10.4161/psb.2.3.4156

Ullah A, Manghwar H, Shaban M, Khan AH, Akbar A, Ali U, Ali E, Fahad S (2018) Phytohormones enhanced drought tolerance in plants: a coping strategy. Environ Sci Pollut Res 25(33):33103–33118. https://doi.org/10.1007/s11356-018-3364-5

Upadhyay SK, Singh JS, Singh DP (2011) Exopolysaccharide-producing plant growth-promoting Rhizobacteria under salinity condition. Pedosphere 21(2):214–222. https://doi.org/10.1016/S1002-0160(11)60120-3

Waqas M, Khan AL, Kamran M, Hamayun M, Kang SM, Kim YH, Lee IJ (2012) Endophytic fungi produce gibberellins and indoleacetic acid and promotes host-plant growth during stress. Molecules 17(9):10754–10773. https://doi.org/10.3390/molecules170910754

Waqas M, Khan AL, Shahzad R, Ullah I, Khan AR, Lee IJ (2015) Mutualistic fungal endophytes produce phytohormones and organic acids that promote japonica rice plant growth under prolonged heat stress. J Zhejiang Univ Sci B 16(12):1011–1018. https://doi.org/10.1631/jzus.B1500081

Wei H, Li Y, Yan J, Peng S, Wei S, Yin Y, Li K, Cheng X (2023) Root cell wall remodeling: a way for exopolysaccharides to mitigate cadmium toxicity in rice seedling. J Hazard Mater 443:130186. https://doi.org/10.1016/j.jhazmat.2022.130186

Xue Y, Gao Y, Liu C, Liu S (2020) A styrene antioxidant NFA from riparian endophytic fungi enhances flooding tolerance in Arabidopsis. J Plant Interact 15(1):111–116. https://doi.org/10.1080/17429145.2020.1761467

Xue Y, Liu C, Bai X, Cheng F, Chen J, Liu S (2021) Antioxidant metabolites from riparian fungal endophytes improve the tolerance of rice seedlings to flooding. Chemoecology 31(5):277–287. https://doi.org/10.1007/s00049-021-00351-7

Yamaji K, Watanabe Y, Masuya H, Shigeto A, Yui H, Haruma T (2016) Root fungal endophytes enhance heavy-metal stress tolerance of *Clethra barbinervis* growing naturally at mining sites via growth enhancement, promotion of nutrient uptake and decrease of heavy-metal concentration. PLoS One 11(12):e0169089. https://doi.org/10.1371/journal.pone.0169089

Zhang H, Zhu J, Gong Z, Zhu J-K (2022) Abiotic stress responses in plants. Nat Rev Genet 23(2):104–119. https://doi.org/10.1038/s41576-021-00413-0

Zhao L, Wang F, Zhang Y, Zhang J (2014) Involvement of *Trichoderma asperellum* strain T6 in regulating iron acquisition in plants. J Basic Microbiol 54(Suppl 1):S115–S124. https://doi.org/10.1002/jobm.201400148

Zhou XR, Dai L, Xu GF, Wang HS (2021) A strain of Phoma species improves drought tolerance of *Pinus tabulaeformis*. Sci Rep 11(1):7637. https://doi.org/10.1038/s41598-021-87105-1

Fungal Metabolites as Crop Pathogen Controllers

Gayathri Segaran, Pearl Shah, Janani Gurumurthy, and Mythili Sathiavelu

Abstract Fungal infections that attack plants cause the vast majority of illnesses encountered in agricultural and horticultural environments. These phytopathogens have evolved strategies for invading plants, obtaining nutrients, and for its growth and development. The biological control of plant pathogens refers to the reduction of disease-causing agents by microorganisms and their products. Endophytes are plant-associated microorganisms that may survive within plant tissues without infecting or harming the host plant. Many endophytes may produce a wide range of active metabolites, which can be used directly or indirectly in a number of therapeutic applications. Secondary compounds are frequently used as chemical defenses against microbes, insects, higher predators, and even other plants, as pollinator attractants, or as chemical responses to environmental challenges. Secondary metabolites (SM) are often higher-value, lower-volume products than primary metabolites, which are employed commercially as physiologically active substances and are generated in specific cell types and at discrete developmental phases. Fungi inhibited by VOCs and SMs generated by Cupressaceae endophytes that includes *Diplodia seriata, Spencermartinsia viticola, Pyricularia oryzae, and Phaeobotryon cupressi. Alternaria, Aspergillus, Penicillia,* and *Trichoderma.* Acids, alcohols, esters, lipids, and ketones are among the five types of chemicals identified in these endophytes. Antimicrobial metabolites are low-molecular-weight chemicals that are effective at low dosages against pathogenic microbes. Antibiosis is vital in the bio control of plant disease, and it is frequently supplemented by competition and parasitism. Trichoderma is an excellent example of a biocontrol agent that generates a variety of secondary metabolites, including butenolides, gliotoxin, gliovirin, pyrones, and volatile terpenes.

Keywords Antimicrobial compounds · Biocontrol agents · Endophytic fungi · Secondary metabolite · Terpenoids

---

G. Segaran · P. Shah · J. Gurumurthy · M. Sathiavelu (✉)
School of Biosciences and Technology, Vellore Institute of Technology, Vellore, Tamil Nadu, India
e-mail: smythili@vit.ac.in

J. Poveda et al. (eds.), *Fungal Metabolites for Agricultural Applications*, Fungal Biology, https://doi.org/10.1007/978-3-031-76587-2_6

1 Introduction

Food production must be expanded to meet the world's fast-growing population (Yu et al. 2010). Experts predict that the demand for food will continue to escalate over the next four decades, fueled by a burgeoning global population. A staggering 70% uptick in food production will be necessary to satisfy this demand by 2050. Tragically, over 1 billion people currently suffer from malnutrition as a result of inadequate food supplies, while twice that number lack access to essential nutrients and vitamins. This predicament is largely due to the shrinking availability of arable land, resulting in decreased productivity (Fang and Ramasamy 2015). Unfortunately, considerable grain loss remains unavoidable as a result of plant diseases, notably those triggered by phytopathogens (Yu et al. 2010). The majority of diseases found in agricultural and horticultural settings are caused by fungal pathogens that attack plants. These phytopathogens have developed methods to forcefully enter plants, source nutrients, and grow and develop. They are capable of reproducing asexually and/or sexually and can overpower the immune defenses of plants. This negatively impacts plant health, homeostasis, and physiology, and can even cause systemic harm in some cases (Shuping and Eloff 2017). Research indicates that pathogen infections can cause crop yields to decrease by 20% to 40%. Maize, barley, rice, and soybeans can experience losses as high as 12%, while groundnuts and potatoes can suffer losses of up to 24%. Wheat and cotton are particularly vulnerable, with estimated losses reaching 50% and 80%, respectively. It is also noteworthy that post-cultivation losses, which are caused by plant diseases and lower quality, can reach up to 30% to 40% (Fang and Ramasamy 2015). Fungi and oomycetes have played significant roles in causing food crises throughout the history of plant diseases. The potato late blight, established by the oomycete Phytophthora infestans, had catastrophic effects in Europe and Ireland in 1845. It poses an unprecedented threat to agriculture (De Silva et al. 2019). To cause fungal diseases in the plant, the fungal phytopathogens need a suitable host where they can germinate on the surface. However, the spores of the plant fungal pathogen can only grow if favorable circumstances include a suitable host, sufficient humidity from rain or dew, and low molecular mass nutrients. These fungal spores can remain viable for years by using self-inhibitors until the conditions become favorable. Once the conditions are advantageous, plant fungal diseases generate infection structures such as the infection peg and the appressorium, which aid in hyphae penetration. Some pathogens, like *Colletotrichum gloeosporioides,* affect avocados by using host waxes to infect the host (Shuping and Eloff 2017). Dealing with plant diseases is a major expense in crop production. Traditionally, agricultural ecosystems have used several approaches to combat diseases, including breeding resistant crop varieties, utilizing fungicides to eradicate any fungal diseases, and maintaining cleanliness to avoid the transmission of tainted soil or seed (O'Brien 2017).

Chemical-manufactured medications with numerous negative effects are being employed to treat various issues. Both human pathogens and phytopathogens are prone to developing "drug resistances,' 'which significantly reduce the efficacy of previous antibiotics (Yu et al. 2010). Historically, the use of fungicides dates back to the discovery of the Bordeaux mixture in 1885, the first generation of fungicides, which was used to control mildew in grapes. This was followed by the second generation, dithiocarbamates, which act on plant surfaces without penetrating tissues. The third generation, developed in the 1970s–1980s, was capable of penetrating the tissues of plants and suppressing infections. Finally, the fourth generation of fungicides, which has been in use since the 1980s, interferes with fungal penetration and activates host-plant resistance (De Silva et al. 2019). In the 1970s, chemical pesticides greatly contributed to improving agricultural production and minimizing damage caused by pests and diseases. However, the persistent application of pesticides since the development of high-yielding cultivars during the Green Revolution has raised a number of concerns. These include residues of pesticides, their harmful effects on plants, environmental pollution, and potential health risks for humans (Raguchander et al. 2011). Repetitive use of certain chemicals can leave harmful residues behind and even contribute to the development of resistance to pathogens. Accordingly, there has been an increasing interest in identifying safer, non-chemical methods for controlling diseases that are both effective and present lower risks to human health and the environment (Talapatra et al. 2017). One such alternative is the implementation of bio-control agents, which can replace fungicides, and aid in the control of plant pathogens, promote the production of safe food, and decrease environmental pollution (Talapatra et al. 2017).

Biological control of plant pathogens refers to the reduction of disease-causing agents by organisms other than humans (Arwiyanto 2014). Biological control has been extensively researched as a substitute for chemical disease control, especially in recent years (Collinge et al. 2022). Plants exist in a variety of habitats, interacting with a wide range of microorganisms. The discovery that microbes and plants communicate as a dynamic entity to develop a combined response to external challenges imposed on them gave rise to the 'holobiont' concept. Endophytes are microorganisms linked to plants that may live within plant tissues without causing infection or other harm to the host plant. Plant endophytes are mostly bacteria and fungi, although archaebacteria, algae, protozoa, and nematodes are also found as endophytes and have significant consequences for plants (Burragoni and Jeon 2021). It has been known since the late 1800s that fungi can exist inside plants, and De Bary coined the word "endophyte" in 1866. Endophytic fungi have been connected with plant tissues for over 400 million years and are frequently employed for defining the mycota found within live plants. They are increasingly recognized as vital components of biodiversity. Endophytic fungi lack a teleomorphic state, with the majority being ascomycetes, and are highly diverse (Rajamanikyam et al. 2017). These organisms can reside within a plant for their entire lifespan or just during specific stages. The interaction between plants and endophytes can range from mutualism to commensalism, amensalism, and even parasitism (Collinge et al. 2022). Endophytes colonize the interior tissues of healthy plants, such as the leaves, stem, and root, without generating any visible signs in the host plant. Endophytes are so important to fungal biodiversity that they not only influence the composition and diversity of the plant kingdom but are additionally impacted by host plant components

(Erfandoust et al. 2020). Endophytes can be found in a range of plants across various families and classes, growing under different geographical and ecological circumstances. The dynamics between the host and endophyte can differ from one host to another. Research indicates that the association between the host plant and endophyte can help regulate pathogen-host interactions, though it may not necessarily be a symbiotic one (Jalgaonwala et al. 2011).

Endophytic fungi use a wide range of enzymes, such as pectinase, cellulase, xylanase, and laccase, to break down the cell walls of host plants. This allows them to enter, inhabit, and multiply within the plant tissue by changing the structural and physical properties of the cell walls. When the plant's cell walls are degraded, the plant activates its natural defense mechanisms, such as pattern-recognition receptors (PRRs), to protect itself. To survive inside the host plant, endophytes need to avoid detection by the plant's immune system. One way to do this is by hiding chitin, which can be done by modifying it or using chitin deacetylases to create chitin-derived oligomers (Jha et al. 2023). Endophytes are classified as systemic/true endophytes or non-systemic/transient endophytes depending on genetics, biology, and transmission methods. Systemic endophytes are generally mutualistic, invade the plant's intercellular compartments systemically, and are transferred vertically by seeds. Non-systemic endophytes, on the other hand, are a taxonomically varied collection of microorganisms that are not systemic in host plant colonization and are not vertically transferred (Burragoni and Jeon 2021). The dynamic between endophytes and plants can exhibit a wide range of outcomes, from being adversarial to cooperative. There is a prevailing theory that the cooperative relationship developed from previous parasitic behavior. Several factors, such as evolutionary background, geographic location, interactions with other species, and environmental conditions, can influence the level of antagonism or mutualism exhibited between the two species (Rajamanikyam et al. 2017). Endophytes are classified into two types: obligatory endophytes, which rely on the plant's metabolism for survival, and facultative endophytes, which spend a portion of their life cycle outside the host and are typically found in surrounding soil conditions (Jha et al. 2023).

Fungal endophytes are a diverse group of fungi, mainly consisting of ascomycetes such as *Alternaria, Aspergillus, Colletotrichum, Cladopsorium, Fusarium, Penicillium, Talaromyces,* and *Trichoderma* species. They produce metabolites that are rich in various properties,beneficial to biotechnology, such as antifungal, antimicrobial, anti-insect, antioxidative, and anticancer compounds. These metabolites are a valuable source of novel and potent compounds (Sharma et al. 2023). These endophytes not only contribute to the adaptation of plants to biotic stresses but also to abiotic stress by increasing tolerance to conditions such as drought, water stress, higher temperatures, and salinity (Lugtenberg et al. 2016). Approximately 300,000 plant species are on the planet, and there are still many plants that need to be investigated for their fungal endophytes and novel bioactive compounds that they may generate (Amirzakariya and Shakeri 2022).

2 Antagonistic Impact of Fungal Endophytes on Phytopathogens

2.1 Direct Mechanisms Involved in Plant Protection

Antibiosis

Antimicrobial metabolites are low-molecular-weight compounds that are potent against pathogenic microbes at low doses (Baker and Satish 2012). Many endophytes may synthesize a variety of active metabolites, which can be employed directly or indirectly for a variety of medicinal applications. These secondary metabolites are postulated to perform the same role as antibiotics in reducing drug use and may contribute to the treatment of malignancies or subsequent infections that are difficult to treat with existing treatments (Erfandoust et al. 2020). Endophytic fungi are able to generate metabolic substances that can effectively control potential plant pathogens. This is achieved through an antagonistic relationship known as antibiosis (Adeleke et al. 2022). Antibiosis plays a crucial role in the biocontrol of plant disease, often complemented by competition and parasitism. *Trichoderma* is a prime example of a biocontrol agent that produces a range of secondary metabolites, such as butenolides, gliotoxin, gliovirin, pyrones, and volatile terpenes, with variations depending on the strain. Exploring these extracellular secondary metabolites associated with biocontrol offers a chance to clone genes that encode their biosynthesis and potentially bolster the efficacy of a given biocontrol agent. When it comes to antibiosis, biocontrol can be enhanced in two ways: by giving it genes that produce antibiotics typically found in other bacteria and fungi or by increasing the amount of antibiotics the biocontrol agent creates through genetic recombination (Raguchander et al. 2011). Alamethicins, flocculosin, glisoprenins, gliovirin, heptelidic acid, harzianic acid, massoilactone, polyketide, terpenoid/steroid, 6-penthyl-α-pyrone, peptaibol, peptaibols, tricholin, and viridin are the toxic substances involved in the process of antibiosis (Olowe et al. 2022).

Endophytic microbes employ a combination of direct and indirect techniques to bolster plant growth and safeguard against harmful pathogens. When utilized as bioinoculants, they can induce a range of physiological and phenotypic modifications in plants, heightening their capacity to withstand both biotic and abiotic stressors. Moreover, endophytic fungi produce highly beneficial metabolites that can stimulate antibiosis against phytopathogens, rendering them indispensable for effective plant disease management. Yet, the complete range of potential inherent in these metabolites has yet to be fully realized (Adeleke et al. 2022). Several investigations have documented that endophytes develop unique antimicrobial secondary metabolites as a defense strategy against pathogenic invasion. Currently, a large number of antimicrobial compounds have been discovered in a fraction of the 1 million diverse endophyte species. It is speculated that searching for natural chemicals created by endophytes might be a viable strategy to overcome the problem of microbes growing resistant to several regularly used medications (Yu et al. 2010).

Five categories of compounds have been identified in these endophytes, including acids, alcohols, esters, lipids, and ketones. Based on previous research, esters are found to be the most potent class of inhibitory chemicals, with isoamyl acetate being the most physiologically active. One specific endophytic fungus, *Pestalotiopsis microspora*, has been detected in various rainforests and produces a compound known as ambuic acid. This particular compound acts as an antifungal and anti-oomycete agent that is active against numerous *Fusarium* species as well as *Pythium ultimum* (Lugtenberg et al. 2016). *Diplodia seriata, Spencermartinsia viticola, Pyricularia oryzae*, and *Phaeobotryon cupressi* are the fungi inhibited by the VOCs and SMs produced by the endophytes of Cupressaceae. The most potent fungus was from the genera *Alternaria, Aspergillus, Penicillia,* and *Trichoderma* (Erfandoust et al. 2020). Nine guaiane sesquiterpenoids (296–304) were identified in *Xylaria* sp. YM 311647 obtained from *Azadirachta indica.* These compounds were the most active against *Candida albicans*, with MICs of 32 g/mL. The antibacterial activity of all metabolites against *Aspergillus niger, Pyricularia oryzae*, and *Hormodendrum compactum* was moderate to poor. Seiricardine D (33), a biscyclic sesquiterpene derived from *Cytospora* sp. of *Ceriops tagal,* inhibited *Magnaporthe oryzae* with a MIC of 839.6 M (Amirzakariya and Shakeri 2022). *Phomopis cassiae*, an endophytic fungus discovered within *Cassia spectabilis*, has been found to yield five derivatives of cadinane sesquiterpenes. One of these derivatives, known as 3,11,12-trihydroxycadalene, has exhibited a notably high level of antifungal activity against both *Cladosporium cladsporioides* and *Cladosporium sphaerospermum.* Volatile oil can have antibiotic properties. *Muscodor albus*, a fungus found in tropical trees, produces various organic compounds, including aciphyllene, 2-butanone, tetrohydofuran, and 2-methyl furan, which are known to have antibiotic effects (Gao et al. 2010). VOCs are low molecular weight hydrophobic compounds with high vapor pressure. The production of VOCs (volatile organic compounds) depends on the availability of nitrogen, carbon, sulfur, and energy obtained from primary metabolic pathways such as the methylerythritol phosphate (MEP), mevalonic acid (MVA), shikimate, and lipoxygenase (LOX) pathways. These pathways use precursors such as pyruvate, acetyl-CoA, E4P, and PEP. Fungal endophytes secrete VOCs that significantly affect plant growth and defense mechanisms, making them useful as "green chemicals" (Sharma et al. 2023).

Mycoparasitism

Mycoparasitism is a process where *Trichoderma* parasitizes or grows on plant fungal pathogens to prevent their growth. This mechanism has been extensively studied for the biocontrol of numerous phytopathogens. Trichoderma fungi attack the pathogens directly by penetrating their cells and acquiring nutrients from them (Olowe et al. 2022). Mycoparasitism is a phenomenon whereby one microorganism parasitizes another organism. In certain strains of *Trichoderma*, this parasitic relationship is observed, as it parasitizes *Fusarium oxysporum*. *Trichoderma* achieves this through the development of haustoria and the generation of secondary metabolites or enzymes to obtain nutrients from the fungal pathogens (De Silva et al. 2019). The process can be divided into four distinct stages. Initially, the fungal pathogens are detected, followed by the chemotropic growth of the endophytic fungal mycelium towards the pathogenic fungi. In the third step, the endophytic fungi establish physical contact with the pathogenic fungi. Finally, in the fourth step, the endophytic fungi infiltrate the target fungal cells, leading to the breakdown of the cell wall (Akram et al. 2023).

Competition for Nutrients and Space

Competition among microbes plays a crucial role in determining which ones inhabit plant tissue, and endophytes may help prevent pathogens from colonizing. Endophytic fungi can colonize plant tissue both within and outside of the tissues, allowing them to rapidly inhabit and feed on available nutrients. By occupying the space that could have been filled by potential pathogens, they may help inhibit their growth (Adeleke et al. 2022). Fungal endophytes have various roles, such as shielding plants from plant parasitic nematodes, phytopathogens, and herbivorous insects. They are natural adversaries of plant pathogens and function as microbial biocontrol agents (MBCAs) to manage plant diseases without causing harm to the host plant, humans, or the environment. Additionally, the risk of pathogen resistance evolving is highly unlikely compared to agrochemicals, according to Shazia Akram. Observations have revealed that the endophytic fungus, *T. viride* demonstrates antagonistic behavior toward various phytopathogens. However, it has been discovered that the hyphae of *T. viride* tend to overtake the pathogen mycelium (Talapatra et al. 2017).

2.2 *Indirect Mechanisms for Plant Protection*

Induced Plant Defence Mechanism

Fungal endophytes can help plants compete against pathogens by boosting their immune status, increasing stress tolerance, and providing protection against diseases. However, achieving stable growth and development of these biocontrol agents under field conditions remains a challenge due to unfavorable environmental factors (Akram et al. 2023). According to the "balanced antagonism" hypothesis, an endophyte can effectively regulate its host's defense mechanisms, withstand the toxicity of the host's metabolites, and thrive within the host without any discernible signs of infection or illness. This mechanism allows the endophyte to develop self-protection and remain resilient against the host's harmful toxins (Rajamanikyam et al. 2017). To assess whether an endophyte activates a resistance response towards a particular pathogen, two standards are employed to assess and classify plant reactions. The initial standard requires that the induced responses efficiently manage the targeted phytopathogens. It must be established that the fungal-induced response can efficiently exterminate the pathogens, and the observed expressions should be correlated with impeding pathogen infection. The second standard entails that the elimination of the phytopathogen should align with Koch's postulate. This can be verified by monitoring the defense response expression following the introduction of the pathogen to the plant (Adeleke et al. 2022). Along with their antibiotic properties, lipopeptides also serve as crucial signaling molecules. They have a significant impact on various processes, including motility, host plant defense mechanisms, and the formation of microbial biofilms on plant surfaces. When inducing ISR, it is preferable to establish a primed state rather than full induction to ensure an enhanced reaction to biotic or abiotic stimuli (O'Brien 2017). Various tactics are employed by fungal pathogens to invade plants. Mechanical pressure and chemical reactions are commonly used, as well as exploiting wounds and stomata. To survive, these pathogens must overcome the plant's defense mechanisms, which include effector-triggered immunity (ETI), pathogen-associated molecular patterns (PAMP) defenses, and PAMP-triggered immunity (PTI). Pathogens that effectively beat PTI and ETI, particularly by altering their avirulent genes to overcome plant R-resistance genes, have the potential to produce severe plant illnesses that might lead to epidemics (Shuping and Eloff 2017). In the context of plant pathology, a pathogen's invasion of a host plant can elicit a response known as systemic acquired resistance (SAR). This response is facilitated by salicylic acid and results in the accumulation of pathogenesis-related (PR) proteins. Conversely, when non-pathogenic rhizobacteria evoke a response known as induced systemic resistance (ISR), it is mediated by either jasmonic acid or ethylene and does not result in the accumulation of PR proteins. Lipopolysaccharides, polysaccharides, and glycoproteins are elicitors that stimulate plant defense and the production of secondary metabolites, effectively suppressing pathogen attacks (Gao et al. 2010). Host-produced microbial metabolites are crucial in providing resistance to invading pathogens. Fungi-specific metabolite inducers can trigger the plant defense response. Plants can easily recognize microbial-associated molecular patterns (MAMPs) and pathogen-associated molecular patterns (PAMPs) as compounds of microbial origin. Chitin and β-glucans, which are components of the fungal cell wall, are important MAMPs that plants can recognize, according to Shazia Akram. There are genes in endophytic microbes that create Microbe-Associated Molecular Patterns (MAMPs) inhibitors. These molecules stop MAMPs from being recognized by plants and trigger their immune response (Jha et al. 2023).

Plant Growth Promoting Characteristics

Plants have the remarkable ability to form symbiotic relationships with microbes, which offer a multitude of benefits. These associations can boost plant growth, enhance drought tolerance, repel insect herbivores, and ward off fungal pathogens. Additionally, microbes produce bioactive compounds that aid in the protection and growth of their plant hosts. For instance, they might synthesize indole-3-acetic acid

(IAA) to promote cell division, elongation, and fruit development, or siderophores to transport essential iron back to the microbial cell. These substances can either be stored intracellularly for future use or released into the environment for immediate impact (Aramsirirujiwet et al. 2016). *Trichoderma* fungi also compete with the pathogens by starving or depriving the plant of its basic life needs. Several strains of *Trichoderma* fungi have been found to secrete substances like siderophores, which starve the pathogens of essential elements. They also produce cellulose, glucan, and chitin to decrease carbon availability and proteins to enhance root colonization, providing protection against pathogens (Olowe et al. 2022). Endophytes actively benefit plants by assisting in the acquisition of nutrients such as phosphorus, iron, and nitrogen from their surroundings or by generating phytohormones, notably gibberellic acid, auxins, ethylene, and IAA (Burragoni and Jeon 2021). Indole-3-acetic acid (IAA), is an essential type of auxin that contributes to plant growth. It plays a significant role in the development and growth of both shoots and roots. Endophytic and soil fungi are responsible for producing plant growth-promoting compounds, including IAA and gibberellins. Trichoderma isolate found in the rhizosphere region is particularly effective in producing IAA. The IAA-producing fungal endophytes are found in various plant species, such as *Solanum nigrum, Withenia somnifera, Camellia sinensis, Coffea arabica, Piper nigrum, Curalluma acutangula*, and *Teucrium polium* (Segaran and Sathiavelu 2023). The endophyte B06, which has been identified as *Lasiodiplodia pseudotheobromae,* demonstrated the most significant production of IAA at a rate of 37.034 μg/ml. Additionally, the endophyte exhibited a noticeable orange halo that is indicative of the production of siderophores (Aramsirirujiwet et al. 2016).

Endophytic microorganisms perform an important function in plant development and maintaining plant health through their involvement in metabolic pathways. These microbes are capable of producing hydrogen cyanide and exopolysaccharide, colonizing plant tissues, and encouraging the expression of certain genes involved in secretion systems and secondary metabolite release, thereby contributing to their biocontrol potential (Adeleke et al. 2022). The utilization of *Trichoderma harzianum, T. asperellum,* and *Paecilomyces formosus* has been shown to significantly improve both the shoot length and plant biomass of *Capsicum chinense* seedlings. Similarly, *Beauveria brongniartii*, derived from *Carica papaya* has been found to increase fruit diameter. These findings suggest that the use of these specific fungi may have practical applications in the cultivation of these particular plant species (Segaran and Sathiavelu 2023). Endophytic microorganisms solubilize organic and inorganic forms of phosphorous by processes such as chelation and exchange, pH reduction, and acidification via the production of organic acids. The main process is the generation of organic acids like acetic acid and lactic acid that chelate the cations associated with phosphate and eventually convert insoluble phosphorous to soluble form (Burragoni and Jeon 2021). Fungal endophytes that solubilize phosphorus (P-solubilizing) are usually found in the genera *Piriformospora, Aspergillus, Curvularia,* and *Penicillium,* while endophytic arbuscular mycorrhizal (AM) fungi are found in the genera *Acaulospora, Entrophospora, Gigaspora, Glomus, Scutellospora, and Sclerocystis.* The efficiency of phosphate solubilization by five endophytic fungi extracted from *Taxus wallichiana* roots was studied. All isolates were capable of solubilizing phosphate using substrates such as phosphate, aluminum, calcium, and iron, in addition to the release of phosphatases and phytases (Sharma et al. 2023).

Hydrolytic Enzyme Production

The *Trichoderma* strain has a remarkable ability to suppress phytopathogenic fungi, its unique lytic enzymes work synergistically to enhance its antagonistic ability. In addition to producing antibiotics, *Trichoderma* excretes cell wall hydrolytic enzymes that infiltrate cells and dissolve fragments of host cells. This approach increases the synthesis of enzymes and triggers a chain of physiological changes, leading to rapid and efficient growth of *Trichoderma* (Talapatra et al. 2017). Numerous microorganisms are known to produce lytic enzymes that catalyze the breakdown of DNA, chitin, cellulose, proteins, and hemicellulose (Gao et al. 2010). The release of enzymes that can effectively break down cell walls, including chitinases, cellulases, β-1,3-glucanases, and proteases, is a critical factor in the control of plant pathogenic fungi through the use of fungal and bacterial antagonists. These biocontrol agents can produce multiple enzymes from a single gene, simplifying the process of extracting and transferring these genes to other biocontrol agents (Raguchander et al. 2011).

3 Fungal Metabolite Classification

3.1 Primary Metabolites

Primary metabolites are crucial chemicals that are necessary for growth to take place. These include proteins, carbohydrates, nucleic acids, and lipids (Pott et al. 2019). If the primary products cannot be acquired from the growing medium, it is necessary to synthesize them. The growth of the fungus is dependent upon the crucial and readily apparent functions performed by these main metabolites. Primary metabolites are commonly linked to the rapid early growth period of the organism, with maximum production occurring towards the conclusion of this phase (Sanchez and Demain 2008). Upon entering the stationary phase of growth, it is noteworthy that the fungus has the potential to undergo further metabolism of primary metabolites. Several primary metabolites are known to be created in significant quantities, including enzymes, lipids, alcohol, organic acids, and low molecular weight molecules (Maeda 2019) (Fig. 1).

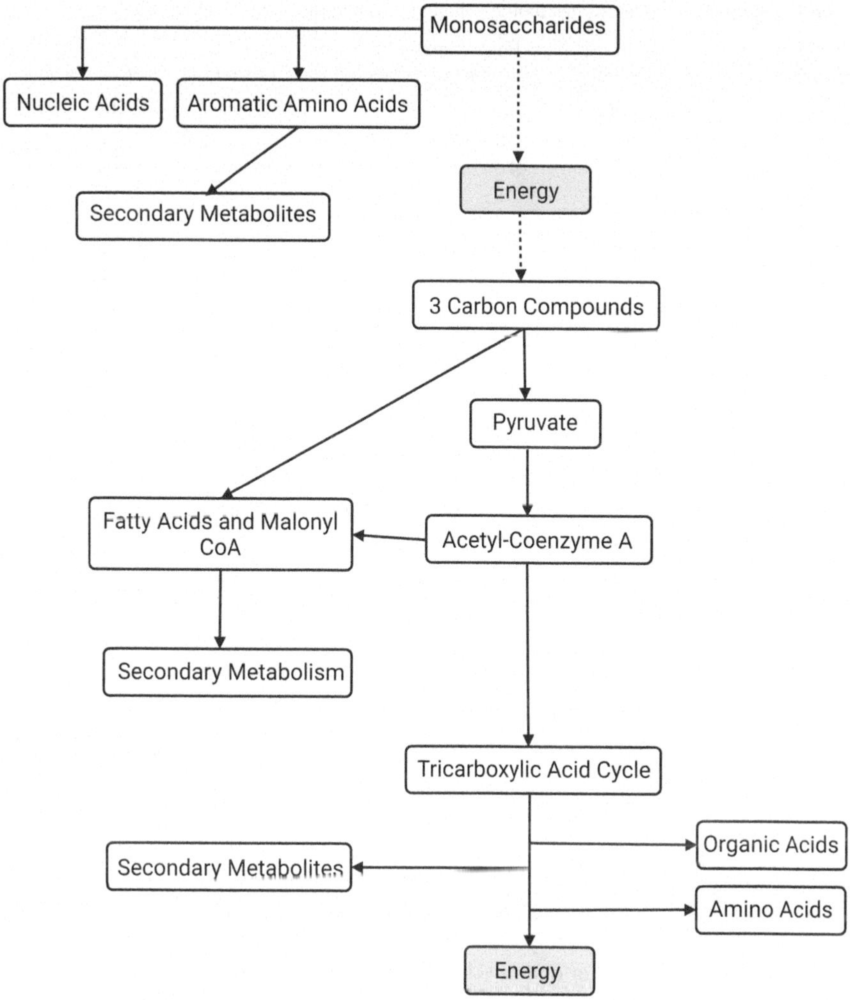

Fig. 1 An overview of central metabolic pathways of fungi showing energy-yielding pathways providing the precursors for biosynthesis of various metabolic products

3.2 Secondary Metabolites

Secondary metabolites (SMs) are chemical compounds produced by particular biosynthetic pathways that are not required for regular fungal growth and development in the laboratory. However, they are found in a wide range of taxa, implying that their persistence in evolution implies a competitive advantage in nature (Avalos and Limón 2021). Secondary compounds are often chemical defenses against microorganisms, insects, higher predators, and even other plants, or they are pollinator

Table 1 Examples of different metabolites obtained from various chemical families

Chemical Family	Metabolite	Fungal family producing metabolite	Function	References
Polyketides	Aflatoxin	*Aspergillus flavus* *Aspergillus parasiticus*	Mycotoxin	Klich (2007)
	Lovastatin	*Aspergillus terreus*	HMG-CoA reductase inhibitor	Mulder et al. (2015)
	Bikaverin	*Fusarium spp.*	Antibiotic	Limón et al. (2010)
	Zearalenone	*Fusarium graminearum*	Mycotoxin	Kim et al. (2005)
	Fumonisin B1	*Fusarium verticillioides* and related Fusarium species	Mycotoxin	Nelson et al. (1993)
Non ribosomal peptides (NRPs)	Apicidin	*Fusarium incarnatum*	Histone deacetylase inhibitor	Jin et al. (2010)
	Enniatin B	*Fusarium spp.*	Mycotoxin (cytotoxic)	Prosperini et al. (2017)
Terpenoids	Gibberellic acid (GA3)	*Gibberella fujikuroi*	Plant hormone	Studt et al. (2012)
	Helvolic acid	*Aspergillus spp.*	Antibiotic (bacteria)	Mitsuguchi et al. (2009)
	Deoxynivalenol	*Fusarium graminearum*	Mycotoxin	Chen et al. (2019)

attractants or chemical adaptations to environmental stresses. Although they have no obvious role in a plant's primary metabolism, secondary compounds frequently play an ecological role (Teoh 2015). Typically, plants accumulate secondary metabolites at a lower rate than primary metabolites. Secondary metabolites are also challenging to obtain. Consequently, secondary metabolites, which are synthesized in specialized cell types and at distinct developmental stages, necessitating extraction and purification, metabolites are typically higher-value, lower-volume products than primary metabolites, which are used commercially as biologically active compounds (pharmaceuticals, flavors, fragrances, and pesticides). As a result, many secondary metabolites might be classified as specialty materials or fine chemicals when compared to primary metabolites (bulk chemicals) (Guerriero et al. 2018). Secondary metabolites are frequently huge chemical compounds that must be produced using a significant number of enzyme processes (Table 1).

Polyketides

Polyketides (PKs) represent the most extensive and prolific group of secondary metabolites. Their biosynthetic pathways commence with the incorporation of acetyl-CoA units, resulting in diverse chemical structures generated through a sequence of enzymatic reactions (Staunton and Weissman 2001). These pathways are initiated by a specialized enzyme known as polyketide synthase (PKS), which was initially identified in bacteria. PKSs can be categorized into three primary types based on their structural characteristics and mechanisms: types I, II, and III (Shen 2003). Type I PKSs are substantial, multifunctional enzymes that bear resemblance to fatty acid synthetases (Herbst et al. 2018). They typically feature essential domains, such as acyltransferase (AT), which recognizes the monomer for synthesis; ketosynthase (KS), responsible for linking it to the growing polyketide chain, and acyl carrier protein (ACP), which houses a phosphopantetheine prosthetic group, serving as a covalent binding site for intermediates (Avalos and Limón 2021). Type II PKSs are extensive, multi-enzymatic complexes predominantly found in prokaryotes, notably absent in fungi based on genome studies (Hertweck et al. 2007). They execute repetitive catalytic processes, typically yielding aromatic molecules (Avalos and Limón 2021). In contrast, Type III PKSs are simpler homodimeric enzymes with fewer functional domains. They are well-known for their involvement in chalcone biosynthesis in plants, contributing to protective functions (Dao et al. 2011). Type III PKSs can directly interact with acyl CoA substrates, distinguishing them from Type I PKSs, which rely on acyl transporter proteins for substrate delivery. Although Type III PKSs exist in fungal genomes, they are far less prevalent compared to Type I PKSs (Avalos and Limón 2021). Polyketides are generally soluble molecules with well-defined chemical structures. However, some exceptions exist, such as melanins, which are insoluble polymeric pigments containing indole and benzene rings (Nosanchuk et al. 2015). Melanin synthesis can occur via two distinct pathways, with the most common one initiated by a type I PKS. In summary, PKs are a highly diverse group of secondary metabolites, with PKS enzymes playing crucial roles in their synthesis, and their classification into types I, II, and III based on structural and functional characteristics highlights their versatility in nature (Eisenman and Casadevall 2011).

Terpenoids

Isoprenoids, also referred to as terpenoids, are a group of natural compounds that have various forms and arise from molecules of terpene with modified carbon skeletons, often incorporating oxygen or other functional groups (Heddergott et al. 2014). They are secondary metabolites produced by both plants and fungi. Fungal terpenoids are derived from acetyl CoA and are formed through two intermediates: isopentyl diphosphate (IPP) and dimethylallyl diphosphate (DMAPP) (Avalos and Limón 2021). The intermediates generated can either be by the mevalonate pathway or the non-mevalonate methyl D erythritol 4-phosphate (MEP) pathway. In the fungal kingdom, terpenoids serve various functions (Matsuda et al. 2016). Some, like botrydial produced by Botrytis cinerea, act as mycotoxins, while others, such as gibberellic acid and carotenoids, function as potent hormones and pigments. Gibberellic acid (GA3), a diterpenoid phytohormone initially isolated from *Gibberella fujikuroi*, is particularly significant in industrial applications. *Gibberella fujikuroi* acts as a growth hormone for plants and can be produced from a fungal strain of *G. fujikuroi* (Fusarium moniliforme) industrially using a SSF (high-output solid-state fermentation) platform (Azizan et al. 2016). This method is preferred due to its cost-effectiveness, higher yield compared to submerged fermentation, and the fungus's ability to produce more GA3 than plants. Due to this, the commercial value of GA3 has increased exponentially and has become a sought-after product in the agricultural sector, notably for enhancing fruit harvests. The advancement of mycological biotechnology has led to the intensive industrialization of GA3 production from fungal sources in recent times (Azizan et al. 2016).

Shikimic Acid

The shikimic acid pathway, also known as the shikimate pathway, is a metabolic system consisting of seven steps. It is utilized by a variety of organisms, including bacteria, archaea, fungi, algae, certain protozoans, and plants. Production of folates and aromatic amino acids (such as tryptophan and phenylalanine) is the primary purpose of this pathway (Santos-Sánchez et al. 2019). The nomenclature of shikimic acid finds its origin in its initial discovery of the highly toxic Japanese shikimi (Illicium anisatum) flower. The metabolic pathway discussed in this context serves as a vital link connecting primary and secondary metabolism in higher plants (Macheroux et al. 1999). In microorganisms, the shikimate pathway plays a pivotal role in the synthesis of essential aromatic amino acids, namely L-phenylalanine (L-Phe), L-tyrosine (L-Tyr), and L-tryptophan (L-Trp), which are fundamental building blocks for protein biosynthesis (Herrmann and Weaver 1999). However, within the realm of plants, these aromatic amino acids play a vital role not only in the synthesis of proteins but also in their capacity as precursors for a wide range of secondary metabolites that hold significant importance in facilitating plant growth.

Non-ribosomal Peptides (NRPs)

Non-ribosomal peptides (NRPs) represent a category of low-molecular-weight peptides renowned for their remarkable chemical diversity (Süssmuth and Mainz 2017). Differing from peptides synthesized on ribosomes during the conventional protein synthesis process, NRPs are generated through a distinctive mechanism. They distinguish themselves from typical proteins not only by their smaller size but also by their frequently cyclic structures and the incorporation of unconventional amino acids, including hydroxylated or methylated variants and D forms. In contrast to proteins, which undergo modifications solely after their synthesis, NRPs experience chemical alterations to their amino acids during their assembly or through the action of other enzymes subsequent to their release. Similar to polyketides (PKs), non-ribosomal peptides (NRPs) are synthesized by substantial multi-enzyme complexes known as NRP synthetases (NRPSs). These complexes are structured into modules, each housing numerous catalytic domains that operate in a coordinated manner

(Finking and Marahiel 2004). A typical module includes an adenylation (A) domain, which is responsible for selecting and activating the substrate. Subsequently, the substrate is transferred to a thiolation domain (T), where it forms a covalent bond with a phosphopantetheine unit. Each module contributes an aminoacyl or aryl residue, culminating in the formation of the final NRP. Other enzymatic domains within the modules can introduce modifications to these residues, such as epimerization (E) or methylation (M), or other chemical alterations. The condensation domains (C) facilitate the formation of peptide bonds between substrates linked to adjacent phosphopantetheine units. These reactions persist until the complete NRP is assembled. The release of the NRP typically involves a cyclic reaction performed by a thioester domain situated at the carboxyl end of the multi-enzyme complex.

After release, many NRPs undergo further chemical modifications by other enzymes, such as oxidation, halogenation, or glycosylation, which further increase their structural diversity and biological activities. The unique and varied chemical properties of NRPs make them valuable in various biological processes and have drawn significant attention in pharmaceutical and biotechnological research (Avalos and Limón 2021). Certain secondary metabolites (SMs) are generated through the synergistic cooperation of polyketide synthases (PKS) and non-ribosomal peptide synthetases (NRPS) (Gil-Serna et al. 2018). These enzymatic complexes are often encoded by distinct genes, collaborating within the same biosynthetic pathway. An example of this phenomenon can be observed in the biosynthetic clusters responsible for ochratoxin A production in various Aspergillus and Penicillium species. The modular structures of Type I PKS and NRPS systems exhibit similarities, facilitating the potential evolution of chimeric genes. These chimeric genes combine PKS and NRPS modules and can adopt two orientations regarding the amino-carboxyl distribution: PKS–NRPS and NRPS–PKS as described by Theobald et al. in 2019. Within these complex multifunctional assemblies, typically comprising a single NRPS module alongside one or more PKS modules, substrates traverse different modules to incorporate corresponding polyketide (PK) or non-ribosomal peptide (NRP) extensions into the final compound. This innovative combinatorial approach significantly enhances the diversity of the resulting compounds (Fisch 2013).

4 Factors Affecting Metabolite Production

4.1 Factors Influencing the Metabolite Production in Plants

Plants face a broad spectrum of environmental stresses, which encompass factors such as extreme temperatures, drought, high salinity, alkalinity, exposure to UV radiation, and susceptibility to pathogen attacks (Fig. 2). These stressors possess the potential to adversely affect plant well-being (Akula and Ravishankar 2011). To amplify productivity or stimulate the biosynthesis of secondary metabolites in plant cell cultures conducted in controlled environments, the technique of elicitation has

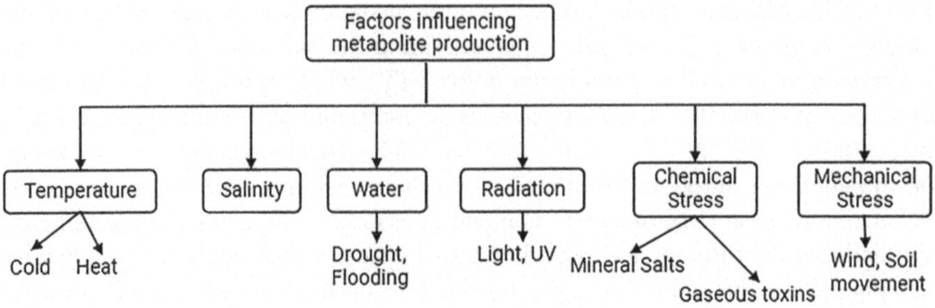

Fig. 2 Factors influencing metabolic production in crops

been extensively employed. Numerous researchers have applied various elicitors to augment the production of secondary metabolites in diverse plant cell, tissue, and organ cultures (Karuppusamy 2009). The production of phenylpropanoids is frequently elevated in response to various environmental stressors, such as pathogen assaults, exposure to UV radiation, intense light, physical damage, nutrient deficiencies, temperature fluctuations, and herbicide exposure (Akula and Ravishankar 2011). Nutrient stress notably influences the levels of phenolic compounds within plant tissues.

Salt Stress: In a saline environment, the high salt concentration causes cellular dehydration, resulting in osmotic stress and subsequent loss of water from the cytoplasm. This process ultimately results in a decrease in both cytosolic and the vacuolar volumes (Mahajan and Tuteja 2005). The exposure of plants to salt stress frequently induces both ionic and osmotic stress, leading to alterations in the levels of certain secondary metabolites in plants, either through their accumulation or depletion (Akula and Ravishankar 2011). The production of various secondary metabolites has been proven to increase due to salt stress.

Drought stress: Drought stress is a prominent abiotic stressor that exerts a substantial impact on the physiological processes and overall growth trajectory of plants (Xu et al. 2010). When the quantity of water present in the soil reaches critically low levels, a phenomenon referred to as drought stress arises. It is exacerbated by meteorological conditions that contribute to ongoing water loss. Drought stress tolerance is observed in all plant species; however, the degree of tolerance differs among different species. Drought stress occurs as a result of a deficiency in water supply, typically coupled with elevated temperatures and solar radiation (Xu et al. 2010).

Influence of heavy metals: The formation of secondary metabolites can also be regulated by metal ions such as lanthanum, europium, silver, and cadmium, as well as oxalate. Nickel (Ni), an essential trace metal, serves a pivotal function as a key component of the urease enzyme and is vital for plant growth and development (Akula and Ravishankar 2011).

Cold stress: The adverse impact of low temperatures on temperate plants is considered to be one of the most detrimental abiotic stressors. These species have demonstrated the ability to adapt to temperature fluctuations by modulating their metabolic processes during the fall season. This adaptation involves an increase in the production of various cryo-protective chemicals, which enhances their ability to withstand cold temperatures (Janska et al. 2010). The cryopreservation technique exposes plants to various environmental stressors, including osmotic stress, desiccation, and low temperatures.

Temperature variations: The influence of temperature on metabolic activity and plant development is significant, with elevated temperatures having the potential to trigger premature leaf senescence. Extensive scholarly research endeavors have been undertaken to investigate the influence of elevated temperatures on the production of secondary metabolites in plants (Akula and Ravishankar 2011). Temperature fluctuations have diverse impacts on the regulation of metabolism, permeability, and the pace of intracellular responses in plant cell cultures (Isah 2019). Modulating the culture temperature has the potential to induce alterations in the physiological and metabolic characteristics of cultured cells, exerting an influence on both their growth patterns and the synthesis of secondary metabolites.

Nutrient stress: The production of secondary metabolites in plants may be heightened in response to stress, as growth is typically more impeded than photosynthesis, resulting in a greater allocation of fixed carbon towards secondary metabolites. The presence of nutrient stress significantly impacts the amounts of phenolic compounds in plant tissues (Isah 2019).

Climate change: Climate change poses a significant risk to biodiversity and is a prominent determinant of well-being as well as human health in the forthcoming decades (Akula and Ravishankar 2011). It is anticipated that cold season crops such as rye, oats, wheat, and apples may experience a decrease in their yield. Plants exhibit a high degree of sensitivity to these alterations and typically do not demonstrate rapid adaptability. Previous studies have demonstrated that exposure to ozone has the capacity to elevate the concentrations of phenolic compounds in conifer species. However, it has been observed that low levels of ozone exposure do not exert any discernible influence on the concentrations of monoterpene and resin acid compounds (Isah 2019). The impact of ozone exposure on crop quality has been investigated in a restricted range of crops.

Environmental factors: The makeup of secondary metabolites is subject to a range of factors, encompassing the local geoclimate, seasonal fluctuations, and external conditions like light, temperature, and humidity (Akula and Ravishankar 2011). The generation of secondary metabolites, like saponins, is contingent upon environmental circumstances and serves as an adaptive mechanism to bolster resilience against abiotic stressors (Isah 2019).

4.2 *Factors Influencing the Metabolite Production in Fungi*

There are various factors that also affect fungal secondary metabolite production (Calvo et al. 2002). The major primary factors include:

Strain selection: This is fundamental. Different strains of same fungi may produce different secondary metabolites.

Nutrient availability: The composition and availability of nutrients in the growth medium significantly impact secondary metabolite production. Fungi require specific carbon and nitrogen sources, as well as minerals and vitamins, to produce secondary metabolites effectively.

Oxygen levels: Oxygen availability affects fungal metabolism and secondary metabolite production. Some secondary metabolites are produced under aerobic conditions, while others are generated under anaerobic or microaerophilic conditions (Akula and Ravishankar 2011).

Genetic modification: The genetic engineering and manipulation of fungal strains offer the potential to augment the production of particular secondary metabolites or activate dormant biosynthetic gene clusters.

Environmental and regulatory factors: Environmental factors, such as light and humidity, can influence the production of certain secondary metabolites in fungi (Lin et al. 2020). Regulatory mechanisms within the fungal cells, including transcription factors and epigenetic modifications, play a role in controlling secondary metabolite production.

Cultivation method: The choice of cultivation method, such as submerged fermentation (in liquid media) or solid-state fermentation (on solid substrates), can impact secondary metabolite production. Different fungi may respond better to specific cultivation methods.

pH and temperature: Fungal growth and secondary metabolite production are often influenced by pH and temperature. The ideal conditions for cultivation differ based on the particular fungal species and the specific secondary metabolite under investigation (Calvo et al. 2002).

5 Applications of Fungal Metabolites in Crop Pathogen Control

5.1 *Fungal Phytopathogens*

Fungi are regarded as some of the most dangerous plant infections due to the fact that they have the capacity to generate mycotoxins, which both people and animals may subsequently ingest. An example is *Aspergillus flavus*, which causes kernel rot in maize and creates aflatoxins. The metabolites produced by endophytes that act antagonistically on fungal pathogen growth can be grouped into a few subcategories based on species (Mousa and Raizada 2013). *Trichoderma* remains among the most prolific secondary metabolite sources with anti-fungal activity. Trichoderma *spp.* derived metabolites include peptaibols, harzianic acid gliotoxins, ETPs (Epipolythiodioxopiperazines), pyrones, pyridines, and trichodermin. The toxicity of ETPs may be traced back to their disulfide bridges, which are responsible for binding to thiol groups, producing reactive oxygen species through redox cycles, and deactivating proteins as a result. Peptaibols are linear peptides that are amphipathic and may self-assemble in membranes to produce ion channels that are voltage-dependent. The antibiotic effects of these substances can be attributed, in large part, to their capacity for this. Trichokonins VI isolated from *Trichoderma koningii* showed antimicrobial efficacy against a range of significant plant diseases, including *Rhizoctonia solani, Fusarium oxysporum, Verticillium dahliae*, and *Botrytis cinerea,* exhibiting broad-spectrum activity. Harzianopyridone (harzianic acid) is derived from *T. harzianum*. It possesses a pyridine ring structure that is arranged in a 2,3-dimethoxy-4-pyridinol configuration. The racemic form of harzianopyridone exhibited potent antifungal action against a plethora of plant-pathogenic fungi, including *Pythium ultimum* and *Gaeumannaomyces graminis* var. *tritici* (Guo et al. 2022). Aside from secondary metabolites alone, some complementary mechanisms are behind *Trichoderma's* success against fungal phytopathogens. One such example is the combination activity of the secondary metabolites and hydrolytic enzymes. *Trichoderma* fungi also secrete chemicals that are similar to elicitors and cause a localized or systemic resistance response in plants. In addition, the role that secondary metabolites play in the struggle for resources is yet another method by which they inhibit phytopathogenic fungi. *Trichoderma spp.* has the capacity to compete with other organisms for space and resources because of its rapid growth rate. In order to prevent other microbes from using iron, *Trichoderma* spp. also secretes siderophores, which are responsible for collecting iron from the surrounding environment. It has been demonstrated that competition for iron is a key factor in the antagonistic action of *T. asperellum* against *F. oxysporum* (causative agent of Fusariam wilt) (Khan et al. 2020, Vinale et al. 2012). Another example is *Purpureocillium lilacinum,* which generates leucinostatins exhibiting a wide array of bioactivity against both bacteria and fungi. The capacity of these chemicals to limit ATP generation in the mitochondria, in addition to several phosphorylation pathways, is the mechanism that underpins the antibacterial and antifungal action that they have. Leucinostatins have been found to inhibit oomycetes growth, preventing the growth of species like *Phytophthora infestans* (Potato late blight) and *P. capsici* (Tomato foliar blight) (Wang et al. 2016).

Epicoccum nigrum is regarded as a saprophytic fungus and can display an endophytic lifestyle, despite the fact that it is classified as a mild plant pathogen in a variety of plants. It has been used as an agent for biocontrol, as it generates metabolites that are effective against *Candida albicans* and bacterial species like *Bacillus subtilis* and *Escherichia coli.* Examples of these metabolites are 2-methyl-3-nonyl prodiginine, and bis (2-ethylhexyl) phthalate, epicorazines, and flavipin. *Epicoccum nigrum* is particularly known for preventing the wheat pathogen *Fusarium graminearum* through metabolite action and upregulating pathogen-related genes in wheat (Lugtenberg et al. 2016; Li et al. 2022). Another example is the metabolite cryptocandin, which is obtained from the endophytic fungus *Cryptosporiopsis quercina*, and the endophyte *Tripterigeum wilfordii*. It shows anti-mycotic action in both human (*C. albicans*) and plant fungi (*Sclerotinia sclerotiorum*). For instance, the tetramic acid known as cryptocin was discovered in the endophytic fungus *Cryptosporiopsis quercina* which is an endophyte of *Tripterigeum wilfordii.* Cryptocandin is only one of the several anti-mycotics produced by *C. quercina.* Cryptocin, a tetramic acid, is another, showing sensitivity to *P. oryzae*, the causative agent of the rice-blast fungus and one of the most pernicious crop pathogen diseases (Li et al. 2000; Strobel et al. 1999). Cordycepsidone, a depsidone metabolite derived from the endophytic fungus *Cordyceps dipterigena* was found to impede the mycelial growth of the pathogenic fungus *Gibberella fujikuroi*, which causes bakanae disease in rice seedlings (Varughese et al. 2012).

5.2 Bacterial Pathogens

Fungal metabolites against bacteria have been extensively exploited for human benefits, with penicillin serving as the foremost and most popular example. In the context of plant health, fungal metabolites with antibacterial properties have a crucial role in defending plants against harmful bacterial pathogens. These metabolites are produced by various fungi, including species of the genus *Trichoderma*, *Rhizobium,* and others. Different strategies can be used by fungus metabolites to prevent bacterial development. Some metabolites, for example, damage bacterial cell membranes, obstruct important cellular functions, or block vital enzymes required for bacterial survival and reproduction. By managing or reducing bacterial infections that might lead to diseases like blight, wilt, or other harmful plant conditions, these fungal metabolites' capacity to inhibit bacterial growth and survival can benefit plants (Dow et al. 2023). The main secondary metabolites generated by *T. harzianum* and other *Trichoderma spp*. that have been found to have an antibacterial action against plant pathogenic bacteria, including *C. michiganensis* and *E. amylovora,* include lysosime, viridiofungin, and trichokonin. These metabolites obtained from *Trichoderma harzianum* have also been effective against *Ralstonia solanacearum*, the cause of bacterial blight in tomatoes. *In vitro* studies have supported the direct bacterial cell death of these metabolites. The antibacterial substances work against the population of bacteria through a number of methods, including the degradation of membranes, DNA, and proteins (Yan and Khan 2021). Colletonoic acid is a substance that was extracted from the genus Colletotrichum. It has antibacterial action against both the fungus *Helminthosporium sativum* and various bacteria (Lugtenberg et al. 2016).

The kiwifruit is easily susceptible to the bacterium *Pseudomonas syringae pv actinidae* (Psa), which causes canker disease. Current methods to control the disease involve copper-based chemicals and the application of streptomycin, both of which have turned out to be ineffective and phytotoxic. However, certain endophytic kiwi-associated fungi, such as *Zopfiella sp.* and *Paraphaeosphaeria sporulosa,* have been identified to produce anti-psa compounds, namely zofielliamides and the antibiotic diketopiperazine. A study by Chen et al., (2022) identified a couple more anti-PSA compounds from *P. sporulosa* belonging to the isocoumarin and isobenzofuranones groups. *Pichia guilliermondii* is an endophytic fungus that is found in the plant *Paris polyphylla yunnanensis*, and among the metabolites isolated, helvolic acid was found to have the most potent anti-bacterial activity on both plant and human pathogens. It was strongest against the bacteria *A. tumefaciens* and *R. solanacearum*. As for fungi, antimycotic activity was observed with *M. oryzae, Fusarium graminearum, Colletotirchum gloeosporioides,* and *Phytophthora capsici*, to name a few (Zhao et al. 2010).

5.3 Nematodes

Nematodes that parasitize plants interact with their hosts in a variety of ways. All of them have projecting, hollow stylets that are frequently referred to as "mouth spears," which are used to penetrate cells to allow feeding and, in the case of endoparasitic species, entry into the host. The parasitic nature of the nematodes causes mechanical damage and nutrient deficiency. Plant parasitic nematodes are classified into several groups, including root-knot nematodes, cyst nematodes, burrowing nematodes, and root lesion nematodes. Among these, the most harmful is root-knot nematodes (*Meloidogyne incognita*), which are known to burrow into the roots as larvae, causing the roots to form "knots" (Jones et al. 2013). *Trichoderma spp.* has been shown to display nematicidal activity, with the secondary metabolites generated by *T. virens* showing the strongest nematicidal action, resulting in the greatest suppression of egg hatch and juvenile mortality. Gliotoxin has been isolated from a wide variety of fungi, including a strain of *T. virens* that demonstrated nematicidal activity along with acetic acid from *T. longibrachiatum*. An amino acid cyclosporin from *T. polysporum* also produced nematicidal activity against *M. incognita* These secondary metabolites were also reported to cause disintegration of internal tissues in nematodes (Khan et al. 2020). Other metabolites from Trichoderma include Serine protease and PRA1 trypsin-like protease, which showed nematicidal action against *M. javanica* juveniles. Various microorganisms have produced volatile organic compounds (VOCs), which have been used as bioagents to control nematodes. *Fusarium oxysporum* produces a number of VOCs, including 2-methyl butyl acetate, ethyl acetate, and 2-methylpropyl acetate, which have nematocidal properties against plant parasitic nematodes. Chaetoglobosin A, an inhibitor of motility and mammalian cell growth, is produced by the fungus *Chaetomium globosum*. It was found to have nematicidal action against *M. incognita* (Bhat et al. 2023).

One unusual source of nematacidal metabolites is *F. oxysporum*, as the genus is well-known for being a plant pathogenic species. However, *Fusarium* species or specific isolates of these species have been found to remain as endophytes inside plants without causing disease symptoms. Numerous studies have shown that the

Fusarium genus of fungus is a rich source of physiologically active secondary metabolites. Fo162 is able to produce a variety of distinct active chemicals, some of which have the potential for dual action, influencing both plant physiology and the nematode at the same time. Some of these metabolites include gibepyrone D, fusarinolic acid, and picolinic acid, with the latter two showing poor nematicidal activity (Bogner et al. 2017). Research conducted in Kenya showed that the inoculation of banana crops with non-pathogenic *F. oxysporum* affected by the nematode species *Helicotylenchus multicinctus* and *Pratylenchus goodeyi* resulted in a reduction of the nematode population and banana root damage. The ability of *Fusarium* to inhibit nematode activity was dependent on the species, as *Meloidogyne* was not affected (Waweru et al. 2014). Aside from the action of secondary metabolites, there also exist nematophagous fungi that kill nematodes by spore production or causing disease. *Drechmeria coniospora* is an example of this, as it produces adhesive spores that affix to and penetrate the cuticle of nematodes. *D. coniospora* also produced 13 metabolites, some of which exhibited nematacidal activity. 5-hydroxymet hylfuran-2-carboxylic acid was among the compounds produced and is a known furan that is found across many fungi, such as *Aspergillus*, where it showed efficient nematicidal activity against *C. elegans* and the pine wood worm *Bursaphelenchus xylophilus*. Furans isolated from the fungus *Coprinus comatus* also had nematicidal effects on *M. arenaria* and *P. redivivus*. This can lead to the inference that furans are a significant source of nematicidal metabolites (Wan et al. 2021). Carbo-furans are very potent commercial nematicides that are toxic to the human body, and their usage is strictly regulated in European territories (Bogner et al. 2017). Other nematophagous fungi include *Harposporium anguillulae,* which produces conidiospores that infect the nematodes. Conidia with a crescent form are abundantly produced, and these spores septate and branching hyphae absorb the worm's nutrients after germinating inside the nematode. *H. anguillulae* was also found to have some metabolites that assist in its infection of nematodes, including a polyketide metabolite with nematode-attracting activity and an indole terpene that has nematicidal activity (Dai et al. 2022).

6 Implementation of Secondary Metabolites for Agricultural Usage and Future Perspectives

As evidenced in this article, there have been numerous research studies dedicated to the biocontrol potential of fungal secondary metabolites; however, comprehensive in-field use on a wide scale is a rather untapped area, despite the abundance of these metabolites and their potential. The usage of biopesticides itself is an area of contention among the beliefs of most conventional farmers. There are very few fungal species that are currently approved for use against fungal phytopathogens, despite the large number of studies (Table 2). However, most, if not all, of currently available products are not made from metabolites, taking the example of *Trichoderma*

Table 2 Fungal metabolites as biocontrol agents against phytopathogens

Fungal Species	Metabolite Produced	Diseases	Pathogenic Species	References
Beauvaria bassiana	Beauvaricin and Bassicridin	Prevents *lepidopteran* insect infestation	Insect species: *Lepidoptera*	Singh et al. (2015)
Trichoderma polysporum	Cyclosporin	Root-knot disease	Nematodal species: *Meloidogyne incognita*	Khan et al. (2020)
Paraphaeosphaeria sporulosa	Zofielliamides and Diketopiperazine	Kiwi-fruit canker disease	Bacterial species: *Pseudomonas syringae pv actinidiae*	Chen et al. (2022)
Pichia guilliermondii	Helvolic acid	Crown gall disease, bacterial wilt, Rice blast fungus, blight	Bacterial species: *A. tumefaciens* and *R. solanacearum* Fungal species: *M. oryzae* and *Fusarium graminearum*	Zhao et al. (2010)
Xylaria sp.	Griseofulvin	Leaf blotch, Fusarium wilt, blast disease	Fungal species: *Alternaria mali, Fusarium oxysporum* and *Magnaporthe grisea*	Macías-Rubalcava and Sánchez-Fernández (2016)
Stachybotrys levispora	Griseofulvin	White mold	Fungal species: *Sclerotinia sclerotiorum*	Ribeiro et al. (2018)
Cordyceps dipterigena	Cordycepsidone	Bakanae disease	Fungal species: *Gibberella fujikuroi*	Varughese et al. (2012)
Trichoderma harzianum	Harzianolide	Stalk rot disease in maize	Fungal species: *Fusarium verticillioides* and *F. proliferatum*	Yassin et al (2021)
Trichoderma brevicompactum	Trichodermin	Seed damping off, root rot, and Fusarium wilt	Fungal species: *Rhizoctum solani* and *Fusarium oxysporum*	Shentu et al. (2014)

(continued)

Table 2 (continued)

Fungal Species	Metabolite Produced	Diseases	Pathogenic Species	References
Cryptosporiopsis quercina	Cryptocandin and Cryptocin	White mold, gray mould, Rice blast disease	Fungal species: *Sclerotinia sclerotiorum*, *Botrytis cinerea*, and *Pyricularia oryzae*	Li et al. (2000), Strobel et al. (1999)
Epicoccum nigrum	Epicorazines and Flavipin	Fusarium head blight	Fungal species: *Fusarium graminearum*	Lugtenberg et al. (2016)
Purpureocillium lilacinum	Leucinostatins	Crown infections, leaf spot, potato blight	Fungal species: *P. capsici* and *P. infestans* Nematodes: Acts as endoparasite of most plant nematodes	Wang et al. (2016)
Pezicula sp.	Ramulosin	Grey mould and tomato leaf mold	Fungal species: *Botrytis cinerea* and *Fulvia fulva*	Wang et al. (2014)
Paraconiothyrium sp.	Globosuxanthone A	Fusarium blight, Grey mould and damping off	Fungal species: *Fusarium graminearum*, *Fusarium solani*, and *Botrytis cinerea*	Miao et al. (2020)
Chaetomium globosum	Flavipin	Fusarium blight	Fungal species: *Fusarium graminearum* various nematodes	Ye et al. (2013)
Aspergillus sp. SPH2	Neo Aspergillic acid and Mellein	Fusarium blight, Grey Mould, black spot	Fungal species: *F. oxysporum, A. alternata*, and *B. cinerea*	Morales-Sánchez et al. (2021)
Aspergillus fumigatus	Verruculogen, Cyclotryprostatin	Grey mould and leaf spot	Fungal species: *B. Cinerea* and *A. Alternata*	Li et al. (2012)
Penicillium canescens	Griseofulvin	Grey mould, Anthracnose,Gummy stem blight, and white mold	Fungal species: *Botrytis cinerea, Colletotrichum orbiculare, Didymella bryoniae* and *Sclerotinia sclerotiorum*	Wang et al. (2010)

spp. There are currently numerous products made from Trichoderma, utilized as biopesticides, biofungicides, and biofertilizers. Most of these formulations consist of *T. harzianum* forming the majority of combintheations (83%), *T. viride* following the second (55%), and then *T. koningi* (28%) (Woo et al. 2014). There is currently no avenue for metabolite-formulated products, as secondary metabolites are still considered a research area, and have not been economically implemented on a wide scale in the agricultural sector. According to Rodrigo et al. (2021), the issue lies in the inability to transfer from *in-vitro* to in-field, as there are many uncontrolled variables. Most of the data obtained has been from a controlled environment that supported endophyte growth and metabolite production. With regards to endophytic fungi, environmental factors may be crucial in establishing the endophyte-plant host link, but also for producing the endophyte's defense response (metabolites) after the association has been made. However, the fungus may also produce other substances in addition to those that have antifungal activity. This multiple production may not occur when the metabolite is used directly in place of the living organism as a bypass, utilizing the metabolite or bioactive compound directly without the plant-fungal intermediary. However, this comes with its own set of problems to consider. It is extremely important to assess the toxicity of the multiple metabolite types across a variety of environments and hosts, and the effectiveness of the metabolite independent from the fungus (Rodrigo et al. 2021). When it comes to endophytes, due to their intracellular lives and development inside healthy plant tissue, biological issues are frequently disregarded or not given adequate consideration. No matter how many control measures are in place when handling a biological entity, they may vary during the process of adapting to different fermentation settings (Ganeshan et al. 2021).

Although there are not any commercially available metabolite-formulated products, there is a prominent case study of utilizing endophytes as an anti-fungal agent. *Epichloe* and *Acremonium spp. endophytes* were used to infect the grasses, and anti-fungal activity was observed against 2 plant pathogens, *Drechslera erythrospila* and *Rhizoctonia zeae*, by the action of certain diffusible substances (Christensen 1996). Fungal endophytes not only display anti-microbial activity but also aid the plant in stress tolerance and adaptivity, enabling better growth. An example of this is *Piriformospora indica*, a mycorrhiza-like endophyte that aids the plant in enhanced root proliferation, bolsters anti-fungal defenses, and helps plantlets overcome transplantation shock (Gill et al. 2016). Following this, *Epichloe* endophytes have been effectively commercialized in perennial ryegrass and tall fescue in South America, Australia, New Zealand, and the USA. Novel *Epichloe* endophytes have been discovered, leading to the development and breeding of grasses with animal-safe characteristics. They can be used in agriculture because they can increase plant survival by shielding them from biotic and abiotic challenges (Lugtenberg et al. 2016). Research focus should be redirected to the efficacy of secondary metabolites as a biocontrol agent, as this is a point that has been proven through decades of studies. The question that must be delved into now is how to implement these products into current agricultural practices and the feasibility of scale-up production. The use of these substances on crops presents an opportunity for the creation of low-cost, sustainable, and eco-friendly agricultural methods.

7 Conclusion

Fungi present a significant hurdle to plant production and can result in severe ramifications for food production. Phytopathogens impede food and floricultural production and also have the potential to impact the safety and yield of medicinal plants post-harvest. The existing measures for controlling plant fungal pathogens are inadequate for managing their emergence and outbreaks. Consequently, more research is vital, including the investigation of plant-based options to develop cost-effective, less hazardous, and efficacious biological products. Endophytes have evolved into an important biological resource repository. Endophytes have the potential to give at least partial, if not total, responses to current issues such as rising strain on the global food supply, climate change, environmental degradation, and the growing incidence of antibiotic-resistant illnesses. New technologies have provided opportunities for endophytic research, which has yet to be fully explored. This source has been found to contain bioactive compounds, as briefly outlined in this review. Researchers are eager to find sustainable and economically viable sources of bioactive compounds, and endophytes have garnered attention as independent biofactories with vast biodiversity. These endophytes are presumed to yield novel natural products that can advance drug discovery. Experts have proposed that incorporating microorganisms as biological control agents in integrated pest management is the most viable and eco-friendly solution in the long run. The isolation of natural substances presents a promising alternative to synthetic chemicals that may result in pathogen resistance and pollute secure habitats like endophytic fungi. The wide range of fungal endophytes presents a significant opportunity to discover novel natural compounds characterized by unique chemical structures that have been perfected through evolution with higher plants.

References

Adeleke BS, Ayilara MS, Akinola SA, Babalola OO (2022) Biocontrol mechanisms of endophytic fungi. Egypt J Biol Pest Control 32:1–17
Akram S, Ahmed A, He P, He P, Liu Y, Wu Y, Munir S, He Y (2023) Uniting the role of endophytic fungi against plant pathogens and their interaction. J Fungi 9:1–23
Akula R, Ravishankar GA (2011) Influence of abiotic stress signals on secondary metabolites in plants. Plant Signal Behav 6:1720–1731
Amirzakariya BZ, Shakeri A (2022) Bioactive terpenoids derived from plant endophytic fungi: an updated review (2011–2020). Phytochemistry 197:1–36

Aramsirirujiwet Y, Gumlangmak C, Kitpreechavanich V (2016) Studies on antagonistic effect against plant pathogenic fungi from endophytic fungi isolated from *Hottuynia cordata* thunb. And screening for siderophore and indole-3-acetic acid production. KKU Resj 21:55–66
Arwiyanto T (2014) Biological control of plant diseases caused by bacteria. J Perlind Tanam Indones 18:1–12
Avalos J, Limón MC (2021) Fungal secondary metabolism. Encyclopedia 2:1–13
Azizan MS, Zamani AI, Stahmann KP, Ng CL (2016) Fungal metabolites and their industrial importance: a brief review. Malays J Biochem Mol Biol 19:15–23
Baker S, Satish S (2012) Endophytes: natural warehouse of bioactive compounds. Drug Invent Today 4:548–553
Bhat AA, Shakeel A, Waqar S, Handoo ZA, Khan AA (2023) Microbes vs. nematodes: insights into biocontrol through antagonistic organisms to control root-knot nematodes. Plan Theory 12(3):451
Bogner CW, Ramsay ST, Kamdem SG, Matthäus C, Hölscher D, Popp J, Proksch P, Florian SAP (2017) Bioactive secondary metabolites with multiple activities from a fungal endophyte. Microb Biotechnol 10:175–188
Burragoni SG, Jeon J (2021) Applications of endophytic microbes in agriculture, biotechnology, medicine, and beyond. Microbiol Res 245:1–16
Calvo AM, Wilson RA, Bok JW, Keller NP (2002) Relationship between secondary metabolism and fungal development. Microbiol MolBiol 66:447–459
Chen Y, Kistler HC, Ma Z (2019) *Fusarium graminearum* trichothecene mycotoxins: biosynthesis, regulation, and management. Annu Rev Phytopathol 57:15–39
Chen Q, Yu JJ, He J, Feng T, Liu JK (2022) Isobenzofuranones and isocoumarins from kiwi endophytic fungus *Paraphaeosphaeria sporulosa* and their antibacterial activity against *Pseudomonas syringae* pv. actinidiae. Phytochemistry 195:113050
Christensen MJ (1996) Antifungal activity in grasses infected with Acremonium and Epichloë endophytes. Australas Plant Pathol 25:186
Collinge DB, Jensen B, Jørgensen HJL (2022) Fungal endophytes in plants and their relationship to plant disease. Curr Opin Microbiol 69:1–14
Dai Z, Gao Y, Zhao P, Li G (2022) Secondary metabolites from the endoparasitic nematophagous fungus *Harposporium anguillulae* YMF 1.01751. Microorganisms 10:1553–1553
Dao TTH, Linthorst HJM, Verpoorte R (2011) Chalcone synthase and its functions in plant resistance. Phytochem Rev 10:397–412
De Silva NI, Brooks S, Lumyong S, Hyde KD (2019) Use of endophytes as biocontrol agents. Fungal Biol Rev 33:131–148
Dow L, Gallart M, Ramarajan M, Law SR, Thatcher LF (2023) Streptomyces and their specialised metabolites for phytopathogen control -comparative in vitro and in planta metabolic approaches. Front Plant, Sci, p 14
Eisenman HC, Casadevall A (2011) Synthesis and assembly of fungal melanin. Appl Microbiol Biotechnol 93:931–940
Erfandoust R, Habibipour R, Soltani J (2020) Antifungal activity of endophytic fungi from Cupressaceae against human pathogenic *Aspergillus fumigatus* and *Aspergillus Niger*. J Mycol Med 30:1–6
Fang Y, Ramasamy RP (2015) Current and prospective methods for plant disease detection. Biosensors 4:537–561
Finking R, Marahiel MA (2004) Biosynthesis of nonribosomal peptides1. Ann Rev Microbiol 58:453–488
Fisch KM (2013) Biosynthesis of natural products by microbial iterative hybrid PKS-NRPS. RSC Adv 3:18228
Ganeshan S, Kim SH, Vujanovic V (2021) Scaling-up production of plant endophytes in bioreactors: concepts, challenges and perspectives. Bioresour Bioprocess 8
Gao F, Dai C, Liu X (2010) Mechanisms of fungal endophytes in plant protection against pathogens. Afr J Microbiol Res 4:1346–1351

Gill SS, Gill R, Trivedi DK, Anjum NA, Sharma KK, Ansari MW, Ansari AA, Johri AK, Prasad R, Pereira E, Varma A, Tuteja N (2016) *Piriformospora indica*: potential and significance in plant stress tolerance. Front, Microbiol, p 7

Gil-Serna J, García-Díaz M, GonzálezJaén MT, Vázquez C, Patiño B (2018) Description of an orthologous cluster of ochratoxin a biosynthetic genes in aspergillus and Penicillium species. A comparative analysis. Int J Food Microbiol 268:35–43

Guerriero G, Berni R, Muñoz-Sanchez J, Apone F, Abdel-Salam E, Qahtan A, Alatar A, Cantini C, Cai G, Hausman JF, Siddiqui K, Hernández-Sotomayor S, Faisal M (2018) Production of plant secondary metabolites: examples, tips and suggestions for biotechnologists. Genes 9:309

Guo R, Li G, Zhang Z, Peng X (2022) Structures and biological activities of secondary metabolites from *Trichoderma harzianum*. Mar Drugs 20:701

Heddergott C, Calvo AM, Latgé JP (2014) The volatome of *Aspergillus fumigatus*. Eukaryot Cell 13:1014–1025

Herbst AD, Townsend AC, Maier T (2018) The architectures of iterative type I PKS and FAS. Nat Prod Rep 35:1046–1069

Herrmann KM, Weaver LM (1999) The shikimate pathway. Annu Rev Plant Biol 50:473–503

Hertweck C, Luzhetskyy A, Rebets Y, Bechthold A (2007) Type II polyketide synthases: gaining a deeper insight into enzymatic teamwork. Nat Prod Rep 24:162–190

Isah T (2019) Stress and defense responses in plant secondary metabolites production. Biol Res 52

Jalgaonwala RE, Mohite BV, Mahajan RT (2011) A review: natural products from plant associated endophytic fungi. J Microbiol Biotech Res 1:21–32

Janská A, Maršík P, Zelenková S, Ovesná J (2010) Cold stress and acclimation-what is important for metabolic adjustment. Plant Biol 12:395–405

Jha P, Kaur T, Chhabra I, Panja A, Paul S, Kumar V, Malik T (2023) Endophytic fungi: hidden treasure chest of antimicrobial metabolites interrelationship of endophytes and metabolites. Front Microbiol 14:1–16

Jin JM, Lee S, Lee J, Baek SR, Kim JC, Yun SH, Park SY, Kang S, Lee YW (2010) Functional characterization and manipulation of the apicidin biosynthetic pathway in *Fusarium semitectum*. Mol Microbiol 76:456–466

Jones JT, Haegeman A, Danchin EGJ, Gaur HS, Helder J, Jones MGK, Kikuchi T, Manzanilla-López R, Palomares-Rius JE, Wesemael WML, Perry RN (2013) Top 10 plant-parasitic nematodes in molecular plant pathology. Mol Plant Pathol 14:946–961

Karuppusamy S (2009) A review on trends in the production of secondary metabolites from higher plants by in vitro tissue, organ and cell cultures. J Med Plant Res 3:1222–1239

Khan RAA, Najeeb S, Mao Z, Ling J, Yang Y, Li Y, Xie B (2020) Bioactive secondary metabolites from Trichoderma spp. against phytopathogenic bacteria and root-knot nematode. Microorganisms 8:401

Kim YT, Lee YR, Jin J, Han KH, Kim H, Kim JC, Lee T, Yun SH, Lee YW (2005) Two different polyketide synthase genes are required for synthesis of zearalenone in *Gibberella zeae*. Mol Microbiol 58:1102–1113

Klich MA (2007) Aspergillus flavus: the major producer of aflatoxin. Mol Plant Pathol 8:713–722

Li JY, Strobel G, Harper J, Lobkovsky E, Clardy J (2000) Cryptocin, a potent tetramic acid antimycotic from the endophytic fungus *Cryptosporiopsis cf.quercina*. Org Lett 2:767–770

Li XJ, Zhang Q, Zhang AL, Gao J-M (2012) Metabolites from aspergillus fumigatus, an endophytic fungus associated with Melia azedarach, and their antifungal, antifeedant, and toxic activities. J Agric Food Chem 60:3424–3431

Li T, Im JH, Lee J (2022) Genetic diversity of *Epicoccum nigrum* and its effects on *Fusarium graminearum*. Mycobiology 50:457–466

Limón MC, Rodríguez-Ortiz R, Avalos J (2010) Bikaverin production and applications. Appl Microbiol Biotechnol 87:21–29

Lin JT, Chen SL, Liu SC, Yang DJ (2020) Effect of harvest time on saponins in yam (*Dioscorea pseudojaponica* Yamamoto). J Food Drug Anal 17:1–8

Lugtenberg BJJ, Caradus JR, Johnson LJ (2016) Fungal endophytes for sustainable crop production. FEMS Microbiol Ecol 92:1–17
Macheroux P, Schmid J, Amrhein N, Schaller A (1999) A unique reaction in a common pathway: mechanism and function of chorismate synthase in the shikimate pathway. Planta 207:325–334
Macías-Rubalcava ML, Sánchez-Fernández RE (2016) Secondary metabolites of endophytic Xylaria species with potential applications in medicine and agriculture. World J Microbiol Biotechnol:33
Maeda HA (2019) Evolutionary diversification of primary metabolism and its contribution to plant chemical diversity. Front Plant Sci 10
Mahajan S, Tuteja N (2005) Cold, salinity and drought stresses: an overview. Arch Biochem Biophys 444:139–158
Matsuda Y, Awakawa T, Mori T, Abe I (2016) Unusual chemistries in fungal meroterpenoid biosynthesis. Curr Opin Chem Biol 31:1–7
Miao C, Wang J, Huang R, Liu S, Zheng K, Liu C, Lu Y, Wu S (2020) Antifungal xanthones produced by the endophytic fungus *Paraconionthyrium* sp. YM 311593. Folia Microbiol 65:567–572
Mitsuguchi H, Seshime Y, Fujii I, Shibuya M, Ebizuka Y, Kushiro T (2009) Biosynthesis of steroidal antibiotic fusidanes: functional analysis of oxidosqualene cyclase and subsequent tailoring enzymes from aspergillus fumigatus. J Am Chem Soc 131:6402–6411
Morales-Sánchez V, Díaz CE, Trujillo E, Olmeda SA, Valcarcel F, Muñoz R, Andrés MF, González-Coloma A (2021) Bioactive metabolites from the endophytic fungus *Aspergillus* sp. SPH2. J Fungus 7:109
Mousa WK, Raizada MN (2013) The diversity of anti-microbial secondary metabolites produced by fungal endophytes: an interdisciplinary perspective. Front Microbiol:4
Mulder KCL, Mulinari F, Franco OL, Soares MSF, Magalhães BS, Parachin NS (2015) Lovastatin production: from molecular basis to industrial process optimization. Biotechnol Adv 33:648–665
Nelson BE, Rosenfield AT, Schwartz PE (1993) Preoperative abdominopelvic computed tomographic prediction of optimal cytoreduction in epithelial ovarian carcinoma. J Clin Oncol 11:166–172
Nosanchuk JD, Stark RE, Casadevall A (2015) Fungal melanin: what do we know about structure? Front Microbiol:6
O'Brien PA (2017) Biological control of plant diseases. Australasian Plant Pathol 46:293–304
Olowe OM, Nicola L, Asemoloye MD, Akanmu AO, Babalola OO (2022) Trichoderma: potential bio-resource for the management of tomato root rot diseases in Africa. Microbiol Res 257:1–4
Pott DM, Osorio S, Vallarino JG (2019) From central to specialized metabolism: an overview of some secondary compounds derived from the primary metabolism for their role in conferring nutritional and organoleptic characteristics to fruit. Front Plant Sci 10
Prosperini A, Berrada H, Ruiz MJ, Caloni F, Coccini T, Spicer LJ, Perego MC, Lafranconi A (2017) A review of the mycotoxin Enniatin B. Front Public Health 5
Raguchander T, Saravanakumar D, Balasubramanian P (2011) Molecular approaches to improvement of biocontrol agents of plant diseases. J Biol Control 25:71–84
Rajamanikyam M, Vadlapudi V, Amanchy R, Upadhyayula SM (2017) Endophytic fungi as novel resources of natural therapeutics. Braz Arch Biol Technol 60:1–26
Ribeiro AI, Costa ES, Thomasi SS, Ribeiro F, Vieira P, Fernandes JB, Forim MR, Ferreira AG, Pascholati SF, Pascholati F, Fátima M (2018) Biological and chemical control of *Sclerotinia sclerotiorum* using *Stachybotrys levispora and* its secondary metabolite Griseofulvin. J Agric Food Chem 66
Rodrigo S, García-Latorre C, Santamaria O (2021) Metabolites produced by fungi against fungal phytopathogens: review, implementation and perspectives. Plan Theory 11:81
Sanchez S, Demain AL (2008) Metabolic regulation and overproduction of primary metabolites. Microb Biotechnol 1:283–319

Santos-Sánchez NF, Salas-Coronado R, Hernández-Carlos B, Villanueva-Cañongo C (2019) Shikimic acid pathway in biosynthesis of phenolic compounds. In: Plant physiological aspects of phenolic compounds. IntechOpen

Segaran G, Sathiavelu M (2023) Fungicidal and plant growth promoting traits of *Lasiodiplodia pseudotheobromae*, an endophyte from *Andrographis paniculata*. Front Plant Sci 14:1–16

Sharma I, Raina A, Choudhary M, Apra KS, Dhar MK (2023) Fungal endophyte bioinoculants as a green alternative towards sustainable agriculture. Heliyon 9:1–17

Shen B (2003) Polyketide biosynthesis beyond the type I, II and III polyketide synthase paradigms. Curr Opin Chem Biol 7:285–295

Shentu X, Zhan X, Ma Z, Yu X, Zhang C (2014) Antifungal activity of metabolites of the endophytic fungus *Trichoderma brevicompactum* from garlic. Braz J Microbiol 45:248–254

Shuping DSS, Eloff JN (2017) The use of plants to protect plants and food against fungal pathogens: a review. Afr J Tradit Complement Altern Med 14:120–127

Singh HB, Keswani C, Ray S, Yadav SK, Singh SP, Singh S, Sarma BK (2015) *Beauveria bassiana*: biocontrol beyond lepidopteran pests. Biocontrol Lepidopteran Pests 43:219–235

Staunton J, Weissman KJ (2001) Polyketide biosynthesis: a millennium review. Nat Prod Rep 18:380–416

Strobel GA, Miller RV, Martinez-Miller C, Condron MM, Teplow DB, Hess WM (1999) Cryptocandin, a potent antimycotic from the endophytic fungus *Cryptosporiopsis cf. quercina*. Microbiol. 145:1919–1926

Studt L, Wiemann P, Kleigrewe K, Humpf HU, Tudzynski B (2012) Biosynthesis of fusarubins accounts for pigmentation of Fusarium fujikuroi perithecia. Appl Environ Microbiol 78(12):4468–4480

Süssmuth RD, Mainz A (2017) Nonribosomal peptide synthesis-principles and prospects. Angew Chem Int Ed 56:3770–3821

Talapatra KA, Das R, Saha AK, Das P (2017) *In vitro* antagonistic activity of a root endophytic fungus towards plant pathogenic fungi. J Appl Biol Biotechnol 5:68–71

Teoh ES (2015) Secondary metabolites of plants. Medicinal Orchids of Asia:59–73

Varughese T, Rios N, Higginbotham S, Elizabeth Arnold A, Coley PD, Kursar TA, Gerwick WH, Rios LC (2012) Antifungal depsidone metabolites from *Cordyceps dipterigena*, an endophytic fungus antagonistic to the phytopathogen *Gibberella fujikuroi*. Tetrahedron Lett 53:1624–1626

Vinale F, Sivasithamparam K, Ghisalberti EL, Ruocco M, Woo S, Lorito M (2012) Trichoderma secondary metabolites that affect plant metabolism. Nat Prod Commun 7:1934578X1200701

Wan J, Dai Z, Zhang K, Li G, Zhao P (2021) Pathogenicity and metabolites of endoparasitic nematophagous fungus Drechmeria coniospora YMF1.01759 against nematodes. Microorganisms 9:1735

Wang Y, Wang G, Wang L, Xu X, Xia J, Huang X, Wu Y, Zhang C (2010) Isolation and identification of an endophytic fungus of *Polygonatum cyrtonema* and its antifungal metabolites. Acta Microbiol Sin 50:1036–1043

Wang J, Wang G, Zhang Y, Zheng B, Zhang C, Wang L (2014) Isolation and identification of an endophytic fungus Pezicula sp. in *Forsythia viridissima* and its secondary metabolites. World J Microbiol Biotechnol 30:2639–2644

Wang G, Liu Z, Lin R, Li E, Mao Z, Ling J, Yang Y, Yin WB, Xie BH (2016) Biosynthesis of antibiotic leucinostatins in bio-control fungus *Purpureocillium lilacinum* and their inhibition on phytophthora revealed by genome mining. PLoS Pathog 12:1005685–e1005685

Waweru B, Turoop L, Kahangi E, Coyne D, Dubois T (2014) Non-pathogenic *Fusarium oxysporum* endophytes provide field control of nematodes, improving yield of banana (Musa sp.). Biol Control 74:82–88

Woo S, Ruocco M, Vinale F, Nigro M, Marra R, Lombardi N, Pascale A, Lanzuise S, Manganiello G, Lorito M (2014) Trichoderma-based products and their widespread use in agriculture. Open Mycol J 8:71–126

Xu Z, Zhou G, Shimizu H (2010) Plant responses to drought and rewatering. Plant Signal Behav 5:649–654

Yan L, Khan RAA (2021) Biological control of bacterial wilt in tomato through the metabolites produced by the biocontrol fungus, *Trichoderma harzianum*. Egypt J Biol Pest Control:31

Yassin MT, Mostafa AAF, Al-Askar AA, Sayed SRM, Rady AM (2021) Antagonistic activity of *Trichoderma harzianum* and *Trichoderma viride* strains against some fusarial pathogens causing stalk rot disease of maize, in vitro. King Saud Univ Sci 33:101363

Ye Y, Xiao Y, Ma L, Li H, Xie Z, Wang M, Ma H, Tang H, Liu J (2013) Flavipin in *Chaetomium globosum* CDW7, an endophytic fungus from *Ginkgo biloba*, contributes to antioxidant activity. Appl Microbiol Biotechnol 97:7131–7139

Yu H, Zhang L, Li L, Zheng C, Guo L, Li W, Sun P, Qin L (2010) Recent developments and future prospects of antimicrobial metabolites produced by endophytes. Microbiol Res 165:437–449

Zhao J, Mou Y, Shan T, Li Y, Zhou L, Wang M, Wang J (2010) Antimicrobial metabolites from the Endophytic fungus *Pichia guilliermondii* isolated from *Paris polyphylla* var. yunnanensis. Molecules 15:7961–7970

Fungal Metabolites as Crop Pest Controllers

Panagiotis A. Eliopoulos and Spyridon Mantzoukas

Abstract Entomopathogens produce a plethora of substances, called secondary metabolites, to facilitate insect infection and maintain their vital functions. These metabolites not only suppress the insects' immune system, but can also have direct negative effects on insects such as contact toxicity, antifeedant, egg-laying deterrence, decrease of developmental rate and weight gain and others. Despite decades of intensive research, many aspects of their biosynthesis pathways, modes of action and ecological roles remain poorly understood or even unknown. This chapter describes the current knowledge on the use of secondary metabolites as biopesticides in the context of Integrated Pest Management. The latest studies evaluating their insecticidal activity in laboratory bioassays and field trials are presented. Future challenges and considerations for the full exploitation of secondary metabolites in pest management strategies are discussed.

Keywords Entomopathogenic fungi · Secondary metabolites · Biopesticides · Biological control · Integrated Pest management

1 Introduction

Entomopathogenic fungi (EPF) are valuable biological control agents for the strategies of Integrated Pest Management (IPM) (Litwin et al. 2020; Mantzoukas an Eliopoulos 2020; Deka et al. 2021). The various modes of their insecticidal action have been extensively studied and it has been well documented that they secrete abundant organic compounds called secondary metabolites (SMs), that are essential

P. A. Eliopoulos (✉)
Laboratory of Plant Health Management, Department of Agrotechnology, University of Thessaly, Volos, Greece
e-mail: eliopoulos@uth.gr

S. Mantzoukas
Department of Agriculture, University of Ioannina, Ioannina, Greece

J. Poveda et al. (eds.), *Fungal Metabolites for Agricultural Applications*, Fungal Biology, https://doi.org/10.1007/978-3-031-76587-2_7

both for maintaining vital fungal functions and for suppressing the immune system of the insect host (Ortiz-Urquiza and Keyhani 2013; Donzelli and Krasnoff 2016; Rodríguez-Romero et al. 2023). The most important effects that SMs exert on insects are mortality, anti-feedant, reduced fecundity and developmental rate, etc., which however appear varying degrees of virulence, and yet in many cases their role remains poorly understood or even unknown (Berestetskiy and Hu 2021; Mannino et al. 2021; Roy et al. 2021). This knowledge is crucial for enhancing the production of SMs and their practical application for pest control.

Based on their chemical compositions, SMs can be categorized into several groups, including cyclic polyketides, amino acid derivatives, peptides, non-ribosomal peptides, depsipeptides, peptide hybrids, and terpenoids (Wang et al. 2018; Litwin et al. 2020). The wide prevalence and distinctive characteristics of many compounds of SMs enable researchers to distinguish and study them extensively. Capitalizing on this, scientists continue to investigate the mechanisms underlying the biosynthesis of SMs and their impact on insects and the environment. More importantly, these compounds can be manipulated to promote insect infection so that they can become useful tools for IPM, which represents a holistic and sustainable approach to pest control in agriculture, forestry, and urban environments (Baker et al. 2020).

The biosynthesis of insecticidal SMs in fungi is a complex and highly regulated process, involving numerous enzymatic reactions and genetic pathways. Transcription factors, enzymes, and regulatory proteins control the expression of genes involved in secondary metabolism. Genes clusters are involved in the biosynthesis of SMs (Wang et al. 2016). Environmental cues, such as nutrient availability and insect presence, can influence the activation of these biosynthetic pathways (Brakhage 2013). Understanding them is essential for harnessing the potential of fungal SMs in IPM. EPF have developed a plethora of mechanisms to produce a great variety of SMs. For example, Polyketides, that constitute a large and structurally diverse group of fungal SMs with insecticidal properties, are biosynthesized by polyketide synthases (PKSs), enzymes that assemble small carbon units into complex polyketide structures (Cox et al. 2018). Another significant SMs group, terpenoids derived from the condensation of isoprene units. Fungi biosynthesize terpenoids via the mevalonate or non-mevalonate pathways (Schmidt-Dannert 2015). Fungi can also produce a diverse array of alkaloids, some of which have potent insecticidal properties, exerting toxic effects on insects by interfering with neurotransmission (Glenn and Bacon 1997). Moreover, certain fungi can synthesize insecticidal peptides that disrupt insect physiology or act as neurotoxins (Hu and Dong 2015). For instance, *Metarhizium anisopliae* produces insecticidal peptides, such as destruxins, which modulate ion transport and induce apoptosis in insect cells (Wang et al. 2012). Apart from that, some SMs (e.g. hybrid peptides, polyketides, and terpenoids) are the result of hybrid biosynthetic pathways, where multiple classes of compounds are combined (Lackner et al. 2012).

The mode of action of many SMs produced by EPF has been reported but remains unclear for many others. It has been documented that SMs can exert negative effects on the permeability of cell membranes, the molting and metamorphosis process, insect's fertility, and they usually cause damages on insect's cellular immune system, malpighian tubules and muscles, resulting in problematic excretion and inability to move and feed (Peng et al. 2005; Pal et al. 2007; Gibson et al. 2014).

This chapter reviews the use of SMs in IPM, focusing on the various types of SMs, their modes of action, their insecticidal action tested on field trials and lab bioassays, their mass production and commercial formulation, and possibilities and challenges of their practical application for pest control (Fig. 1).

2 Major Groups of Fungal SMs

Fungal SMs encompass a broad spectrum of bioactive compounds with diverse chemical structures and functions. When it comes to pest control, these metabolites have proven to be a treasure trove of potential solutions. Here, we dive deeper into the various classes of fungal metabolites and their specific effects on pests, shedding light on the mechanisms behind their pesticidal properties.

2.1 Destruxins

Destruxins (DTXs) are cyclic hexadepsipeptides produced by many important EPF like *Beauveria bassiana* and *M. anisopliae* (Bagheri et al. 2021; Islam et al. 2021; Paschapur et al. 2021). More than 40 types of DTXs have been reported so far, with A, B, and E being the most important (Ravindran et al. 2016; Arroyo-Manzanares et al. 2017). Their mode of action is mainly by inhibiting the activity of V-ATPase proton pumps in insect epithelia (Liu and Tzeng 2012). The use of these SMs disrupts the cellular immune system of insects, hinders the formation of nodules, harms both the muscular system and malpighian tubules, thereby impacting excretion. This disruption results in difficulties related to feeding and mobility (Pal et al. 2007; Peng et al. 2005). Type A of DTX caused inhibition of fluids secretion in the Malpighian tubules of *Schistocerca gregaria* (Skrobek et al. 2008). Although, the complete mechanism of DTXs remains unclear, Research has indicated that the primary function of DTXs is to support the pathogen's establishment in host tissues. This occurs through alterations in the ion channels, potentially leading to insect mortality post-infection due to the actions of these toxins (Islam et al. 2021). DTXs are strongly associated with EPF pathogenicity, with the strains producing higher levels of DTXs being the most virulent (Ravindran et al. 2016; Mukherjee and Vilcinskas 2018). Nevertheless, the existence of these SMs in food chain raises concerns, as they have the potential to harm human health and the ecosystem in high doses (Taibon et al. 2015).

There is variability among insects in their susceptibility to DTXs, with certain Lepidoptera species being the most sensitive (Kershaw et al. 1999, Sree et al. 2008; Tiang-Yan et al. 2016). When DTXs were added in the larval diet the larvae of

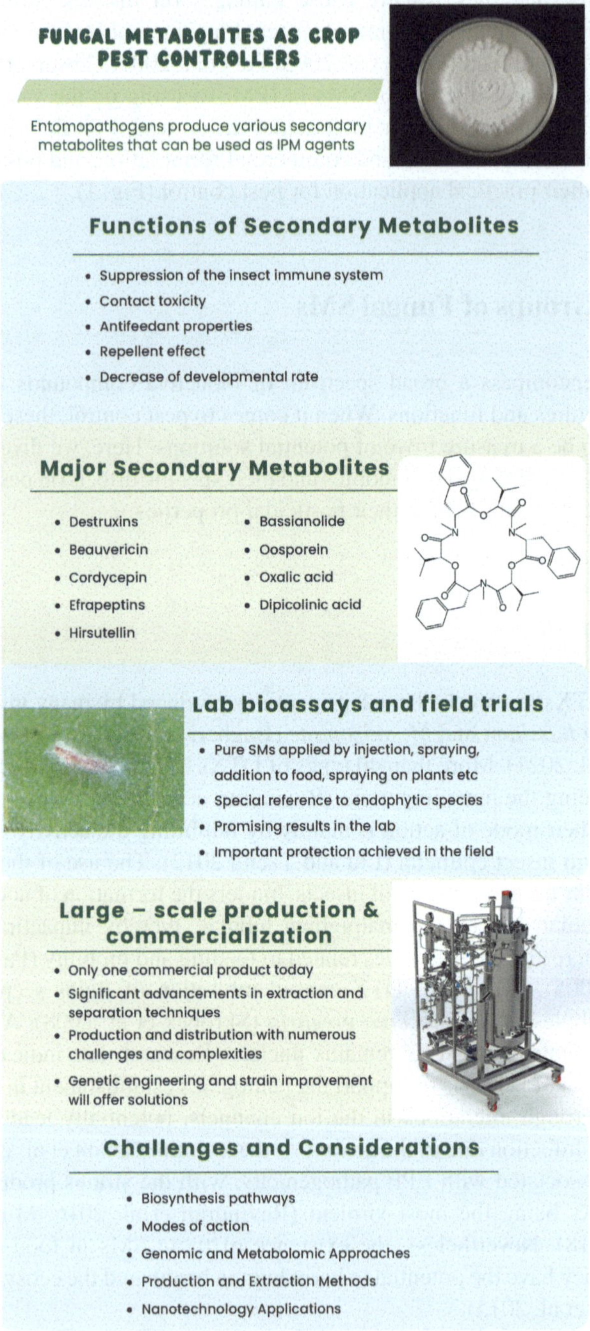

Fig. 1 Main points of the chapter

Agrotis segetum starved to death (Thomsen et al. 1996). They were also very toxic to *Spodoptera litura* (Sree et al. 2008) and *Plutella xylostella* larvae (Tiang-Yan et al. 2016). DTXs also affect insect growth, decreasing dramatically the developmental rate, since larvae grew slower when exposed to leaves treated with DTXs (Brousseau et al. 1996; Amiri et al. 1999).

The potential of DTXs to induce contact toxicity in insects remains uncertain. Reports suggest that when applied to the integument, DTXs do not exhibit contact toxicity (Vey et al. 2001), but topical toxicity to the *P. xylostella* and *Phaedon cochleareae* larvae (Amiri et al. 1999), *Empoasca vitis* nymphs (Poprawski et al. 1994) is documented. The mechanism by which these SMs overcome the barrier of insect cuticle is still unclear.

There is also some evidence that some DTXs, such as DTX E, can act systemically inside plant tissues. The aphids *Brevicoryne brassicae* and *Myzus persicae* were repelled by leaves soaked in DTX E solution (Robert and Riba 1989).

Although, production of DTXs has been improved through biotechnological advancements, the yield in the fungal cultures remains relatively low reaching 200–500 mg/l (Hu et al. 2006; Kim et al. 2019), making the practical application of DTXs as biopesticides unlikely.

2.2 Beauvericin and Other B. Bassiana SMs

The EPF *B. bassiana* produces a plethora of SMs. The most important of them are beauvericin, bassianolides, bassiacridin, bassianin, beauverolides, bassianolone, oosporein and others (Wang and Xu 2012; Cheong 2015; Mallebrera et al. 2018; Canassa et al. 2019; Kim et al. 2023).

Among these, beauvericin exhibits extensive and noteworthy bioactivities, and it can also be synthesized by various other entomopathogenic fungal genera, including *Fusarium*, *Isaria*, *Paecilomyces*, and others (Jaber and Ownley 2018; Weng et al. 2019). Its biosynthesis involves the enzyme enniatin synthetase (Mallebrera et al. 2018). Similarly to DTXs, they are cyclic hexadepsipeptides with broad pesticidal properties as it has caused high mortality against mosquitos, butterflies and beetles mainly by cell disruption (Wang and Xu 2012; Daniel et al. 2017; Luz et al. 2017). They have also demonstrated immunosuppressive properties, as most EPF SMs, playing crucial role in the progression of insect mycoses (Berestetskiy and Hu 2021). Moreover, this SM is capable of increasing the permeability of membranes, inducing dehydration of tissues leading to the death of the host. It has also been reported that it interferes with molting and fertility (Mallebrera et al. 2018). The production of beauvericin reached 400 mg/L of medium, under optimum conditions, which appears inadequate for its potential commercialization as a bioinsecticide (Wang and Xu 2012).

Another important *B. bassiana* SM is bassianolide, also found in *Lecanicillium* sp. (Xu et al. 2009; Ravindran et al. 2018). Bassianolide is known for its antifeedant and insecticidal effects. Injection of 2 μg of pure bassianolide to *Bombyx mori* and

Helicoverpa zea larvae caused atony, whereas it was lethal at higher doses (>5 μg/larva) (Champlin and Grula 1979; Roberts et al. 1992). Antifeedant effect was recorded when larvae fed on diet with 13 ppm of bassianolide died after 4 days (Kanaoka et al. 1978). A concentration of 7.5 μg proved to be effective in killing aphids upon direct contact, that was likely attributed to its interior being relatively hydrophilic and its exterior being hydrophobic, enabling it to permeate the lipophilic cuticle of insects. (Cheong 2015). It has also been documented that it can affect acetylcholine receptors of insect muscles (Ali et al. 2017; Villegas-Mendoza et al. 2019). The importance of Bassianolide as a virulence factor was confirmed by observing the loss of insect pathogenicity when the bassianolide synthetase gene cluster was removed (Xu et al. 2009).

Oosporein is another SM, isolated from *Beauveria*, known to accelerate insect mortality primarily through its potent immunosuppressive properties rather than direct toxicity (Abendstein et al. 2003; Feng et al. 2015; McNamara et al. 2019). In feeding bioassays, pure oosporein did not exhibit any repellent or insecticidal activity (Abendstein and Strasser 2000; Abendstein et al. 2003). Likewise, leaf-dip assays indicated that oosporein did not have any repellent or antifeedant impact on *P. xylostella* and *P. cochleariae* larvae at concentrations below 2 mg/l (Abendstein and Strasser 2000).

2.3 Cordycepin

Cordycepin is produced by *Cordyceps* and *Isaria* and belongs to octadepsipeptides. Its insecticidal potential has been well documented by several reports, like for *P. xylostella* (Kim et al. 2002) and *Leptinotarsa decemlineata* (Kryukov et al. 2014). Complete mortality was recorded on *P. xylostella* larvae after 5 days on treated leaves (0.3–0.5 mg/ml), while it did not cause any topical toxicity (Kim et al. 2002). While cordycepin can be manufactured on an industrial scale (Chen et al. 2020), its application as a biopesticide is improbable due to its phytotoxic characteristics (Quy et al. 2019) and potential health safety concerns associated with high doses (Roberts 1981; Radhi et al. 2021).

2.4 Efrapeptins

These SMs are isolated by *Tolypocladium* and *Beauveria nivea* (Krasnoff and Gupta 1991; Bandani et al. 2000; Brückner and Degenkolb 2020) and exert lethal action against arthropod pests (e.g. spidermites, flies, mosquitoes, beetles and moths) (Matha et al. 1988; Krasnoff et al. 1991; Bandani et al. 2000). Significant (>50%) antifeedant and insect growth inhibition was recorded when *P. xylostella* and

P. cochleariae consumed leaf discs treated with efrapeptins. Apart from reduced larval and pupal weights, the rate of development was significantly retarded (Bandani and Butt 1999). In injection assays, the average LD_{50} values for mature larvae of *Galleria mellonella* and *Manduca sexta* were 30 and 47 ng, respectively, for the purified efrapeptin (Bandani et al. 2000).

2.5 Hirsutellin

Hirsutella thompsonii, another common EPF, synthesizes the insecticidal SM hirsutellin, which has been assessed for its lethal effects on a range of insects. When ingested in *G. melonella* larvae it was highly toxic at very low concentrations (70 pmol, 1 mg/gr) (Liu et al. 1995; Mazet and Vey 1995). It was also effective against the mite *Phyllocoptruta oleivora* causing 100% adult mortality at 100 μg/ml on leaf assay tests after 3 days, while fecundity prior to death was also dramatically reduced (Omoto and McCoy 1998).

2.6 Organic Acids

Many EPF genera (e.g., *Beauveria*, *Paecilomyces*) produce organic acids like oxalic acid and dipicolinic acid. Following topical applications, ticks and whitefly nymphs have shown susceptibility to the insecticidal effects of these SMs (Kirkland et al. 2004; Asaff et al. 2005). Oxalic acid has proven effective in dissolving cuticle tissues of grasshoppers (Bidochka and Khachatourians 1991) and has managed to control *Varroa* mites in bee hives. Notably, it exhibits a toxic effect solely on the mites, without adversely affecting the bees (Aliano and Ellis 2009; Maggi et al. 2017).

Dipicolinic acid has been isolated by *Isaria*, *M. anisopliae*, *B. bassiana*, and *L. lecanii* (Watanabe et al. 2006; Han et al. 2014). It showed lethal effects against *Culliphora erythrocephala* (Claydon and Grove 1982), nymphs of *Bemisia tabaci* and *B. argentifolii.* (Asaff et al. 2005) and *B. mori* pupae (Watanabe et al. 2006).

3 Field Trials and Lab Tests with Fungal SMs Against Certain Insect Pests

It has been already mentioned above that SMs have been evaluated for their insecticidal effect in various lab bioassays and field tests. In Table 1, we present relevant studies where pure SMs were applied to insect pests by injection, spraying on insect body, addition to food, spraying on plants etc.

Table 1 Studies evaluating the insecticidal activity of major fungal secondary metabolites via lab tests and/or field trials against plant pests (2000–2023)

Metabolite	Fungal sources	Target pests	Conclusion	References
Agglutinin	*Rhizoctonia solani*	*Spodoptera littoralis*	Weight of grown larvae was reduced 89% when 1% was added in the diet. Afterwards, 82% mortality was recorded.	Hamshou et al. (2010a)
	Sclerotinia sclerotiorum	*Acyrthosiphon pisum*	High mortality was recorded when fed on artificial diet containing various doses (LC_{50} = 66 mg/ml)	Hamshou et al. (2010b)
Bassiacridine	*Beauveria bassiana*	*Locusta migratoria*	Injection at very low dose (3.3 μg/gr) caused 50% mortality on fourth instar nymphs	Quesada-Moraga and Vey (2004)
Beauvericin	*Beauveria, Cordyceps, Isaria, Fusarium*	*Spodoptera frugiperda*	After 48 h of exposure, it exhibited significant toxic activity on both insect cell lines	Valencia et al. (2011)
		Schizaphis graminum	Longevity was not significantly decreased when feeding treated leaves. However, offspring production was significantly smaller with large number of abortive embryos.	Ganassi et al. (2002)
			In leaf contact bioassays, complete mortality was observed at a dose of 0.5 mg/mL	Salimova et al. (2021)
		Tetranychus urticae	Doses from 10 to 1000 μg/g caused high mortality (84–100%)	Al Khoury et al. (2019)
			In leaf contact bioassays, complete mortality was observed at a dose of 0.5 mg/mL	Salimova et al. (2021)
Bislongiquinolide	*Trichodedrma citrinoviride*	*Schizaphis graminum*	Aphids that settled on treated leaves were significantly fewer (0,34 and 0,68 mg/ml)	Evidente et al. (2009)
Citrantifidiene Citrantifidiol	*Trichodedrma citrinoviride*	*Schizaphis graminum*	Aphids demonstrated strong antifeedant effect on treated leaves (dipped in solution of 0.57 mg/mL citrantifidiene and 0.91 mg/mL citrantifidiol)	Evidente et al. (2008)
Cordycepin	*Cordyceps militaris*	*Plutella xylostella*	Complete mortality was recorded on larvae after 5 days on treated leaves (0.3–0.5 mg/ml), while it did not cause any topical toxicity	Kim et al. (2002)
		Leptinotarsa decemlineata	When third instar larvae were fed with aqueous cordycepin solution (7 mg/g) significant decrease in weight was observed, but not in larval survival (82%)	Kryukov et al. (2014)

Cyclosporin A	*Aspergillus terreus, Cladosporium columbinum, Penicillium fellutanum, Cylindrocarpon lucidum, Trichoderma sp.*	*Mythimna separata*	When larvae fed on treated artificial diet significant insecticidal action was recorded (LC_{50} = 19.23 μg/g for newly hatched, 78.36 for third-instar larvae). Sublethal effects like decrease in offspring production, body size, weight, pupation rate, emergence rate, as well as increase of developmental time were also observed.	Sun et al. (2023)
Destruxins	*Metarhizium anisopliae*	*Ceratitis capitata*	High adult mortality (70–100%) after 48 h when fed on treated artificial diet	Lozano-Tovar et al. (2015)
		Myzus persicae	High aphid mortality (LC_{50}: 58 and 66 ppm). During field tests, treated plants had significantly lower aphid infestation.	Sabbour (2019)
		Pegomya mixta	High adult mortality (LC_{50}: 76 and 141 ppm). Fertility was was significantly decreased. under field conditions, treatments with destruxins in the field resulted in significant increase in sugar beet production.	Sabbour et al. (2020)
		Spodoptera litura	Ingestion or/and topical application caused significant larval mortality. Young larvae (<6 days) were much more susceptible than the old ones (>12 days).	Sree et al. (2008)
			There was high antifeedant effect (dose >100 mg/l) and contact toxicity (LC_{50}: 113.99–292.00 mg/l) 48 h after treatment.	Hu et al. (2007)
Dihydrotrichodimerol	*Trichodedrma citrinoviride*	*Schizaphis graminum*	Mean number of aphids on treated leaves (dipped in solution with doses 2.5 and 1.25 mg/ml) was significantly reduced compared with the control	Evidente et al. (2009)
Dipicolinic acid	*Penicillium, Beauveria, Paecilomyces, Cordyceps, Lecanicilium, Verticillium etc*	*Bemisia sp.*	Moderate lethal action (LD_{50}: 44.5 mg/l) was recorded, when it was sprayed on whitefly nymphs.	Asaff et al. (2005)

(continued)

Table 1 (continued)

Metabolite	Fungal sources	Target pests	Conclusion	References
Efrapeptins	*Tolypocladium, Beauveria nivea*	*Manduca sexta*	Extreme lethal toxicity (LD_{50}: 47 ng) when injected in larvae.	Bandani et al. (2000)
Fumonisins	*Fusarium verticillioides*	*Sitophilus zeamais.*	All effects (contact toxicity, antifeedant action, grain damage) were insignificant.	Usseglio et al. (2020)
Gliotoxin	*Neosartorya pseudofischeri*	*Acyrthosiphon pisum*	Gliotoxin demonstrated high antifeedant effect (FRI: 43–54%) buy low mortality and great phytotoxicity.	Aznar-Fernández et al. (2019)
Hirsutellin A	*Hirsutella thompsonii*	*Locusta migratoria*	Hirsutellin A caused >30–40% nymphal mortality	Quesada-Moraga and Vey (2004)
Ergosterol	*Moelleriella (Hypocrella) raciborskii*	*Tetranychus urticae*	Ergosterol induced 75% mite mortality, after mites being fed with mulberry leaf discs treated with 0.5% concentration, after 3 days.	Buttachon et al. (2013)
Dustanin			Dustanin caused 53% mortality, after mites being fed with mulberry leaf discs treated with 0.5% concentration, after 3 days. Direct spraying reduced mite egg production by 70%.	
3β-acetoxy-15α,22-dihydroxyhopane			46% mortality was recorded, after mites being fed with mulberry leaf discs treated with 0.5% concentration, after 3 days. Direct spraying reduced mite egg production by 71%.	
		Tetranychus truncatus	It was the most virulent compound (LC_{50}: 0.232% w/v) 72 h post treatment. It also caused showed strong repellent effect (70.2%)	Buttachon et al. (2022)
Lectin	*Sclerotium rolfsii*	*Spodoptera litura*	Feeding bioassays with various lectin doses, showed low inhibition of larval growth but significantly increased mortality rate (68.5% at 600 mg/g). LC_{50} value: 430 mg/g of diet.	Vishwanathreddy et al. (2014)

Oosporein	*Beauveria*	*Bemisia tabaci*	Oosporein caused low toxicity (18%) to nymphs. Synergistic effect was recorded when oosporein was mixed with *B. bassiana* conidia (92% mortality).	Amin et al. (2010)
		Capua reticulana	Ingestion of Oosporein into larvae did not cause any insecticidal, antifeedant or growth inhibitory effect	Abendstein et al. (2003)
		Cydia pomonella		
		Mamestra brassicae		
		Plutella xylostella	Oosporein did not have any repellent or antifeedant impact on larvae at doses <2 mg/l	Abendstein and Strasser (2000)
		Phaedon cochleariae		
Penitrem	*Penicillium crustosum*	*Oncopeltus fasciatus*	Percentage of mortality after 72 h of exposure, at the dose of 20 µg/cm^2 ranged from 6,6-31,1% depending on the Penitrem type (A-G)	Carmen González et al. (2003)
		Ceratitis capitata	Topical application (10 µg/fly) resulted in 41,7-73,3% mortality and 77,1–100% fecundity reduction, after 10 days.	
Pyripyropene A	*Aspergillus fumigatus*	*Myzus persicae*	>80% mortality rate at 100 ppm (spraying) (LC_{50} = 0,98 mg/L). When 1 mg per pot was applied by soil drenching the numbers of aphids decreased remarkably after 24 h.	Horikoshi et al. (2017, 2018)
		Bemisia tabaci	100% mortality rate at 5 ppm (spraying) after 6 days (LC_{50} = 1,8 mg/L)	Horikoshi et al. (2018)
Seiridin	*Seiridium cardinale*	*Acyrthosiphon pisum*	High aphid feeding deterrence at 0.5 mM	Aznar-Fernández et al. (2019)

(continued)

Table 1 (continued)

Metabolite	Fungal sources	Target pests	Conclusion	References
Tenuazonic acid	*Alternaria alternata*	*Macrosiphum rosivorum*	Low aphid mortality was found in bioassays with the crude toxin. Spraying with the pure SM (0.06 μg/ml) great inhibitory effect was observed on aphid population (83.60%) after 15 days. It was proved that plant's resistance against aphids was significantly induced by pure tenuazonic acid.	Yang et al. (2015)
		Galleria mellonella	In injection bioassays, the dose 50 μg/larva was lethal (LT50: 6 days). Addition in food (250 μg/g) inhibited larval growth	Salimova et al. (2021)
		Zophobas morio	In injection bioassays, the dose 50 μg/larva was lethal (LT_{50}: 2 days)	
		Acheta domesticus	Addition in food (250 μg/g) caused adult mortality ($LT_{50:}$ 7 days)	
		Tetranychus urticae	Low contact toxicity was recorded at a dose of 1 mg/mL (mortality 12%) and 2 mg/mL (mortality 19%)	
		Schizaphis graminum	Moderate contact toxicity was recorded at a dose of 1 mg/mL (mortality 27%) and 2 mg/mL (mortality 43%)	
Vertlecanins	*Verticillium lecanii*	*Helicoverpa zea*	When added in larval diet (1500 ppm) 49% reduction in weight gain relative to controls was recorded.	Soman et al. (2001)

3.1 SMs from Endophytic EPF

Endophytic EPF grow asymptomatically inside plant tissues (Vega 2018; Mantzoukas and Eliopoulos 2020). Likewise other EPF, they produce various SMs helping the host plant to inhibit the infestation by insect pests (Zhang et al. 2012; Schulz et al. 2015). For instance, the genera *Neotyphodium* and *Epichloë* in grasses produce numerous alkaloids (ergoalkaloids, lolitrems, lolines), with insecticidal properties, while others (e.g. peramine) cause antifeedant effects (Bush et al. 1997; Potter et al. 2008; Schardl et al. 2013). Specifically, the concentration of lolins in the tissues of fescue *Festuca* sp. was strongly correlated with the decrease in aphid populations (Wilkinson et al. 2000), while their colonization in the stems of ryegrass *Neotyphodium uncinatum* inhibited the growth and development of the larvae of the weevil *Listronotus bonariensis* (Popay et al. 2009). Similar studies about the insecticidal effect of endophytic EPF SMs are presented in Table 2.

4 SMs Large: Scale Production and Commercialization

Secondary metabolites from plants (Divekar et al. 2022) and bacteria (Aggarwal et al. 2016) have been commercially available for pest control for many years. However, there is currently only one commercial product on the market that utilizes EPF SMs as its active ingredient (Koradin et al. 2021) (Table 3).

The large-scale cultivation of EPF and the extraction of their SMs are essential steps for their practical application for pest control. To achieve this difficult task a plethora of methods and techniques have been developed, consisted of certain steps. The first step in SMs mass production is the careful selection of EPF strains with desirable characteristics. Different EPF species and strains may produce specific SMs with distinct pesticidal properties. Researchers and manufacturers choose strains based on their desired bioactivity, production yields, and compatibility with certain target pests (Muñiz-Paredes et al. 2017). Except strain selection, EPF require suitable growth substrates and culture media. These can include natural materials such as grains (e.g., rice, wheat) or synthetic media with defined nutrient compositions. The choice of substrate and medium depends on the fungal species and the intended application (Jaronski 2023). Fermentation is the most common method for cultivating EPF to produce SMs. Submerged liquid fermentation and solid-state fermentation are two primary techniques used. In submerged fermentation, fungi grow in a liquid medium, while solid-state fermentation involves fungal growth on solid substrates (Jaronski and Mascarin 2017; Jaronski 2023). Various fermentation systems have been developed and are able to provide high yields of SMs (Velmurugan et al. 2010). To meet commercial production demands, EPF cultivation often involves scale-up from laboratory or small-scale bioreactors to larger industrial bioreactors. These bioreactors provide controlled conditions (aeration, agitation, temperature, pH) for optimal fungal growth and metabolite production.

Table 2 Endophytic EPF species with insecticidal metabolites

Metabolite	Endophytic EPF	Host Plant	Insecticidal Action	References
3-O-methylfunicone	*Talaromyces pinophilus*	*Arbutus unedo*	Moderate toxic effects against the pea aphid *A. pisum* were displayed (26,2% mortality after 72 h)	Vinale et al. (2017)
Aspergide	*Aspergillus fumigatus*	*Acrostichum specioum*	Strong insecticidal action (mortality, reduced larval development and pupa weight) on larvae of *S. litura*. Larval mortality 76.67% after 20 days.	Guo et al. (2017)
Ergot alkaloids	*Neotyphodium lolii × Epichloë typhina*	*Lolium perenne*	Lethal effects were recorded when larvae of *Agrotis ipsilon* were fed on colonized plants	Potter et al. (2008)
Lolines	*Neotyphodium coenophialum*	*Lolium arundinaceum*	Severity of damage by the weevil *Listronotus bonariensis* was less on colonized plants. Larval development was decreased by 36%.	Popay et al. (2009)
	Epichloë festucae	*Lolium pratense*	Complete mortality of the aphids *Schizaphis graminum* and *Rhopalosiphum padi* when loline levels were > 67 ppm in plant tissues.	Wilkinson et al. (2000)
	Neotyphodium uncinatum & *N. occultans*	*Lolium pratense*	Negative effects on survival of the rice leaf bug *Trigonotylus caelestialium*, even on plants with the lowest concentration (50 μg/g).	Shiba and Sugawara (2009)
	Neotyphodium occultans	*Lolium multiflorum*	Strong deterrent effect and lower survival of *T. caelestialium* on plants infected with the loline N-formylloline. The effects were strongly correlated with the loline concentration.	Shiba et al. (2011)

(continued)

Table 2 (continued)

Metabolite	Endophytic EPF	Host Plant	Insecticidal Action	References
	Neotyphodium uncinatum	*Lolium multiflorum*	Infestation of *Stenotus rubrovittatus* was decreased in loline-infected plants. Presence of *Laodelphax striatellus* was not affected.	Matsukura et al. (2014)
			Survival of *Laodelphax striatellus, Sogatella furcifera and Cicadulina bipunctata* was significantly decreased when fed with the phloem-sap of loline-infected ryegrass. No effect was observed on *Nephotettix cincticeps* which is a xylem-sap feeder.	Matsukura et al. (2012)
Naphthalene	*Muscodor vitigenus*	*Paullinia paullinioides*	Strong repellent effect on the adult sawfly, *Cephus cinctus*, in olfactory tests.	Daisy et al. (2002)
Peramine	*Acremonium Ioli*	*Lolium perenne*	Antifeedant effect on adults and larvae of *L. bonariensis*	Rowan et al. (1990)
Phyllostin	*Diaporthe miriciae*	*Cyperus iria*	Significant contact toxicity antifeedant and oviposition deterrent was observed against *P. xylostella*. Complete antifeedant effect and oviposition deterrence at 50 μg/cm^2 in no-choice leaf disc bioassays. The LC_{50} was estimated 4.38 μg/larva in contact toxicity bioassays.	Ratnaweera et al. (2020)
Rugulosin	*Phialocephala sp.*	*Picea glauca*	Significant decrease of body size and weight for *Choristoneura fumiferana, Zeiraphera canadensis* and *Lambdina fiscellaria.*	Sumarah et al. (2008)
	Phialocephala scopiformis		Significant inhibition of larval development of *C. fumiferana* when fed with colonized needles.	Miller et al. (2002)

Table 3 Commercial biopesticides based on fungal secondary metabolites

Trade name	Active ingredients	Fungal source	Manufacturer	Target pests
Inscalis®	Afidopyropen, a semisynthetic product of pyripyropen A	*Aspergillus fumigatus*, *Penicillium coprobium*	BASF, Germany	Aphids, whiteflies, leafhoppers, psyllids, scale insects

Despite significant advancements in extraction and separation techniques, isolating SMs from EPF remains a challenging task. The method of harvesting SMs depends on the nature of the metabolites and the cultivation system. Filtration, centrifugation, solvent extraction, ionic liquids, ultrasound-assisted extraction, distillation and chromatography are common techniques for separating fungal biomass from the culture medium and isolating the metabolites (Berestetskiy and Hu 2021). Nevertheless, addressing several considerations is essential when extracting the desired compounds from the non-soluble matrix in which they are embedded. In fungal cultures, secondary metabolites (SMs) are typically intracellular. Therefore, enhancing yield involves grinding the culture, breaking tissue, and compromising cell integrity before extraction.

It should be mentioned that the production and distribution of EPF SMs present numerous challenges and complexities. Moving from laboratory-scale fermentation to large-scale production often involves challenges related to scaling up. Ensuring optimal conditions, mixing, and oxygen supply in industrial bioreactors can be addressed by working with specialized engineers and researchers with expertise in bioprocess engineering (Rodríguez-Durán et al. 2023). The cost of providing suitable nutrients for fungal growth can be a significant challenge. Utilizing low-cost, sustainable substrates and optimizing nutrient utilization can help mitigate these expenses. Agro-industrial waste materials and recycling of nutrients are potential solutions (Sala et al. 2019). Achieving high yields of EPF SMs is often required to meet market demands. Genetic engineering and strain improvement programs can enhance the productivity of selected fungal strains (Reyes Haro 2018).

5 Conclusion and Future Directions

The use of EPF SMs in IPM is an evolving field with a growing body of research and innovation. As outlined in this chapter, SMs have multiple roles to play in IPM, but our knowledge and technological status about their biosynthetic pathways, ecological functions, regulatory mechanisms, modes of action inside the insect body, mass production and extraction methods needs to be improved. Emerging trends and research areas that are shaping the future of SMs as IPM agents, offering novel strategies and solutions, are mentioned below.

New strains: One of the key trends in EPF SMs research is the continuous exploration of new EPF strains for their potential in pest control. Researchers are isolating and characterizing SMs from previously unexplored strains to discover novel compounds (Rajula et al. 2021). Different conditions of humidity, temperature, host, habitat, competition, etc. are forcing the creation of novel isolates or strains of EPF that may possibly differ in pathogenicity (Al-Ani 2019).

Genomic and Metabolomic Approaches: Advancements in genomics and metabolomics have revolutionized the study of fungal SMs. Researchers are using these techniques to identify, predict, and engineer the biosynthesis of SMs for enhanced virulence against pests (Harith-Fadzilah et al. 2021). Genomic investigations into *Metarrhizium* and *Beauveria* (Gao et al. 2011; Xiao et al. 2012) indicate the presence of distinctive and extensive gene pools linked to SMs production. Unraveling the dynamics of these interactions between pathogens and hosts not only enhances our comprehension but also contributes to the development of new IPM strategies, effectively addressing the escalating challenges posed by insect pests. Moreover, Biotechnology enables the modification of fungal strains to enhance secondary metabolite production. Genetic engineering techniques can be used to overexpress key genes involved in secondary metabolite biosynthesis, resulting in increased metabolite yields (Vidhate et al. 2023).

Enhanced Production and Extraction Methods: Research is focused on improving the production efficiency of fungal SMs, which includes optimizing fermentation processes, scaling up EPF production and improving extraction methods and SMs yield to reduce production costs and make biopesticides more economically viable (Michaliski et al. 2023).

Nanotechnology Applications: The use of nanotechnology to enhance the delivery and effectiveness of biopesticides is a growing trend, given that nanoformulations can improve their stability and targeted delivery. Nanoparticles can protect and enhance the stability of SMs, ensuring their effectiveness. Recent studies with nanoparticles-loaded with fungal SMs revealed that their pesticidal action was significantly increased (Namasivayam and Arvind Bharani 2021).

In conclusion, despite the impressive scientific progress, we have not yet fully decoded how SMs suppress insect defense mechanisms and counteract the host immune system. Many questions about the exact link between SMs and EPF pathogenicity remain to be answered. With the expanding understanding of molecular genetic mechanisms governing the regulation of SMs biosynthesis and the vast diversity of EPF in nature, there remains considerable potential to comprehend the significance of fungal SMs and advance their application in IPM.

References

Abendstein D, Strasser H (2000) Considerations on toxic metabolites produced by *Beauveria brongniartii*, in integrated control of soil Pest. IOBC WPRS Bull 23:99–105

Abendstein D, Schweigkofler W, Strasser H (2003) Study on insecticidal, antifeedant and growth inhibitory properties of oosporein on selected pest organisms. IOBC WPRS Bull 26:103–106

Aggarwal N, Thind SK, Sharma S (2016) Role of secondary metabolites of actinomycetes in crop protection. In: Plant growth promoting actinobacteria: a new avenue for enhancing the productivity and soil fertility of grain legumes. Springer, pp 99–121. https://doi.org/10.1007/978-981-10-0707-1_7

Al Khoury C, Guillot J, Nemer N (2019) Lethal activity of beauvericin, a *Beauveria bassiana* mycotoxin, against the two-spotted spider mites, *Tetranychus urticae* Koch. J Appl Entomol 143:974–983. https://doi.org/10.1111/jen.12684

Al-Ani LKT (2019) Entomopathogenic fungi in IP landscape. In: Intellectual property issues in microbiology. Springer, pp 223–238. https://doi.org/10.1007/978-981-13-7466-1_12

Ali S, Zhang C, Wang Z, Wang XM, Wu JH, Cuthbertson AGS, Shao Z, Qiu BL (2017) Toxicological and biochemical basis of synergism between the entomopathogenic fungus *Lecanicillium muscarium* and the insecticide matrine against *Bemisia tabaci* (Gennadius). Sci Rep 7. https://doi.org/10.1038/srep46558

Aliano NP, Ellis MD (2009) Oxalic acid: a prospective tool for reducing *Varroa mite* populations in package bees. Exp Appl Acarol 48:303–309. https://doi.org/10.1007/s10493-009-9240-8

Amin GA, Youssef NA, Bazaid S, Saleh WD (2010) Assessment of insecticidal activity of red pigment produced by the fungus *Beauveria bassiana*. World J Microbiol Biotechnol 26:2263–2268. https://doi.org/10.1007/s11274-010-0416-5

Amiri B, Ibrahim L, Butt TM (1999) Antifeedant properties of destruxins and their use with the entomogenous fungus *Metarhizium anisopliae* for improved control of crucifer pests. Biocontrol Sci Tech 9:487–498. https://doi.org/10.1080/09583159929451

Arroyo-Manzanares N, Diana Di Mavungu J, Garrido-Jurado I, Arce L, Vanhaecke L, Quesada-Moraga E, Saeger SD (2017) Analytical strategy for determination of known and unknown destruxins using hybrid quadrupole-Orbitrap high resolution mass spectrometry. Anal Bioanal Chem 409:3347–3357. https://doi.org/10.1007/s00216-017-0276-z

Asaff A, Cerda-García-Rojas C, de la Torre M (2005) Isolation of dipicolinic acid as an insecticidal toxin from *Paecilomyces fumosoroseus*. Appl Microbiol Biotechnol 68:542–547. https://doi.org/10.1007/s00253-005-1909-2

Aznar-Fernández T, Cimmino A, Masi M, Rubiales D, Evidente A (2019) Anti-feedant activity of long-chain alcohols, and fungal and plant metabolites against pea aphid (*Acyrthosiphon pisum*) as potential biocontrol strategy. Nat Prod Res 33:2471–2479. https://doi.org/10.1080/14786419.2018.1452013

Bagheri A, Askari Seyahooei M, Fathipour Y (2021) The endophytes. In: Microbial approaches for insect pest management. Springer, pp 151–215. https://doi.org/10.1007/978-981-16-3595-3_4

Baker BP, Green TA, Loker AJ (2020) Biological control and integrated pest management in organic and conventional systems. Biol Control 140:104095. https://doi.org/10.1016/j.biocontrol.2019.104095

Bandani R, Butt TM (1999) Insecticidal, anti-feedant, growth inhibitory activities of efrapeptins, metabolites of entomogenous fungus *Tolypocladium*. Biocontrol Sci Tech 9:499–506. https://doi.org/10.1080/09583159929460

Bandani AR, Khambay BPS, Faull J, Newton R, Deadman M, Butt TM (2000) Production of efrapeptins by *Tolypocladium* species and evaluation of their insecticidal and antimicrobial properties. Mycol Res 104:537–544. https://doi.org/10.1017/s0953756299001859

Berestetskiy A, Hu Q (2021) The chemical ecology approach to reveal fungal metabolites for arthropod pest management. Microorganisms 9:1379. https://doi.org/10.3390/microorganisms9071379

Bidochka MJ, Khachatourians GG (1991) The implication of metabolic acids produced by *Beauveria bassiana* in the pathogenesis of the migratory grasshopper, *Melanoplus sanguinipes*. J Invertebr Pathol 58:106–117. https://doi.org/10.1016/0022-2011(91)90168-p

Brakhage AA (2013) Regulation of fungal secondary metabolism. Nat Rev Microbiol 11:21–32. https://doi.org/10.1038/nrmicro2916

Brousseau C, Charpentier G, Belloncik S (1996) Susceptibility of spruce budworm, *Choristoneura fumiferana* Clemens, to destruxins, cyclodepsipeptidic mycotoxins of *Metarhizium anisopliae*. J Invertebr Pathol 68:180–182. https://doi.org/10.1006/jipa.1996.0079

Brückner H, Degenkolb T (2020) Sequences of tolypins, insecticidal efrapeptin-type peptaibiotics from species of the fungal genus *Tolypocladium*. Chem Biodivers 17:e200027. https://doi.org/10.1002/cbdv.202000276

Bush LP, Wilkinson HH, Schardl CL (1997) Bioprotective alkaloids of grass-fungal endophyte symbioses. Plant Physiol 114:1–7. https://doi.org/10.1104/pp.114.1.1

Buttachon S, Chandrapatya A, Himaman W, Kijjoa A (2013) Acaricidal activity of *Hypocrella raciborskii* Zimm. (Hypocreales: Clavicipitaceae) crude extract and some pure compounds on *Tetranychus urticae* Koch (Acari: Tetranychidae). Afr J Microbiol Res 7:577–585. https://doi.org/10.5897/AJMR12.1727

Buttachon S, Zin WWM, Himaman W, Sinchayakul P, Kijjoa A (2022) Laboratory-based toxicity of scale insect pathogen *Moelleriella raciborskii* (Zimm.) (Hypocreales: Clavicipitaceae) crude extracts and isolated compounds against *Tetranychus truncatus* Ehara (Acari: Tetranychidae). Syst Appl Acarol 27:1152–1165. https://doi.org/10.11158/saa.27.6.13

Canassa F, Tall S, Moral RA, de Lara IA, Delalibera I Jr, Meyling NV (2019) Effects of bean seed treatment by the entomopathogenic fungi *Metarhizium robertsii* and *Beauveria bassiana* on plant growth, spider mite populations, and behavior of predatory mites. Biol Control 132:199–208. https://doi.org/10.1016/j.biocontrol.2019.02.003

Carmen González-Mas MC, Lull C, Moya P, Ayala I, Primo J, Yúfera EP (2003) Insecticidal activity of Penitrems, including Penitrem G, a new member of the family isolated from *Penicillium crustosum*. J Agric Food Chem 51:2156–2160. https://doi.org/10.1021/jf020983e

Champlin FR, Grula EA (1979) Noninvolvement of beauvericin in the entomopathogenicity of *Beauveria bassiana*. Appl Environ Microbiol 37:1122–1126. https://doi.org/10.1128/aem.37.6.1122-1126.1979

Chen B-X, Wei T, Xue L-N, Zheng Q-W, Ye Z-W, Zou Y, Yang Y, Yun F, Guo L-Q, Lin J-F (2020) Transcriptome analysis reveals the flexibility of cordycepin network in *Cordyceps militaris* activated by L-alanine addition. Front Microbiol 11:577. https://doi.org/10.3389/fmicb.2020.00577

Cheong P (2015) Bioactive metabolites of an isolate of the entomopathogenic fungus *Beauveria bassiana*. Doctoral dissertation, Lincoln University, p 282

Claydon N, Grove JF (1982) Insecticidal secondary metabolic products from the entomogenous fungus *Verticillium lecanii*. J Invertebr Pathol 40:413–418. https://doi.org/10.1016/0022-2011(82)90180-x

Cox RJ, Skellam E, Williams K (2018) Biosynthesis of fungal polyketides. In: Physiology and genetics: selected basic and applied aspects. Springer, pp 385–412. https://doi.org/10.1007/978-3-319-71740-1_13

Daisy BH, Strobel GA, Castillo U, Ezra D, Sears J, Weaver DK, Runyon JB (2002) Naphthalene, an insect repellent, is produced by *Muscodor vitigenus*, a novel endophytic fungus. Microbiology 148:3737–3741. https://doi.org/10.1099/00221287-148-11-3737

Daniel J, Silva A, Nakagawa D, de Madeiros L, Carvalho M, Tavares L, Abreu LM, Rodrigues-Filho E (2017) Larvicidal activity of *Beauveria bassiana* extracts against *Aedes aegypti* and identification of beauvericins. J Braz Chem Soc 28:1003–1013. https://doi.org/10.21577/0103-5053.20160253

Deka B, Baruah C, Babu A (2021) Entomopathogenic microorganisms: their role in insect pest management. Egyptian J Biol Pest Control 31:121. https://doi.org/10.1186/s41938-021-00466-7

Divekar PA, Narayana S, Divekar BA, Kumar R, Gadratagi BG, Ray A, Singh AK, Rani V, Singh V, Singh AK, Kumar A, Singh RP, Meena RS, Behera TK (2022) Plant secondary metabolites as defense tools against herbivores for sustainable crop protection. Int J Mol Sci 23:2690. https://doi.org/10.3390/ijms23052690

Donzelli BGG, Krasnoff SB (2016) Molecular genetics of secondary chemistry in *Metarhizium* fungi. Adv Genet 94:365–436. https://doi.org/10.1016/bs.adgen.2016.01.005

Evidente A, Ricciardiello G, Andolfi A, Sabatini MA, Ganassi S, Altomare C, Favilla M, Melck D (2008) Citrantifidiene and citrantifidiol: bioactive metabolites produced by *Trichoderma citrinoviride* with potential antifeedant activity toward aphids. J Agric Food Chem 56:3569–3573. https://doi.org/10.1021/jf073541h

Evidente A, Andolfi A, Cimmino A, Ganassi S, Altomare C, Favilla M, De Cristofaro A, Vitagliano S, Agnese Sabatini M (2009) Bisorbicillinoids produced by the fungus *Trichoderma citrino-viride* affect feeding preference of the aphid *Schizaphis graminum*. J Chem Ecol 35:533–541. https://doi.org/10.1007/s10886-009-9632-6

Feng P, Shang Y, Cen K, Wang C (2015) Fungal biosynthesis of the bibenzoquinone oosporein to evade insect immunity. Proc Natl Acad Sci 112:11365–11370

Ganassi S, Moretti A, Pagliai AMB, Logrieco A, Sabatini MA (2002) Effects of beauvericin on *Schizaphis graminum* (Aphididae). J Invertebr Pathol 80:90–96. https://doi.org/10.1016/s0022-2011(02)00125-8

Gao Q, Jin K, Ying S-H, Zhang Y, Xiao G, Shang Y et al (2011) Genome sequencing and comparative Transcriptomics of the model Entomopathogenic fungi *Metarhizium anisopliae* and *M. acridum*. PLoS Genet 7:e1001264. https://doi.org/10.1371/journal.pgen.1001264

Gibson DM, Donzelli BGG, Krasnoff SB, Keyhani NO (2014) Discovering the secondary metabolite potential encoded within entomopathogenic fungi. Nat Prod Rep 31:1287–1305. https://doi.org/10.1039/C4NP00054D

Glenn AE, Bacon CW (1997) Distribution of ergot alkaloids within the family Clavicipitaceae. In: *Neotyphodium*/grass interactions. Springer, pp 53–56. https://doi.org/10.1007/978-1-4899-0271-9_7

Guo Z, Gai C, Cai C, Chen L, Liu S, Zeng Y et al (2017) Metabolites with insecticidal activity from *Aspergillus fumigatus* JRJ111048 isolated from mangrove plant *Acrostichum specioum* endemic to Hainan Island. Mar Drugs 15:381. https://doi.org/10.3390/md15120381

Hamshou M, Smagghe G, Van Damme EJM (2010a) Entomotoxic effects of fungal lectin from *Rhizoctonia solani* towards *Spodoptera littoralis*. Fungal Biol 114:34–40. https://doi.org/10.1016/j.mycres.2009.10.002

Hamshou M, Smagghe G, Shahidi-Noghabi S, De Geyter E, Lannoo N, Van Damme EJ (2010b) Insecticidal properties of *Sclerotinia sclerotiorum* agglutinin and its interaction with insect tissues and cells. Insect Biochem Mol Biol 40:883–890. https://doi.org/10.1016/j.ibmb.2010.08.008

Han JH, Jin BR, Kim JJ, Lee SY (2014) Virulence of entomopathogenic fungi *Metarhizium anisopliae* and *Paecilomyces fumosoroseus* for the microbial control of *Spodoptera exigua*. Mycobiology 42:385–390. https://doi.org/10.5941/MYCO.2014.42.4.385

Harith-Fadzilah N, Abd Ghani I, Hassan M (2021) Omics-based approach in characterising mechanisms of entomopathogenic fungi pathogenicity: a case example of *Beauveria bassiana*. J King Saud Univ Sci 33:101332. https://doi.org/10.1016/j.jksus.2020.101332

Horikoshi R, Goto K, Mitomi M, Oyama K, Sunazuka T, Ōmura S (2017) Identification of pyripyropene A as a promising insecticidal compound in a microbial metabolite screening. J Antibiot 70:272–276

Horikoshi R, Goto K, Mitomi M, Oyama K, Sunazuka T, Ōmura S (2018) Insecticidal properties of pyripyropene A, a microbial secondary metabolite, against agricultural pests. J Pestic Sci 43:266–271. https://doi.org/10.1584/jpestics.d18-030

Hu Q, Dong T (2015) Non-ribosomal peptides from entomogenous fungi. In: Biocontrol of lepidopteran pests: use of soil microbes and their metabolites. Springer International Publishing, pp 169–206. https://doi.org/10.1007/978-3-319-14499-3_8

Hu Q-B, Ren S-X, Wu J-H, Chang J-M, Musa PD (2006) Investigation of destruxin A and B from 80 Metarhizium strains in China, and the optimization of cultural conditions for the strain MaQ10. Toxicon 48:491–498. https://doi.org/10.1016/j.toxicon.2006.06.018

Hu QB, Ren SX, An XC, Qian MH (2007) Insecticidal activity influence of destruxins on the pathogenicity of *Paecilomyces javanicus* against *Spodoptera litura*. J Appl Entomol 131:262–268. https://doi.org/10.1111/j.1439-0418.2007.01159.x

Islam W, Adnan M, Shabbir A, Naveed H, Abubakar YS, Qasim M et al (2021) Insect-fungal-interactions: a detailed review on entomopathogenic fungi pathogenicity to combat insect pests. Microb Pathog 159:105122. https://doi.org/10.1016/j.micpath.2021.105122

Jaber LR, Ownley BH (2018) Can we use entomopathogenic fungi as endophytes for dual biological control of insect pests and plant pathogens? Biol Control 116:36–45. https://doi.org/10.1016/j.biocontrol.2017.01.018

Jaronski ST (2023) Mass production of entomopathogenic fungi – state of the art. In: Mass production of beneficial organisms. Elsevier, pp 317–357. https://doi.org/10.1016/b978-0-12-822106-8.00017-8

Jaronski ST, Mascarin GM (2017) Mass production of fungal entomopathogens. In: Microbial control of insect and mite pests. Springer, pp 141–155. https://doi.org/10.1016/b978-0-12-803527-6.00009-3

Kanaoka M, Isogai A, Murakoshi S, Ichinoe M, Suzuki A, Tamura S (1978) Bassianolide, a new insecticidal cyclodepsipeptide from *Beauveria bassiana* and *Verticillium lecanii*. Agric Biol Chem 42:629–635. https://doi.org/10.1080/00021369.1978.10863029

Kershaw MJ, Moorhouse ER, Bateman R, Reynolds SE, Charnley AK (1999) The role of destruxins in the pathogenicity of *Metarhizium anisopliae* for three species of insects. J Invertebr Pathol 74:213–223. https://doi.org/10.1006/jipa.1999.4884

Kim JR, Yeon SH, Kim HS, Ahn YJ (2002) Larvicidal activity against *Plutella xylostella* of cordycepin from the fruiting body of *Cordyceps militaris*. Pest Manag Sci 58:713–717. https://doi.org/10.1002/ps.508

Kim HM, Choi IS, Lee S, Hwang IM, Chun HH, Wi SG, Kim J-C, Shin TY, Kim JC, Kim JS et al (2019) Advanced strategy to produce insecticidal destruxins from lignocellulosic biomass Miscanthus. Biotechnol Biofuels 12:1–10. https://doi.org/10.1186/s13068-019-1530-8

Kim JC, Hwang IM, Kim HM, Kim S, Shin TS, Woo SD, Park HW (2023) Rapid analysis of insecticidal metabolites from the entomopathogenic fungus *Beauveria bassiana* 331R using UPLC-Q-Orbitrap MS. Mycotoxin Res:1–10. https://doi.org/10.1007/s12550-023-00509-y

Kirkland B, Westwood GS, Keyhani NO (2004) Pathogenicity of entomopathogenic fungi *Beauveria bassiana* and *Metarhizium anisopliae* to Ixodidae tick species *Dermacentor variabilis*, *Rhipicephalus sanguineus*, and *Ixodes scapularis*. J Med Entomol 41:705–711. https://doi.org/10.1603/0022-2585-41.4.705

Koradin C, Schröder H, Oyama K, Ōmura S (2021) Chemistry and biology connected: the development of Inscalis. In: Recent highlights in the discovery and optimization of crop protection products. Academic, pp 231–239. https://doi.org/10.1016/b978-0-12-821035-2.00011-5

Krasnoff SB, Gupta S (1991) Identification and biosynthesis of efrapeptins in the fungus *Tolypocladium geodes* gams (Deuteromycotina, Hyphomycetes). J Chem Ecol 17:1953–1960. https://doi.org/10.1007/bf00992580

Krasnoff SB, Gupta S, St. Leger RJ, Renwick JA, Roberts DW (1991) Antifungal and insecticidal properties of efrapeptin metabolites of the fungus *Tolypocladium niveum*. J Invertebr Pathol 58:180–188. https://doi.org/10.1016/0022-2011(91)90062-u

Kryukov VY, Yaroslavtseva ON, Dubovskiy IM, Tyurin MV, Kryukova NA, Glupov VV (2014) Insecticidal and immunosuppressive effect of ascomycete *Cordyceps militaris* on the larvae of the Colorado potato beetle *Leptinotarsa decemlineata*. Biol Bull 41:276–283. https://doi.org/10.1134/s1062359014020046

Lackner G, Misiek M, Braesel J, Hoffmeister D (2012) Genome mining reveals the evolutionary origin and biosynthetic potential of basidiomycete polyketide synthases. Fungal Genet Biol 49:996–1003. https://doi.org/10.1016/j.fgb.2012.09.009

Litwin A, Nowak M, Różalska S (2020) Entomopathogenic fungi: unconventional applications. Rev Environ Sci Bio Technol 19:23–42. https://doi.org/10.1007/s11157-020-09525-1

Liu BL, Tzeng YM (2012) Development and applications of destruxins: a review. Biotechnol Adv 30:1242–1254. https://doi.org/10.1016/j.biotechadv.2011.10.006

Liu WZ, Boucias DG, McCoy CW (1995) Extraction and characterization of the insecticidal toxin hirsutellin A produced by *Hirsutella thompsonii* var. *thompsonii*. Exp Mycol 19:254–262. https://doi.org/10.1006/emyc.1995.1032

Lozano-Tovar MD, Garrido-Jurado I, Lafont F, Quesada-Moraga E (2015) Insecticidal activity of a destruxin-containing extract of *Metarhizium brunneum* against *Ceratitis capitata* (Diptera: Tephritidae). J Econ Entomol 108:462–472. https://doi.org/10.1093/jee/tov041

Luz C, Saladino F, Luciano FB, Mañes J, Meca G (2017) Occurrence, toxicity, bioaccessibility and mitigation strategies of beauvericin, a minor *Fusarium* mycotoxin. Food Chem Toxicol 107:430–439. https://doi.org/10.1016/j.fct.2017.07.032

Maggi MD, Damiani N, Ruffinengo SR, Brasesco MC, Szawarski N, Mitton G et al (2017) The susceptibility of *Varroa* destructor against oxalic acid: a study case. Bull Insectol 70:39–44

Mallebrera B, Prosperini A, Font G, Ruiz MJ (2018) In vitro mechanisms of Beauvericin toxicity: a review. Food Chem Toxicol 111:537–545. https://doi.org/10.1016/j.fct.2017.11.019

Mannino MC, Davyt-Colo B, Pedrini N (2021) Toxic secondary metabolites and virulence factors expression by Entomopathogenic fungi during insect infection and potential impact as a tool for pest management. In: Microbes for sustainable Insect pest management: hydrolytic enzyme & secondary metabolite, vol 2. Springer, pp 121–134. https://doi.org/10.1007/978-3-030-67231-7_6

Mantzoukas S, Eliopoulos PA (2020) Endophytic entomopathogenic fungi: a valuable biological control tool against plant pests. Appl Sci 10:360. https://doi.org/10.3390/app10010360

Matha V, Weiser J, Olejnicek J (1988) The effect of tolypin in *Tolypocladium niveum* crude extract against mosquito and black fly larvae in the laboratory. Folia Parasitol 35:381–383

Matsukura K, Shiba T, Sasaki T, Matsumura M (2012) Enhanced resistance to four species of Clypeorrhynchan pests in *Neotyphodium uncinatum* infected Italian ryegrass. J Econ Entomol 105:129–134. https://doi.org/10.1603/ec11216

Matsukura K, Shiba T, Sasaki T, Yoshida K, Matsumura M (2014) Dynamics of *Neotyphodium uncinatum* and N-formylloline in Italian ryegrass, and their relation to insect resistance in the field. J Appl Microbiol 116:400–407. https://doi.org/10.1111/jam.12374

Mazet I, Vey A (1995) Hirsutellin A, a toxic protein produced in vitro by *Hirsutella thompsonii*. Microbiology 141:1343–1348. https://doi.org/10.1099/13500872-141-6-1343

McNamara L, Dolan SK, Walsh JMD et al (2019) Oosporein, an abundant metabolite in Beauveria caledonica, with a feedback induction mechanism and a role in insect virulence. Fungal Biol 123:601–610. https://doi.org/10.1016/j.funbio.2019.01.004

Michaliski LF, Ióca LP, Oliveira LS, Crnkovic CM, Takaki M, Freire VF, Berlinck RG (2023) Improvement of targeted fungi secondary metabolite production using a systematic experimental design and Chemometrics analysis. Methods Protocols 6:77. https://doi.org/10.1186/s12866-019-1555-y

Miller JD, MacKenzie S, Foto M, Adams GW, Findlay JA (2002) Needles of white spruce inoculated with rugulosin-producing endophytes contain rugulosin reducing spruce budworm growth rate. Mycol Res 106:471–479. https://doi.org/10.1017/s0953756202005671

Mukherjee K, Vilcinskas A (2018) The entomopathogenic fungus *Metarhizium robertsii* communicates with the insect host *Galleria mellonella* during infection. Virulence 9:402–413. https://doi.org/10.1080/21505594.2017.1405190

Muñiz-Paredes F, Miranda-Hernández F, Loera O (2017) Production of conidia by entomopathogenic fungi: from inoculants to final quality tests. World J Microbiol Biotechnol 33:1–9. https://doi.org/10.1007/s11274-017-2229-2

Namasivayam SK, Arvind Bharani RS (2021) Biocompatible silver nanoparticles-loaded fungal metabolites nanoconjugate (agnp–fm) preparation for the noteworthy pesticidal activity. Natl Acad Sci Lett 44:1–7. https://doi.org/10.1007/s40009-021-01044-z

Omoto C, McCoy CW (1998) Toxicity of purified fungal toxin Hirsutellin a to the citrus rust mite *Phyllocoptruta oleivora* (ash.). J Invertebr Pathol 72:319–322. https://doi.org/10.1006/jipa.1998.4813

Ortiz-Urquiza A, Keyhani NO (2013) Stress response signaling and virulence: insights from entomopathogenic fungi. Curr Genet 59:141–163. https://doi.org/10.1007/s00294-014-0439-9

Pal S, Leger RJS, Wu LP (2007) Fungal peptide Destruxin A plays a specific role in suppressing the innate immune response in *Drosophila melanogaster*. J Biol Chem 282:8969–8977. https://doi.org/10.1074/jbc.m605927200

Paschapur A, Subbanna ARNS, Singh AK, Jeevan B, Stanley J, Rajashekhar H, Mishra KK (2021) Unraveling the importance of metabolites from entomopathogen-ic fungi in insect pest management. In: Microbes for sustainable Insect pest management: hydrolytic enzyme & secondary metabolite, vol 2. Springer, pp 89–120. https://doi.org/10.1007/978-3-030-67231-7_5

Peng KC, Huang HS, Tzeng YM, Liu SY (2005) Circular dichroism analysis of destruxins from *Metarhizium anisopliae*. J Biochem Biophys Methods 62:41–50. https://doi.org/10.1016/j.jbbm.2004.08.001

Popay A, Tapper B, Podmore C (2009) Endophyte infected meadow fescue and loline alkaloids affect Argentine stem weevil larvae. N Z Plant Prot 62:19–27. https://doi.org/10.30843/nzpp.2009.62.4801

Poprawski TJ, Robert PH, Maniania NK (1994) Contact toxicity of the mycotoxin destruxin E to *Empoasca vitis* (Göthe)(Hom., Cicadellidae). J Appl Entomol 117:135–143. https://doi.org/10.1111/j.1439-0418.1994.tb00718.x

Potter DA, Stokes JT, Redmond CT, Schardl CL, Panaccione DG (2008) Contribution of ergot alkaloids to suppression of a grass-feeding caterpillar assessed with gene knockout endophytes in perennial ryegrass. Entomol Exp Appl 126:138–147. https://doi.org/10.1111/j.1570-7458.2007.00650.x

Quesada-Moraga E, Vey A (2004) Bassiacridin, a protein toxic for locusts secreted by the entomopathogenic fungus *Beauveria bassiana*. Mycol Res 108:441–452. https://doi.org/10.1017/s0953756204009724

Quy TN, Xuan TD, Andriana Y, Tran HD, Khanh TD, Teschke R, Qu X (2019) Tran Cordycepin isolated from *Cordyceps militaris*: its newly discovered herbicidal property and potential plant-based novel alternative to glyphosate. Molecules 24:2901. https://doi.org/10.3390/molecules24162901

Radhi M, Ashraf S, Lawrence S, Tranholm AA, Wellham PAD, Hafeez A, Khamis AS, Thomas R, McWilliams D, de Moor CH (2021) A systematic review of the biological effects of cordycepin. Molecules 26:5886. https://doi.org/10.3390/molecules26195886

Rajula J, Karthi S, Mumba S, Pittarate S, Thungrabeab M, Krutmuang P (2021) Current status and future prospects of entomopathogenic fungi: a potential source of biopesticides. In: Recent advancement in microbial biotechnology. Elsevier, pp 71–98. https://doi.org/10.1016/b978-0-12-822098-6.00013-6

Ratnaweera PB, Jayasundara M, Herath HSD, Williams DE, Rajapaksha SU, Nishantha KP, de Silva ED, Andersen RJ (2020) Antifeedant, contact toxicity and oviposition deterrent effects of phyllostine acetate and phyllostine isolated from the endophytic fungus *Diaporthe miriciae* against *Plutella xylostella* larvae. Pest Manag Sci 76:1541–1548. https://doi.org/10.1002/ps.5673

Ravindran K, Akutse KS, Sivaramakrishnan S, Wang L (2016) Determination and characterization of destruxin production in *Metarhizium anisopliae* Tk6 and formulations for *Aedes aegypti* mosquitoes control at the field level. Toxicon 120:89–96. https://doi.org/10.1016/j.toxicon.2016.07.016

Ravindran S, Sivaramakrishnan S, Hussain M, Dash CK, Bamisile BS, Qasim M, Lian-de W (2018) Investigation and molecular docking studies of Bassianolide from *Lecanicillium lecanii* against *Plutella xylostella* (Lepidoptera: Plutellidae). Comp Biochem Physiol Part C Toxicol Pharmacol 206–207:65–72. https://doi.org/10.1016/j.cbpc.2018.03.004

Reyes Haro LE (2018) Developing a strain improvement system for the entomopathogenic fungus *Beauveria basssiana*: a way to get better biocontrol agents? Doctoral dissertation, University of Warwick, p. 198

Robert P, Riba G (1989) Toxic and repulsive effects of spray, 'per os' and systemic applications of destruxin E to aphids. Mycopathologia 108:179–183. https://doi.org/10.1007/bf00436223

Roberts DW (1981) Toxins of entomopathogenic fungi. In: Microbial control of pests and plant disease 1970–1980. Academic, pp 441–464

Roberts DW, St Leger RJ, Gupta S (1992) Metabolite production by entomopathogenic fungi. Pesq Agrop Brasileira 27:351–373

Rodríguez-Durán LV, Michel MR, Pichardo A, Aguilar-Zárate P (2023) Microbial bioreactors for secondary metabolite production. In: Microbial bioreactors for industrial molecules. Wiley, pp 275–296. https://doi.org/10.1002/9781119874096.ch13

Rodríguez-Romero H, Rodríguez-Peláez L, Reyes-Castro A, Tlakaele Notario-Rendón O, González-Peréz M, Reza-Salgado J et al (2023) Secondary metabolites of Entomopathogens as biotechnological tools for the biological control of agricultural insect pests. In: Insecticides – advances in insect control and sustainable pest management. IntechOpen. https://doi.org/10.5772/intechopen.111465

Rowan DD, Dymock JJ, Brimble MA (1990) Effect of fungal metabolite peramine and analogs on feeding and development of Argentine stem weevil (*Listronotus bonariensis*). J Chem Ecol 16:1683–1695. https://doi.org/10.1007/bf01014100

Roy A, Ahuja S, Garg S (2021) Fungal secondary metabolites: biological activity and potential applications. In: Recent trends in mycological research. Fungal biology. Springer, pp 159–188. https://doi.org/10.1007/978-3-030-60659-6_7

Sabbour MM (2019) Effect of destruxin on the population reduction of green peach aphid *Myzus persicae* (Hemiptera: Aphididae) and the predator *Coccinella undecimpunctata* (Coleoptera: Coccinellidae) in tomato fields. Bull Natl Res Cent 43:1–7. https://doi.org/10.1186/s42269-019-0170-2

Sabbour MM, Solieman NY, Abdel-Raheem MA (2020) Influence of *Metarhizium anisopliae*-based destruxin A-760 and destruxin-A-724 on the sugar beet fly, *Pegomya mixta* Vill.(Diptera: Anthomyiidae). J Biopest 13:21–27

Sala A, Barrena R, Artola A, Sánchez A (2019) Current developments in the production of fungal biological control agents by solid-state fermentation using organic solid waste. Crit Rev Environ Sci Technol 49:655–694. https://doi.org/10.1080/10643389.2018.1557497

Salimova D, Dalinova A, Dubovik V, Senderskiy I, Stepanycheva E, Tomilova O, Hu Q, Berestetskiy A (2021) Entomotoxic activity of the extracts from the fungus, *Alternaria tenuissima* and its major metabolite, tenuazonic acid. J Fungi 7:774. https://doi.org/10.3390/jof7090774

Schardl CL, Young C, Hesse U, Amyotte SG, Andreeva K, Calie PJ, Fleetwood DJ, Haws DC, Moore N, Oeser B et al (2013) Plant-symbiotic fungi as chemical engineers: multi-genome analysis of the Clavicipitaceae reveals dynamics of alkaloid loci. PLoS Genet 9:e1003323. https://doi.org/10.1371/journal.pgen.1003323

Schmidt-Dannert C (2015) Biosynthesis of terpenoid natural products in fungi. In: Biotechnology of isoprenoids. Springer, pp 19–61. https://doi.org/10.46936/jejc.proj.2013.48096/60005300

Schulz B, Haas S, Junker C, Andrée N, Schobert M (2015) Fungal endophytes are involved in multiple balanced antagonisms. Curr Sci 109:39–45

Shiba T, Sugawara K (2009) Fungal loline alkaloids in grass–endophyte associations confer resistance to the rice leaf bug, *Trigonotylus caelestialium*. Entomol Exp Appl 130:55–62. https://doi.org/10.1111/j.1570-7458.2008.00791.x

Shiba T, Sugawara K, Arakawa A (2011) Evaluating the fungal endophyte *Neotyphodium occultans* for resistance to the rice leaf bug, *Trigonotylus caelestialium*, in Italian ryegrass, *Lolium multiflorum*. Entomol Exp Appl 141:45–51. https://doi.org/10.1111/j.1570-7458.2011.01162.x

Skrobek A, Shah FA, Butt TM (2008) Destruxin production by the entomogenous fungus *Metarhizium anisopliae* in insects and factors influencing their degradation. BioControl 53:361–373. https://doi.org/10.1007/s10526-007-9077-1

Soman AG, Glor JB, Angawi RF, Wicklow DT, Dowd PF (2001) Vertilecanins: new phenopicolinic acid analogues from *Verticillium lecanii*. J Nat Prod 64:189–192. https://doi.org/10.1021/np000094q

Sree KS, Padmaja V, Murthy YLN (2008) Insecticidal activity of destruxin, a mycotoxin *from-Metarhizium anisopliae* (Hypocreales), against *Spodoptera litura* (Lepidoptera: Noctuidae) larval stages. Pest Manag Sci 64:119–125. https://doi.org/10.1002/ps.1480

Sumarah MW, Adams GW, Berghout J, Slack GJ, Wilson AM, Miller JD (2008) Spread and persistence of a rugulosin-producing endophyte in Picea glauca seedlings. Mycol Res 112:731–736. https://doi.org/10.1016/j.mycres.2008.01.007

Sun C, Li X, Xiao Z, Li S, Wang K, Tian C, Feng H, Liu X, Yin X, Wei J, Bai S, An S (2023) Cyclosporin A acts as an insecticide candidate: providing sustainable biocontrol potential for managing Mythimna separata. J Pest Sci 96:1103–1115. https://doi.org/10.1007/s10340-022-01577-2

Taibon J, Sturm S, Seger C, Strasser H, Stuppner H (2015) Quantitative assessment of destruxins from strawberry and maize in the lower parts per billion range: combination of a QuEChERS-based extraction protocol with a fast and selective UHPLCQTOF-MS assay. J Agric Food Chem 63:5707–5713. https://doi.org/10.1021/acs.jafc.5b01562

Thomsen L, Eilenberg J, Esbjerg P (1996) Effects of destruxins on *Pieris brassicae* and *Agrotis segetum*. IOBC WPRS Bull 19:190–195

Tiang-Yan D, Bo-Wen Z, Qun-Fang W, Qiong-Bo H (2016) The production relationship of destruxins and blastospores of *Metarhizium anisopliae* with virulence against *Plutella xylostella*. J Integr Agric 15:1313–1320. https://doi.org/10.1016/S2095 3119(15)61322-3

Usseglio VL, Dambolena JS, Martinez MJ, Zunino MP (2020) The role of Fumonisins in the biological interaction between *Fusarium verticillioides* and *Sitophilus zeamais*. J Chem Ecol 46:1059–1068. https://doi.org/10.1007/s10886-020-01220-3

Valencia JWA, Gaitán Bustamante AL, Jiménez AV, Grossi-de-Sá MF (2011) Cytotoxic activity of fungal metabolites from the pathogenic fungus *Beauveria bassiana*: an intraspecific evaluation of beauvericin production. Curr Microbiol 63:306–312. https://doi.org/10.1007/s00284-011-9977-2

Vega FE (2018) The use of fungal entomopathogens as endophytes in biological control: a review. Mycologia 110:4–30. https://doi.org/10.1080/00275514.2017.1418578

Velmurugan P, Lee YH, Venil CK, Lakshmanaperumalsamy P, Chae JC, Oh BT (2010) Effect of light on growth, intracellular and extracellular pigment production by five pigment-producing filamentous fungi in synthetic medium. J Biosci Bioeng 109:346–350. https://doi.org/10.1016/j.jbiosc.2009.10.003

Vey A, Hoagland RE, Butt TM (2001) Toxic metabolites of fungal biocontrol agents. In: Fungias biocontrol agents. CAB International, pp 311–346. https://doi.org/10.1079/9780851993560.0311

Vidhate RP, Dawkar VV, Punekar SA, Giri AP (2023) Genomic determinants of entomopathogenic fungi and their involvement in pathogenesis. Microb Ecol 85:49–60. https://doi.org/10.1007/s00248-021-01936-z

Villegas-Mendoza JM, Mireles-Martínez M, Rodríguezcastillejos G, Santiago Adame R, Rosas-García NM (2019) Toxicity evaluation of a Methanolic extract from *Beauveria bassiana* Sordariomycetes: Hypocreales against *Spodoptera exigua* and *Spodoptera frugiperda* Lepidoptera: Noctuidae larvae. Southwest Entomol 44:715–722. https://doi.org/10.3958/059.044.0316

Vinale F, Nicoletti R, Lacatena F, Marra R, Sacco A, Lombardi N, D'Errico G, Digilio MC, Lorito M, Woo SL (2017) Secondary metabolites from the endophytic fungus *Talaromyces pinophilus*. Nat Prod Res 31:1778–1785. https://doi.org/10.1080/14786419.2017.1290624

Vishwanathreddy H, Bhat GG, Inamdar SR, Gudihal RK, Swamy BM (2014) *Sclerotium rolfsii* lectin exerts insecticidal activity on *Spodoptera litura* larvae by binding to membrane proteins of midgut epithelial cells and triggering caspase-3-dependent apoptosis. Toxicon 78:47–57. https://doi.org/10.1016/j.toxicon.2013.11.012

Wang Q, Xu L (2012) Beauvericin, a bioactive compound produced by fungi: a short review. Molecules 17:2367–2377. https://doi.org/10.3390/molecules17032367

Wang B, Kang Q, Lu Y, Bai L, Wang C (2012) Unveiling the biosynthetic puzzle of destruxins in *Metarhizium* species. Proc Natl Acad Sci USA 109:1287–1292. https://doi.org/10.1073/pnas.1115983109

Wang JB, St Leger RJ, Wang C (2016) Advances in genomics of entomopathogenic fungi. In: Advances in genetics. Academic, pp 67–105. https://doi.org/10.1016/bs.adgen.2016.01.002

Wang X, Gong X, Li P, Lai D, Zhou L (2018) Structural diversity and biological activities of cyclic depsipeptides from fungi. Molecules 23:169. https://doi.org/10.3390/molecules23010169

Watanabe N, Hattori M, Yokoyama E, Isomura S, Ujita M, Hara A (2006) Entomogenous fungi that produce 2,6-pyridine dicarboxylic acid (dipicolinic acid). J Biosci Bioeng 102:365–368. https://doi.org/10.1263/jbb.102.365

Weng Q, Zhang X, Chen W, Hu Q (2019) Secondary metabolites and the risks of *Isaria fumosorosea* and *Isaria farinosa*. Molecules 24:664. https://doi.org/10.3390/molecules24040664

Wilkinson HH, Siegel MR, Blankenship JD, Mallory AC, Bush LP, Schardl CL (2000) Contribution of fungal Loline alkaloids to protection from aphids in a grass-endophyte mutualism. Mol Plant-Microbe Interact 13:1027–1033. https://doi.org/10.1094/mpmi.2000.13.10.1027

Xiao G, Ying SH, Zheng P, Wang ZL, Zhang S, Xie X-Q, Shang Y, Leger RJ, Zhao G-P, Wang C, Feng M-G (2012) Genomic perspectives on the evolution of fungal entomopathogenicity in *Beauveria bassiana*. Sci Rep 2:483. https://doi.org/10.1038/srep00483

Xu Y, Orozco R, Kithsiri Wijeratne EM, Espinosa-Artiles P, Gunatilaka AAL, Stock SP, Molnár I (2009) Biosynthesis of the cyclooligomer depsipeptide bassianolide, an insecticidal virulence factor of *Beauveria bassiana*. Fungal Genet Biol 46:353–364. https://doi.org/10.1016/j.fgb.2009.03.001

Yang FZ, Yang B, Li BB, Xiao C (2015) Alternaria toxin-induced resistance in rose plants against rose aphid (*Macrosiphum rosivorum*): effect of tenuazonic acid. J Zhejiang Univ Sci B 16:264. https://doi.org/10.1631/jzus.b1400151

Zhang Y, Han T, Ming Q, Wu L, Rahman K, Qin L (2012) Alkaloids produced by Endophytic fungi: a review. Nat Prod Commun 7:963–968. https://doi.org/10.1177/1934578x1200700742

Trichoderma as a Source of Metabolites for Applications in Agriculture

Mario Riolo, Victor D'Opazo, and Santa Olga Cacciola

Abstract In the context of escalating global food demand and the associated threat to food security, the use of synthetic chemicals in agriculture has raised concerns about environmental impact and human health risks. This chapter explores the application of *Trichoderma* species and their metabolites as a sustainable alternative for controlling plant diseases and fungal contaminants in agriculture. *Trichoderma*, a versatile genus of filamentous fungi, has gained prominence as an effective biocontrol agent (BCA) due to its adaptability, symbiotic relationships with crops, and diverse metabolite production. The metabolites produced by *Trichoderma* species play a crucial role in their antagonistic activity against a broad spectrum of fungal and oomycete plant pathogens. This antagonism involves the production of enzymes, such as chitinases and glucanases, which degrade fungal cell walls, highlighting their direct impact on pathogen control. The application of *Trichoderma* species as BCAs offers a sustainable alternative to synthetic fungicides, mitigating environmental pollution and reducing the risk of pathogen resistance. Moreover, *Trichoderma* metabolites contribute to plant growth promotion and induce plant defence mechanisms, presenting a multifaceted approach to agricultural challenges. These metabolites also exhibit anti-mycotoxigenic and detoxification potential, further enhancing their utility. The chapter emphasizes the diverse applications of *Trichoderma* metabolites in agriculture, ranging from antifungal bioproducts to biofertilizers and biostimulants. The ability of *Trichoderma* species to enhance nutrient uptake, particularly nitrogen, phosphorus, and potassium, addresses critical chal-

M. Riolo (✉)
Department of Agriculture, Food and Environment, University of Catania, Catania, Italy

Laboratory of Food Chemistry and Toxicology, Faculty of Pharmacy, University of Valencia, Burjassot, Spain
e-mail: mario.riolo@unict.it

V. D'Opazo
Laboratory of Food Chemistry and Toxicology, Faculty of Pharmacy, University of Valencia, Burjassot, Spain

S. O. Cacciola
Department of Agriculture, Food and Environment, University of Catania, Catania, Italy

J. Poveda et al. (eds.), *Fungal Metabolites for Agricultural Applications*, Fungal Biology, https://doi.org/10.1007/978-3-031-76587-2_8

lenges in modern agriculture related to sustainable nutrient management. In summary, this chapter provides a comprehensive overview of the nature and diversity of metabolites produced by *Trichoderma* species, with a special focus on their applications in sustainable agriculture.

Keywords *Trichoderma* spp. · Secondary metabolites · BCA · Agriculture · Bio-stimulant · Bioremediation

1 Introduction

In a world/global scenario with a growing population and an increasing food demand, crop losses are one of the primary threats to food security (Nunes 2012). Plant diseases caused by fungi and bacteria, microbial contamination of agricultural produce and foods, depletion of soil nutrients and accumulation of toxic substances in the soil are major factors contributing to these losses (Kerchev et al. 2020; Pandey et al. 2022). Microbial contamination has an enormous economic impact on the food industry and is also a concern for the health security of the final consumers. It has been estimated that up to one third of the crop losses are caused by microbial contamination and that more than 1 billion people have suffered diseases related to microbial food contamination (Azevedo et al. 2015). Infections by bacteria and fungi give rise to numerous plant diseases, occurring both in the field and during post-harvest handling. Infectious diseases have an incubation period, whose length depends on the type of disease, the susceptibility level of the host and the environmental conditions. Consequently, symptoms may emerge even several days after the onset of the infection, during product storage and distribution. Moreover, several fungi may contaminate agricultural products and foods by producing mycotoxins (Nazarov et al. 2020; Nazareth et al. 2018; Pitt and Hocking 2009; Rovetto et al. 2023). Contaminated fruits and fruit with latent infections can act as inoculum reservoir and vectors, contributing to the transmission of the infection to healthy fruits and leading to significant losses throughout the supply chain (Dutot et al. 2013). Poor soil quality due to scarce availability of nutrients or pollution lead to substantial yield reductions and require the supply of fertilizers and the application of remediation techniques, respectively (Zin and Badaluddin 2020; Krasilnikov and Taboada 2022; Grifoni et al. 2022).

The traditional solution to reduce losses due to infections and contaminations by fungi, nutrient deficiencies in soil, infestations by insect pests or competition from weeds is the use of synthetic fungicides, fertilizers, insecticides and herbicides, respectively. These compounds, collectively referred to as agrochemicals, are effective, relatively cheap and easy to apply (Brauer et al. 2019). However, the use of synthetic chemicals in horticultural production and post-harvest processes have raised concerns over the polluting effects on the environment and risks to human health and food security (Wightwick et al. 2010). In particular, the excessive use of broad-spectrum synthetic fungicides can alter the delicate balance of the rhizosphere microbiome, affecting beneficial symbiotic fungi and bacteria. These microorganisms play crucial roles in ensuring nutrient availability to the plant, such as nitrogen fixation by rhizobacteria or phosphorous uptake by mycorrhizal fungi (Fasusi et al. 2023; Mendes et al. 2011; Raklami et al. 2019). Furthermore, the alarming reduction in efficacy of some synthetic fungicides due to the emergence of fungicide-resistant pathogen strains has been extensively documented by various studies (Brent and Hollomon 1998; Fisher et al. 2012). The excessive use of fertilizers also lead to problems such as the accumulation of nitrogenated compounds or K particles in the soil (Basak et al. 2017; Vitousek et al. 2009). Moreover, over the years, the application of synthetic fertilizers may lead to the accumulation of heavy metals, which not only affects the soil fertility but also pollutes waters (Briffa et al. 2020).

These are just some of the major problems resultant from the use of synthetic compounds in agriculture. In response to these challenges, national governments and international institutions have recognized the urgency to limit the use of synthetic chemicals and shift their focus to alternative, more sustainable and eco-friendly means. The application of biocontrol agents (BCAs) and their metabolites to control plant diseases and fungal contaminants stands out as a promising solution to mitigate the reliance on synthetic fungicides (La Spada et al. 2020; WHO 2018). Biocontrol is defined as the use of beneficial antagonist microorganisms for protection of plants and agricultural produce from fungal infections and contaminations, soil remediation and plant growth promotion (He et al. 2021; Masmoudi et al. 2017). Several bacteria and fungi have been tested as potential BCAs and plant growth promoters, of which only a limited number have been exploited as commercial products. In particular, among bacteria species of *Bacillus* and lactic acid, bacteria have been proven to be effective as BCAs of a broad spectrum of soil-borne and post-harvest fungal diseases and agricultural produce contaminants (Backer et al. 2018; Riolo et al. 2023; Wilson and Chalutz 1998; Wisniewski et al. 2016), yeasts have been shown to be mainly effective and have been exploited commercially as BCAs of post-harvest fruit diseases (Abrah et al. 2010; Droby et al. 1998; Droby et al. 2009; Hammami et al. 2022; Spadaro and Droby 2016; Spadaro and Gullino 2004) and rhizo-bacteria have been applied to promote the growth of diverse plant species (Etesami and Maheshwari 2018).

In recent decades, the fungal *Trichoderma* genus has emerged as an effective BCA and a promising source of metabolites with potential for agricultural applications (Patil et al. 2016; Zhang et al. 2021; Vinale et al. 2008). Its widespread presence underscores the adaptability to diverse environments. *Trichoderma* species are filamentous fungi that in nature reside prevalently in rhizosphere soil, indicating a remarkable aptitude to interact with plants. Their unique ability to establish symbiotic relationships with various crops has led to an increasing interest in harnessing their metabolites as natural compounds for sustainable agricultural practices (Lorito et al. 2010; Vinale et al. 2008). The establishment of symbiosis between *Trichoderma* and plants relies on the production of an array of active metabolites with multiple functions. These metabolites play a crucial role in interacting with and shaping the plant rhizosphere micro-bioma. The effectiveness of *Trichoderma* species as BCAs of a wide range of fungal and oomycete plant pathogens is due to their antagonistic activity, efficiency in promoting plant growth and ability to induce plant defense mechanisms (La Spada et al. 2020). In turn, the antagonistic activity is the result of different mechanisms, direct and indirect, acting synergistically. *Trichoderma* metabolites are major determinants of this antagonistic activity. For instance, *Trichoderma* species exert direct antagonism by producing enzymes, such as chitinases and glucanases, which degrade the cell walls of fungal plant pathogens. Of all the hydrolytic enzymes of *Trichoderma*, chitinases are the most relevant in mycoparasitism (Osorio-Hernández et al. 2016). The application of *Trichoderma* species such as BCAs reduces the dependence on synthetic chemical fungicides, thereby minimizing environmental pollution and mitigating the risk of pathogen resistance to fungicides. By contrast, the complex mechanisms of biocontrol exerted by *Trichoderma* species minimize the risk of pathogen resistance. Presently, it is estimated that more than 60% of available bio-fungicides are based on strains of *Trichoderma* (Sood et al. 2020). This prevalence can be attributed to several factors, including the capacity of these fungi to generate an array of secondary metabolites and enzymes conferring on them a broad-spectrum antifungal activity, their high rhizosphere competence and ability to adapt to diverse environmental conditions, as well as the straightforward and rapid methods for their isolation and growth. Moreover, some *Trichoderma* species can have anti-mycotoxigenic and detoxification potential (Tian et al.; 2018; Pellan et al. 2021). Finally, it has been demonstrated that *Trichoderma* metabolites can enhance the uptake of essential nutrients such as nitrogen, phosphorus, and potassium and mineral salts (Sharma et al. 2023). This trait is of utmost importance in modern agriculture, where sustainable nutrient management is a critical challenge. Consequently, *Trichoderma* species can have multiple applications such as antifungal bioproducts, biofertilizers and biostimulants on a variety of crops. This chapter highlights the nature and diversity of metabolites produced by *Trichoderma* species with special focus on their application in agriculture (Fig. 1).

2 The *Trichoderma* Genus: A Brief Overview

The genus *Trichoderma* belongs to the domain *Eukaryota*, kingdom *Fungi*, subkingdom *Dikarya*, division *Ascomycota* (Mycobank 2023). It is predominantly asexual (the teleomorphic form is *Hypocrea*) and most species produce large numbers of single-celled conidia and thick-walled chlamydospores. The genus was established in 1794 by Persoon (1794) with the description of *T. reesei* and, since then, more than 375 species have been formally described (Nascimento et al. 2022). In the twentieth century, novel DNA-based methods, including rDNA sequence analysis, random amplified polymorphic DNA (RAPDs) analysis, and PCR fingerprinting, revolutionized the systematics and taxonomy of fungi, including *Trichoderma*. For instance, phylogenetic analysis revealed *T. harzianum* was a species complex comprising four distinct taxa, *T. harzianum sensu stricto, T. atroviride,*

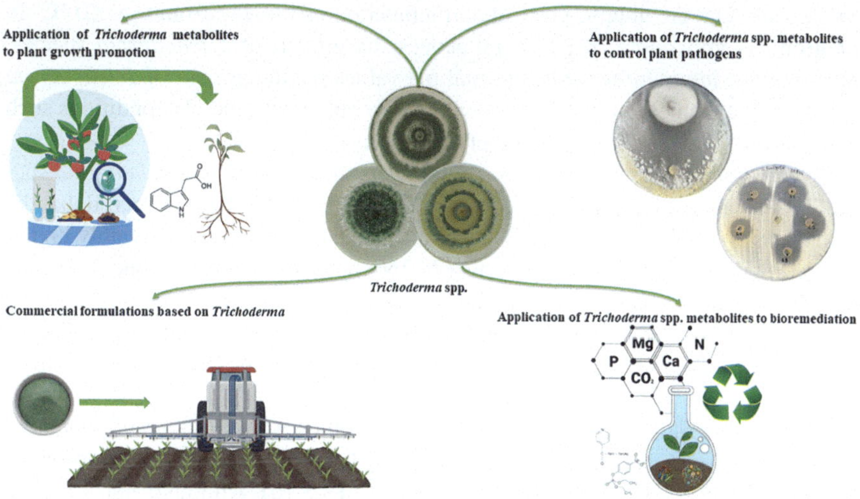

Fig. 1 Graphical representation of the main applications of the genus *Trichoderma* and their secondary metabolites in agriculture

T. longibrachiatum, and *T. asperellum* (Dutta et al. 2022). *Trichoderma* species are characterized by fast mycelium growth and the production of spherical conidia with different colors varying from dark brown (e.g. in *T. melanomagna*) to green (e.g., in *T. harzianum* and *T. asperellum*) (Adnan et al. 2019; Błaszczyk et al. 2014). Conidia are produced by multiple conidiophores that terminate in one or more phialides (Tyśkiewicz et al. 2022).

Trichoderma species are ubiquitous and have a cosmopolitan distribution. They have adapted to a broad range of ecosystems and are major components of the mycobiota of different types of soil, including agricultural, forest, pastureland, organic matter-rich, and even desert soils (Dutta et al. 2022). Moreover, they are able to colonize and decompose dead organic matter such as wood (Manzar et al. 2022; Tyśkiewicz et al. 2022). Most *Trichoderma* species establish with plants a symbiotic relationship and some can even grow as endophytes. However, there are reports of *Trichoderma* species behaving as opportunistic pathogens of mushrooms and plants (Aydoğdu et al. 2020; Hatvani et al. 2007; Li Destri Nicosia et al. 2015; Mazin et al. 2018). Moreover, *T. longibrachiatum* was reported as an emerging human pathogen in immunocompromised patients and some *Trichoderma* species produce mycotoxins. For instance, *T. brevicompactum* produces trichodermin, a mycotoxin in the trichothecine group, which is a food contaminant (Tijerino et al. 2011). The ability of *Trichoderma* species to colonize soil and stimulate plant biological functions is strictly related to ecological parameters including soil temperature, oxygen, pH and organic composition. According to Zehra et al. (2017) high soil temperature and moisture, negatively affect the production of both volatile organic compounds (VOCs) and secondary metabolites and consequently the beneficial effect of *Trichoderma*. However, there are species, e.g. *T. viride* and *T.*

polysporum that are able to grow at soil temperatures ranging from 0 to 40 °C. In addition to temperature, pH also influences the growth of *Trichoderma* species. According to literature records, optimal pH values for the growth of *Trichoderma* species range from 3.5 to 5.6, but growth may occur at extreme pH conditions such as 2.1 (Dutta et al. 2022; Ghazanfar et al. 2018).

It is known that the type of ecosystem affects the diversity of *Trichoderma* species. For instance, in the Xinjiang region of China Ma et al. (2020) observed that the proportion of *Trichoderma* fungi in a soil microbial community was notably higher in forest soils compared to steppe soils (79.26% against 36.55%). Of the 23 identified *Trichoderma* species, only *T. asperellum*, *T. harzianum*, *T. oblongisporum*, *T. paraviridescens*, *T. rossicum*, and *T. saturnisporum* were present across all ecosystem types, including coniferous forests, mixed forests, desert steppes and temperate steppes.

Trichoderma is a valuable reservoir of metabolites with considerable potential for applications in agriculture. It is estimated that *Trichoderma* species produce around 1000 distinct metabolites (Zeilinger et al. 2016). Although not all these metabolites are essential for the growth and basic functions, they play key roles in priming plant defense mechanisms and signaling processes (Sood et al. 2020).

3 Application of *Trichoderma* spp. Metabolites to Control Plant Pathogens

Antibiosis, mycoparasitism, and competitive interactions with plant pathogens for resources and space are the primary mechanisms through which *Trichoderma* species exert their antifungal activity (Kubiak et al. 2023; Manzar et al. 2022; Sood et al. 2020). The synergistic deployment of these strategies confers upon these fungi an exceptional capacity for antagonism against pathogenic microorganisms (Kubiak et al. 2023). Mycoparasitism and competitive interactions involve direct contact between *Trichoderma* and plant pathogens, whereas antibiosis primarily relies on the production of secondary metabolites. These secondary metabolites include antibiotics and toxic compounds that hinder the growth, development, and colonization of pathogens (Kubiak et al. 2023; Riolo et al. 2023; Sood et al. 2020). This occurs through several mechanisms including the deactivation of proteins and enzymes vital for fundamental metabolic processes, the generation of reactive oxygen species and the creation of channels within cell membranes that cause cytoplasm to spill out and the subsequent death of microbes (Khan et al. 2020; Kubiak et al. 2023; Zeilinger et al. 2016). In the study conducted by Khan et al. 2020, up to 11 diverse groups of secondary metabolites synthesized by *Trichoderma* fungi during interactions with pathogens were identified. These groups include anthraquinones, azaphilones, butenolides, epipolythiodioxopiperazines, koninginins, lactones, peptaibols, pyridones, pyrones, steroids, and trichothecenes. Of all the *Trichoderma* metabolites, those with antifungal properties are the most studied. Main compounds

Table 1 *Trichoderma* metabolites with antimicrobial activity against fungal plant pathogens

Strains	Metabolites	Plant pathogens	References
Several *Trichoderma* species	Trichothecenes	Several pathogens	Dutta et al. (2022)
T. harzianum	Azaphilones	*Rhizoctonia solani*, *Phytium ultimum* and *Gaeumannomyces graminis*	Dutta et al. (2022)
T. asperellum	Pyrones	*Magnaporthiopsis maydis*	Degani and Gordani (2022)
Several *Trichoderma* species	6-pentyl-2H-pyran-2-one	*Cylindrocarpon destructans*	Dutta et al. (2022), Ma et al. (2020)
Several *Trichoderma* species	Cyclic peptides	Several pathogens	Dutta et al. (2022)
Several *Trichoderma* species	Chitinases Glucanases	Several pathogens	Saldaña-Mendoza et al. (2023)

with antifungal activity are listed in Table 1 and their characteristics are reported below:

Trichothecenes: Trichotecenes belong to the group of terpenoids and are sesquiterpene-based compounds with a tricyclic 12,13-epoxytrichothec-9-ene skeleton. Structurally, they are classified into different families, nivalenols, neosolaniols, isotrichodermins, calonectrins, trichothecene, and trichobreols, based on the substitution pattern. More than 200 trichothecene derivates have been described and all are toxins, such as trichodermin and its derivates, for instance, which impair plant growth (Dutta et al. 2022).

Azaphilones: Azaphilones are fungal polyketide pigments characterized by a highly oxygenated pyranoquinone bicyclic core and chiral quaternary centre. Currently they are the object of great interest for their applications in the agro-alimentary, printing, cosmetic, and pharmaceutical industries. This class of secondary metabolites produced by *T. harzianum* was reported to inhibit the growth of several fungal plant pathogens, such as *Rhizoctonia solani*, *Phytium ultimum* and *Gaeumannomyces graminis* var. *graminis* (Dutta et al. 2022).

Pyrones: Pyrones are a group of heterocyclic chemical compounds that include several metabolites with antifungal activity. Recently, Degani and Gordani (2022) isolated a new pyrone compound, 6-pentyl-α- pyrone, from *T. asperellum*. This pyrone showed a significant antifungal activity against the late wilt pathogen in maize, *Magnaporthiopsis maydis*. Another pyrone, 6-pentyl-2H-pyran-2-one, produced by several species of *Trichoderma* including *T. atroviridae*, *T. harzianum*, *T. koningii* and *T. viride*, may be an additional example of a secondary metabolite of *Trichoderma* with antifungal activity. It is one of the major volatile organic compounds (VOCs) biosynthesized by *Trichoderma* species that is supposed to be involved in plant-fungal cross-kingdom signalling and was shown to be strongly active against *Cylinfrocarpon destructans*, one of the most destructive pathogens of Chinese ginseng (*Panax notoginseng*) (Dutta et al. 2022; Ma et al. 2020).

Cyclic peptides: These are a group of peptides with antimicrobial properties (bacteriocins and antifungal peptides), the best known of which are suzukacillin, the lipophilic alamethicin, trichopolyns, trichotoxins, and trichorianine, which attack the cell membrane of bacteria and eukaryotes. Various *Trichoderma* species were demonstrated to produce these metabolites to compete with other microorganisms (Dutta et al. 2022).

Chitinases and Glucanases: Of all the enzymes produced by *Trichoderma,* chitinases and glucanases are those mainly responsible for the ability of this group of fungi to parasitize other fungi. They degrade chitin and glucans, major components of fungal cell walls, thereby preventing pathogen adhesion (Saldaña-Mendoza et al. 2023).

4 Application of *Trichoderma* spp. Metabolites to Bioremediation

Diverse *Trichoderma* strains have been shown to have bioremediation potential. Some *Trichoderma* species are particularly resistant to many of the compounds like polycyclic aromatic hydrocarbon, heavy metals and various toxic synthetic active ingredients that accumulate in and pollute soil and water as a consequence of the excessive use of fertilizers and pesticides (Mohsenzadeh and Shahrokhi 2014; Mutlag et al. 2023; Raspanti et al. 2009). These *Trichoderma* species show the ability to grow in polluted media and metabolize toxic compounds into non-toxic metabolites (Maurya et al. 2019; Puglisi et al. 2012; Zin and Badaluddin 2020). In some cases, metabolites responsible for detoxificant capability of *Trichoderma* strains were isolated and used for bioremediation treatments of soils (Zin and Badaluddin 2020). *Trichoderma* metabolites with proved detoxificant properties are listed below and reported in Table 2. However, it is very likely many other *Trichoderma*-derived compounds with similar characteristics are yet to be discovered.

Arylamine N-acetyltransferases. These enzymes are xenobiotic-metabolizers enabling acetyl coenzyme A-dependent degradation of aromatic amines, which are one of the principal residues of 3,4-dichloroaniline, a pesticide largely used in agriculture. An accumulation of these amines also occurs in soil of highly industrialized areas. Some other soil-borne fungi such as *Penicillium* species were proven to be effective in remediating this soil contamination (Umana et al. 2016). There is evidence in the literature indicating the capacity of *T. virens* and *T. reesei* strains to thrive in soils with a high content of aromatic amines and degrade these pollutants

Table 2 Metabolites of *Trichoderma* species with soil bioremediation activity

Strains	Metabolites	Soil pollutants	References
T. virens and *T. reesei*	Arylamine N-acetyltransferases	Aromatic amines	Cocaign et al. (2013)
T. atroviride	Paraxonases	2,2-dichlorovinyl dimethyl phosphate (dichlorvos pesticide)	Sun et al. (2019)
Several *Trichoderma* species	Metal binding proteins	Heavy metals	López Errasquín and Vázquez (2003)
Several *Trichoderma* species	Cyanide hydratase and rhodanese	Cyanide	Ezzi and Lynch (2005)
Several *Trichoderma* species	Peroxidase	Pyrethroids and carbamates	Mendarte-Alquisira et al. (2022)
Several *Trichoderma* species	Laccase (LAC1)	Polycyclic aromatic hydrocarbon	Behera et al. (2023)

(Cocaign et al. 2013). A strict correlation was observed between the production of arylamine N-acetyltransferases by *T. virens* and its ability to degrade aromatic amines indicating these were the primary enzymes involved in the degradation process, while to degrade aromatic amines *T. reesei* used a wider pool of enzymes yet to be determined (Cocaign et al. 2013).

Paraxonases. Paraxonases are a family of proteins known to have a protective effect against lipidic oxidation. Some organophosphate pesticides, such as dichlorvos, are known to produce this type of oxidative stress. The literature evidences the potential of *T. atroviride* to efficiently degrade 2,2-dichlorovinyl dimethyl phosphate, the active ingredient of the largely used dichlorvos pesticide. An accurate study revealed this degradation mechanism was linked to the production of paraxonases, proteins produced by *T. atroviride* that degraded the pesticide by a P-O bond hydrolysis (Sun et al. 2019).

Metal binding proteins. Diverse proteins associated with the cell walls of fungi have been proven to bind heavy metals, like copper, zinc and cadmium, from the environment (El-Gendy et al. 2017; Saiano et al. 2005). These heavy metals are usually in inorganic forms which can cause an increase in the soil toxicity. Their occurrence is related to the use of diverse agricultural compounds and industrial activities and since they are simple molecules, they cannot be further degraded. Consequently, binding to larger and more complex molecules is the best way to reduce their amount in the soil. Numerous studies have demonstrated the capacity of cell wall proteins of diverse *Trichoderma* species to bind heavy metals from the soil and reduce their toxicity. For instance, López Errasquín & Vázquez (2003) showed the heavy metal biosorption capacity of a *T. atroviride* strain under a wide range of environmental conditions. The removal was especially effective after the autolysis of the mycelium, indicating that not only the cell wall binding proteins were involved in this process.

Cyanide hydratase and rhodanese. Both enzymes are known to be able to detoxify cyanide. Cyanide hydratases is a type of hydro-lyase able to transform the cyanide into formamide; while rhodanese, also known as thiosulfate sulfotransferase, is an enzyme that degrades cyanide into sulphite thiocyanate in a two step-reaction. Both enzymes intervene as defense mechanisms against the presence of this contaminant. Cyanide is a natural pollutant, but anthropogenic activity increases its occurrence. Some studies relate the reduction of this contaminant through the production of cyanide hydratase and rhodanese by strains of *Trichoderma* species. Some of these strains not only degraded the cyanide, but use it as the only carbon source for growth (Ezzi & Lynch 2005).

Peroxidases. Peroxidases are a large group of enzymes which play a role in various biological processes; they catalyze several oxidative reactions by using other substrates such as electron donors. The production of diverse peroxidases by *Trichoderma* strains was related to their ability to reduce the amount of pyrethroid and carbamate residues in soil (Mendarte-Alquisira et al. 2022). Pyrethroids and carbamates are widely used as insecticides in agriculture and forestry as well as in wood and textile industries. Both are very stable molecules that persist for a long time and contaminate the soil.

Laccases. These enzymes are widespread in the biosphere and serve as catalyzers for the oxidation of a great variety of phenolic compounds. Each laccase is involved in the metabolization of a specific group of phenolic compounds. Polycyclic aromatic hydrocarbons are contaminants produced by human activities (agricultural, industrial, oil extraction) which are rather persistent in nature and hazardous for humans, plants, bacteria and fungi. Some microorganisms have developed the ability to degrade these molecules. For instance, a *Trichoderma* strain was able to metabolize polycyclic aromatic hydrocarbon compounds into non-toxic molecules by the laccase LAC 1 (Behera et al. 2023).

5 Application of *Trichoderma* Metabolites to Plant Growth Promotion

Trichoderma species are known for their ability to establish symbiotic relations with plants. This symbiosis commonly occurs through colonization of roots and, less frequently, of fruit, flowers and foliage of host plants (Harman et al. 2004). The symbiosis process, at least during the early phases of the interaction, is similar to that by which *Trichoderma* species parasitize other fungi and plants. The ability of *Trichoderma* species to thrive and proliferate in complex associations with plants is the result of a coevolution process. Like in a host-parasite relationship, *Trichoderma* penetrates the plant tissues and, once it has entered the plant, produces an array of metabolites to suppress to the plant defenses, such as reactive oxygen species, antimicrobial peptides and proteases. Then, *Trichoderma* grows inside the plant and starts to produce a series of metabolites, from phytohormones to enzymes. Which interfere with the plant development and exert growth-promoting effects (Saldaña-Mendoza et al. 2023). In the formation of numerous plant-pathogen associations, there is a stimulation of sucrose hydrolysis in plant cells to increase the flow of monosaccharides. Vargas et al. (2009) elaborated on the identification and characterization of an intracellular invertase from *Trichoderma virens* (TvInv), demonstrating its potential to regulate mechanisms involved in symbiosis in the presence of sucrose derived from plants. Some of the *Trichoderma* metabolites with plant growth promoting effects are reported in Table 3 and are briefly described below.

Indole-3-acetic acid. Indole-3-acetic acid, which belongs to the auxin group, is one of phytohormones which occurs most frequently. Several authors related this hormone to different plant gene regulation systems and also to microbe-plant symbiotic interaction. It is known that the production of this metabolite by several microbial species promotes the growth of leaves and roots in plants. Among many examples from the literature, Nieto-Jacobo et al. (2017) related the production of this phytohormone by different *Trichoderma* species to their growth-promoting effect on roots.

6-pentyl-2H-pyran-2-one. This is a volatile organic compound produced by many *Trichoderma* species. It is believed that most of the VOCs produced by *Trichoderma* are involved into the cross-kingdom signaling, particularly between *Trichoderma* and plants. For instance, the stimulation of root genesis observed in *Arabidopsis thaliana* by *T. atroviride* was attributed to an over-expression of genes involved in this activity (Garnica-Vergara et al. 2016).

Chelating and redox molecules. Some heavy metals, such as Fe, Cu, Zn, and Mn, which in low amounts are essential to plant growth, occur in the soil in an insoluble or sparingly soluble state. The production by *Trichoderma* spp. of metabolites with chelating and redox activity, most of which have not yet been identified, has been

Table 3 Metabolites of *Trichoderma* species with plant growth promotion activity

Strains	Metabolites	Type of plant benefit	References
Several *Trichoderma* species	Indole-3-acetic acid	Promote foliar and radical growth	Nieto-Jacobo et al. (2017)
T. atroviride	6-pentyl-2H-pyran-2-one	Stimulates root genesis	Garnica-Vergara et al. (2016)
T. harzianum	Chelating and redox molecules	Favors mineral uptake	Altomare et al. (1999)
Trichoderma spp.	1-aminocyclopropane-1-carboxylate deaminase	Promotes plant growth and increases stress tolerance	Zhang et al. (2019)
Trichoderma spp.	Gibberellin	Increases stem weight and resilience to water stress	Illescas et al. (2021), Jaroszuk-ścisel et al. (2019)
T. asperellum	Abscisic acid	Increases resistance to salt stress	Zhao and Zhang (2015)

proved to increase the solubility of these metals. Altomare et al. (1999) related the plant growth-promoting effect of a *T. harzianum* strain to its chelating and redox activity, which increased the solubility of Fe (III) by transforming it into molecular species with an elevated water solubility, thus increasing the uptake by the plant.

1-aminocyclopropane-1-carboxylate deaminase. This metabolite, also refered to as ACC, is a bacterium-fungus-derived hydrolytic enzyme which regulates the ethylene levels of the plant. The literature links the production of this metabolite by different *Trichoderma* strains to both the plant growth and an increase in resilience to abiotic stresses, such as the tolerance of saline soils (Zhang et al. 2019).

Gibberellins. Gibberellins are a group of phytohormones that regulate important development processes of the plant, such as stem elongation, germination or flowering. Like other metabolites already mentioned in here, the hormones are produced by diverse plant-symbiotic microorganisms. Various studies correlate the production of gibberellins by *Trichoderma* species with both the increase in stem weight and resistance of plants to water stress (Illescas et al. 2021; Jaroszuk-ścisel et al. 2019).

Abscisic acid. Like gibberellins, abscisic acid is a phytohormone that regulates different aspects of the plant physiology. It has a crucial role in the adaptation to biotic and abiotic stress, in the seed development and in regulating the seed dormancy. This hormone was reported to be produced by *T. asperellum* and in the plant stimulated root activity and induced resistance to salt stress (Zhao and Zhang 2015).

6 Commercial Formulations Based on *Trichoderma*

Various *Trichoderma*-based products have been developed for commercial use as biopesticides and biofertilizers. These bioproducts are formulated to ensure the stability and long-term viability of Trichoderma propagules and to facilitate their application on a large scale (Peles et al. 2021). Woo et al. (2014) reported several commercial formulations used to contrast plant pathogens, with most of these employing *Trichoderma harzianum*, *T. viride*, and *T. atroviride* as species. Most of these commercial bioproducts, such as CANNA AkTRivator (UK), Lycomax (UK), Micosat F (Italy), Sani-Root (Spain), T34 Biocontrol (Spain), and TRIANUM-P (EU and New Zealand), are used to counteract soil-borne pathogens and to control root rot and wilt diseases of horticultural crops. Meanwhile, other bioformulations have been patented as bio-stimulants, such as Trichodermas BioFlower (Spain), Trichomic (Spain), and Custom GP Beneficial Fungi (USA), which are effective in the growth of the root system of tree and vegetable species as well as in increasing plant disease resistance mechanisms.

Moreover, some *Trichoderma* metabolites are being tested as biostimulants to enhance plant tolerance to abiotic stresses, such as drought and salinity, and to optimize nutrient uptake and utilization (Fiorentino et al. 2018; Ventorino et al. 2014). Although the application of *Trichoderma* metabolites for sustainable and eco-friendly farming practices is promising, further research is needed to optimize both production and formulation methods.

7 Conclusions

The *Trichoderma* genus is a rich source of active metabolites with several potential applications in agriculture. The activity spectrum of secondary metabolites produced by species in this genus ranges from biocontrol to plant growth promotion but has not yet been fully explored and exploited. Some selected strains have already been developed as successful commercial products, however. Moreover, it can be envisaged that, in the coming years, new *Trichoderma* species and strains will be discovered and selected, thus broadening the spectrum of potential applications of this group of fungi. A new promising research line is the direct application of *Trichoderma* metabolites, which could be a way to bypass the inherent limits and concerns of the use of living organisms as post-harvest antifungal agents and food preservatives. In conclusion, it can be said *Trichoderma* species and their active metabolites are one of the most promising source of technical tools for a more eco-friendly and sustainable agriculture.

Bibliography

Abraha OA, Laing MD, Bower JP (2010) Isolation and in vivo screening of yeast and *bacillus* antagonists for the control of *Penicillium digitatum* of citrus fruit. Biol Control 53:32–38. https://doi.org/10.1016/j.biocontrol.2009.12.009

Adnan M, Islam W, Shabbir A, Khan KA, Ghramh HA, Huang Z, Chen HYH, Lu G (2019) Plant defense against fungal pathogens by antagonistic fungi with *Trichoderma* in focus. Microb Pathog 129:7–18. https://doi.org/10.1016/j.micpath.2019.01.042

Altomare C, Norvell WA, Bjorkman T, Harman GE (1999) Solubilization of phosphates and micronutrients by the plant-growth- promoting and biocontrol fungus *Trichoderma harzianum* Rifai 1295-22. Appl Environ Microbiol 65(7):2926–2933. https://doi.org/10.1128/AEM.65.7.2926-2933.1999

Aydoğdu M, Kurbetli İ, Kitapçı A, Sülü G (2020) Aggressiveness of green mould on cultivated mushroom (*Agaricus bisporus*) in Turkey. J Plant Dis Prot 127:695–708. https://doi.org/10.1007/s41348-020-00328-8

Azevedo MM, Faria-Ramos I, Cruz LC, Pina-Vaz C, Gonçalves Rodrigues A (2015) Genesis of azole antifungal resistance from agriculture to clinical settings. J Agric Food Chem 63(34):7463–7468. https://doi.org/10.1021/acs.jafc.5b02728

Backer R, Rokem JS, Ilangumaran G, Lamont J, Praslickova D, Ricci E, Subramanian S, Smith DL (2018) Plant growth-promoting rhizobacteria: context, mechanisms of action, and roadmap to commercialization of biostimulants for sustainable agriculture. Front Plant Sci 871:1–17. https://doi.org/10.3389/fpls.2018.01473

Basak BB, Sarkar B, Biswas DR, Sarkar S, Sanderson P, Naidu R (2017) Bio-intervention of naturally occurring silicate minerals for alternative source of potassium: challenges and opportunities. Adv Agron 141(115–145). https://doi.org/10.1016/bs.agron.2016.10.016

Behera AD, Chittoria N, Kumari S, Chatterjee S, Das S (2023) Homology modeling of laccase enzyme of filamentous fungus *Trichoderma* sp. CNSC-2 and its role in pyrene degradation. Geomicrobiol J. https://doi.org/10.1080/01490451.2023.2218362

Błaszczyk L, Siwulski M, Sobieralski K, Lisiecka J, Jędryczka M (2014) *Trichoderma* spp. Application and prospects for use in organic farming and industry. J Plant Prot Res 54(4):309–317. https://doi.org/10.2478/jppr-2014-0047

Brauer VS, Rezende CP, Pessoni AM, De Paula RG, Rangappa KS, Nayaka SC, Gupta VK, Almeida F (2019) Antifungal agents in agriculture: friends and foes of public health. Biomol Ther 9(10):1–21

Brent KJ, Hollomon DW (1998). Fungicide resistance the assessment of risk. FRAC, "Global Crop Protection Federation"

Briffa J, Sinagra E, Blundell R (2020) Heavy metal pollution in the environment and their toxicological effects on humans. Heliyon 6(9). https://doi.org/10.1016/j.heliyon.2020.e04691

Cocaign A, Bui LC, Silar P, Chan Ho Tong L, Busi F, Lamouri A, Mougin C, Rodrigues-Lima F, Dupret JM, Dairou J (2013) Biotransformation of *Trichoderma* spp. and their tolerance to aromatic amines, a major class of pollutants. Appl Environ Microbiol 79(15):4719–4726. https://doi.org/10.1128/AEM.00989-13

Degani O, Gordani A (2022) New antifungal compound, 6-pentyl-α-pyrone, against the maize late wilt pathogen, *Magnaporthiopsis maydis*. Agronomy 12(10):2339. https://doi.org/10.3390/agronomy12102339

Droby S, Cohen L, Daus A, Weiss B, Horev B, Chalutz E, Katz H, Keren-Tzur M, Shachnai A (1998) Commercial testing of aspire: a yeast preparation for the biological control of postharvest decay of citrus. Biol Control 12:97–101. https://doi.org/10.1006/bcon.1998.0615

Droby S, Wisniewski M, Macarisin D, Wilson C (2009) Twenty years of postharvest biocontrol research: is it time for a new paradigm? Postharvest Biol Technol 52:137–145. https://doi.org/10.1016/j.postharvbio.2008.11.009

Dutot M, Nelson LM, Tyson RC (2013) Predicting the spread of postharvest disease in stored fruit, with application to apples. Postharvest Biol Technol 85:45–56. https://doi.org/10.1016/j.postharvbio.2013.04.003

Dutta P, Deb L, Pandey AK (2022) *Trichoderma*- from lab bench to field application: looking back over 50 years. Front Agron 4:1–25. https://doi.org/10.3389/fagro.2022.932839

El-Gendy MMAA, Hassanein NM, Abd El-Hay Ibrahim H, Abd El-Baky DH (2017) Heavy metals biosorption from aqueous solution by endophytic *Drechslera hawaiiensis* of *Morus alba* L. derived from heavy metals habitats. Mycobiology 45(2):73–83. https://doi.org/10.5941/MYCO.2017.45.2.73

Etesami H, Maheshwari DK (2018) Use of plant growth promoting rhizobacteria (PGPRs) with multiple plant growth promoting traits in stress agriculture: action mechanisms and future prospects. Ecotoxicol Environ Saf 156:225–246. https://doi.org/10.1016/j.ecoenv.2018.03.013

Ezzi MI, Lynch JM (2005) Biodegradation of cyanide by *Trichoderma* spp. and *Fusarium* spp. Enzym Microb Technol 36(7):849–854. https://doi.org/10.1016/j.enzmictec.2004.03.030

Fasusi OA, Babalola OO, Adejumo TO (2023) Harnessing of plant growth-promoting rhizobacteria and arbuscular mycorrhizal fungi in agroecosystem sustainability. CABI Agric Biosci 4:26. https://doi.org/10.1186/s43170-023-00168-0

Fiorentino N, Ventorino V, Woo SL, Pepe O, De Rosa A, Gioia L, Romano I, Lombardi N, Napolitano M, Colla G, Rouphael Y (2018) *Trichoderma*-based biostimulants modulate rhizosphere microbial populations and improve N uptake efficiency, yield, and nutritional quality of leafy vegetables. Front Plant Sci 9:743. https://doi.org/10.3389/fpls.2018.00743

Fisher MC, Henk DA, Briggs CJ, Brownstein JS, Madoff LC, McCraw SL, Gurr SJ (2012) Emerging fungal threats to animal, plant and ecosystem health. Nature 484(7393). https://doi.org/10.1038/nature10947

Garnica-Vergara A, Barrera-Ortiz S, Muñoz-Parra E, Raya-González J, Méndez-Bravo A, Macías-Rodríguez L, Ruiz-Herrera LF, López-Bucio J (2016) The volatile 6-pentyl-2H-pyran-2-one from *Trichoderma atroviride* regulates *Arabidopsis thaliana* root morphogenesis via auxin signaling and ethylene insensitive 2 functioning. New Phytol 209(4):1496–1512. https://doi.org/10.1111/nph.13725

Ghazanfar MU, Raza M, Raza W (2018) Effect of physiological parameters on mass production of *Trichoderma* species. Pakistan. J Phytopathol 30(1):59–65. https://doi.org/10.33866/phytopathol.030.01.0447

Grifoni M, Franchi E, Fusini D, Vocciante M, Barbafieri M, Pedron F, Rosellini I, Petruzzelli G (2022) Soil remediation: towards a resilient and adaptive approach to deal with the ever-changing environmental challenges. Environments 9:18. https://doi.org/10.3390/environments9020018

Hammami R, Oueslati M, Smiri M, Nefzi S, Ruissi M, Comitini F, Romanazzi G, Cacciola SO, Sadfi Zouaoui N (2022) Epiphytic yeasts and bacteria as candidate biocontrol agents of green and blue molds of citrus fruits. J Fungi 8(8):818. https://doi.org/10.3390/jof8080818

Harman GE, Lorito M, Lynch JM (2004) Uses of *Trichoderma* spp. to alleviate or remediate soil and water pollution. Adv Appl Microbiol 56:313–330. https://doi.org/10.1016/S0065-2164(04)56010-0

Hatvani L, Antal Z, Manczinger L, Szekeres A, Druzhinina IS, Kubicek CP, Nagy A, Nagy E, Vágvölgyi C, Kredics L (2007) Green mold diseases of *Agaricus* and *Pleurotus* spp. are caused by related but phylogenetically different *Trichoderma* species. Phytopathology 97:532–537

He D-C, He MH, Amalin DM, Liu W, Alvindia DG, Zhan J (2021) Biological control of plant diseases: an evolutionary and eco-economic consideration. Pathogens 10:1311. https://doi.org/10.3390/pathogens10101311

Illescas M, Pedrero-Méndez A, Pitorini-Bovolini M, Hermosa R, Monte E (2021) Phytohormone production profiles in *Trichoderma* species and their relationship to wheat plant responses to water stress. Pathogens 10(8). https://doi.org/10.3390/pathogens10080991

Jaroszuk-Ściseł J, Tyśkiewicz R, Nowak A, Ozimek E, Majewska M, Hanaka A, Tyśkiewicz K, Pawlik A, Janusz G (2019) Phytohormones (Auxin, gibberellin) and ACC deaminase in vitro synthesized by the mycoparasitic trichoderma DEMTKZ3A0 strain and changes in the level of auxin and plant resistance markers in wheat seedlings inoculated with this strain conidia. Int J Mol Sci 20(19):4923. https://doi.org/10.3390/ijms20194923

Kerchev P, Van der Meer T, Sujeeth N, Verlee A, Stevens CV, Van Breusegem F, Gechev T (2020) Molecular priming as an approach to induce tolerance against abiotic and oxidative stresses in crop plants. Biotechnol Adv 40(November 2019)

Khan RAA, Najeeb S, Hussain S, Xie B, Li Y (2020) Bioactive secondary metabolites from *Trichoderma* spp. against phytopathogenic fungi. Microorganisms 8(6). https://doi.org/10.3390/microorganisms8030401

Krasilnikov P, Taboada MA (2022) Fertilizer use, soil health and agricultural sustainability. Agriculture 12:462. https://doi.org/10.3390/agriculture12040462

Kubiak A, Wolna-Maruwka A, Pilarska AA, Niewiadomska A, Piotrowska-Cyplik A (2023) Fungi of the *Trichoderma* genus: future perspectives of benefits in sustainable agriculture. Appl Sci 13(11):6434. https://doi.org/10.3390/app13116434

La Spada F, Stracquadanio C, Riolo M, Pane A, Cacciola SA (2020) *Trichoderma* counteracts the challenge of *Phytophthora nicotianae* infections on tomato by modulating plant defense mechanisms and the expression of Crinkler, necrosis-inducing Phytophthora protein 1, and cellulose-binding elicitor lectin pathogenic effector. Front Plant Sci 11:583539. https://doi.org/10.3389/fpls.2020.583539

Li Destri Nicosia MG, Mosca S, Mercurio R, Schena L (2015) Dieback of *Pinus nigra* seedlings caused by a strain of *Trichoderma viride*. Plant Dis 99:44–49. https://doi.org/10.1094/PDIS-04-14-0433-RE

López Errasquín E, Vázquez C (2003) Tolerance and uptake of heavy metals by *Trichoderma atroviride* isolated from sludge. Chemosphere 50(1):137–143. https://doi.org/10.1016/S0045-6535(02)00485-X

Lorito M, Woo SL, Harman GE, Monte E (2010) Translational research on *Trichoderma*: from omics to the field. Annu Rev Phytopathol 48:395–417

Ma J, Tsegaye E, Li M, Wu B, Jiang X (2020) Biodiversity of *Trichoderma* from grassland and forest ecosystems in Northern Xinjiang, China. 3 Biotech 10(8):1–13. https://doi.org/10.1007/s13205-020-02301-6

Manzar N, Kashyap AS, Goutam RS, Rajawat MVS, Sharma PK, Sharma SK, Singh HV (2022) *Trichoderma*: advent of versatile biocontrol agent, its secrets and insights into mechanism of biocontrol potential. Sustain For 14(19):12786. https://doi.org/10.3390/su141912786

Masmoudi F, Ben Khedher S, Kamoun A, Zouari N, Tounsi S, Trigui M (2017) Combinatorial effect of mutagenesis and medium component optimization on *Bacillus amyloliquefaciens* antifungal activity and efficacy in eradicating *Botrytis cinerea*. Microbiol Res 197:29–38. https://doi.org/10.1016/j.micres.2017.01.001

Maurya S, Rashk-E-Eram NSK, Choudhary JS, Kumar S (2019) Heavy metals scavenging potential of *Trichoderma asperellum* and *Hypocrea nigricans* isolated from acid soil of Jharkhand. Indian J Microbiol 59(1):27–38. https://doi.org/10.1007/s12088-018-0756-7

Mazin M, Andreadis SS, Jenkins NE, Rajotte EG (2018) The mushroom sciarid fly, *Lycoriella ingenua*, benefits from its association with green mold disease (*Trichoderma aggressivum*) in commercial mushroom production. J Pest Sci 91:815–822. https://doi.org/10.1007/s10340-017-0930-4

Mendarte-Alquisira C, Alarcón A, Ferrera-Cerrato R (2022) Growth, tolerance, and enzyme activities of *Trichoderma* strains grown in culture media added with a pyrethroid-based insecticide. SSRN Electron J

Mendes R, Kruijt M, de Bruijn I, Dekkers E, van der Voort M, Schneider JH, Piceno YM, DeSantis TZ, Andersen GL, Bakker PA, Raaijmakers JM (2011) Deciphering the rhizosphere microbiome for disease-suppressive bacteria. Science 332(6033):1097–1100. https://doi.org/10.1126/science.1203980

Mohsenzadeh F, Shahrokhi F (2014) Biological removing of Cadmium from contaminated media by fungal biomass of *Trichoderma* species. J Environ Health Sci Eng 12(1):102. https://doi.org/10.1186/2052-336X-12-102

Mutlag NH, Hamoudi SA, Radhi WN (2023) Bioremediation of polluted soil with hydrocarbons by some of *Trichoderma* spp in Al-Najaf-Iraq. Egypt J Hosp Med 90(2):2620–2627

Mycobank (2023) Retrieved September 23, 2023, from https://www.mycobank.org/

Nascimento VC, Rodrigues-Santos KC, Carvalho-Alencar KL, Castro MB, Kruger RH, Lopes FAC (2022) *Trichoderma*: biological control efficiency and perspectives for the Brazilian Midwest states and Tocantins. Braz J Biol 82. https://doi.org/10.1590/1519-6984.260161

Nazareth TM, Corrêa JAF, Pinto ACSM, Palma JB, Meca G, Bordin K, Luciano FB (2018) Evaluation of gaseous allyl isothiocyanate against the growth of mycotoxigenic fungi and mycotoxin production in corn stored for 6 months. J Sci Food Agric 98:5235–5241. https://doi.org/10.1002/jsfa.9061

Nazarov PA, Baleev DN, Ivanova MI, Sokolova LM, Karakozova MV (2020) Infectious plant diseases: etiology, current status, problems and prospects in plant protection. Acta Nat 12(3):46–59. https://doi.org/10.32607/actanaturae.11026

Nieto-Jacobo MF, Steyaert JM, Salazar-Badillo FB, Vi Nguyen D, Braithwaite M, De Souza JT, Jimenez-Bremont JF, Ohkura M, Stewart A, Mendoza-Mendoza A (2017) Environmental growth conditions of trichoderma spp. Affects indole acetic acid derivatives, volatile organic compounds, and plant growth promotion. Front Plant Sci 8:245069. https://doi.org/10.3389/fpls.2017.00102

Nunes CA (2012) Biological control of postharvest diseases of fruit. Eur J Plant Pathol 133:181–196. https://doi.org/10.1007/s10658-011-9919-7

Osorio-Hernández E, Hernandez-Morales J, Conde-Martinez V, Michel-Aceves AC, Lopez-Santillan J, Torres Castillo JA (2016) In vitro activities of *Trichoderma* species against *Phytophthora parasitica* and *Fusarium oxysporum*. Afr J Microbiol Res 10:521–527

Pandey RN, Jaisani P, Singh HB (2022) *Trichoderma*: agricultural applications and beyond. In: Rakshit A, Meena VS, Abhilash PC, Sarma BK, Singh HB, Fraceto L, Parihar M, Singh AK (eds) Advances in bio-inoculant science, biopesticides. Woodhead Publishing, pp 353–381

Patil AS, Patil SR, Paikrao HM (2016) *Trichoderma* secondary metabolites: their biochemistry and possible role in disease management. In: Microbial-mediated induced systemic resistance in plants, pp 69–102

Peles F, Sipos P, Kovács S, Győri Z, Pócsi I, Pusztahelyi T (2021) Biological control and mitigation of aflatoxin contamination in commodities. Toxins 13:104. https://doi.org/10.3390/toxins13020104

Pellan L, Dieye CAT, Durand N, Fontana A, Strub C, Schorr-Galindo S (2021) Biocontrol agents: toolbox for the screening of weapons against mycotoxigenic *Fusarium*. J Fungi 7(6):446. https://doi.org/10.3390/jof7060446

Persoon CH (1794) Observationes mycologicae

Pitt JI, Hocking AD (2009) Fungi and food spoilage, fungi and food spoilage. Springer, New York

Puglisi I, Faedda R, Sanzaro V, Lo Piero AR, Petrone G, Cacciola SO (2012) Identification of differentially expressed genes in response to mercury I and II stress in *Trichoderma harzianum*. Gene 506(2):325–330. https://doi.org/10.1016/j.gene.2012.06.091

Raklami A, Bechtaoui N, Tahiri AI, Anli M, Meddich A, Oufdou K (2019) Use of rhizobacteria and mycorrhizae consortium in the open field as a strategy for improving crop nutrition, productivity and soil fertility. Front Microbiol 10:1106. https://doi.org/10.3389/fmicb.2019.01106

Raspanti E, Cacciola SO, Gotor C, Romero LC, García I (2009) Implications of cysteine metabolism in the heavy metal response in *Trichoderma harzianum* and in three *Fusarium* species. Chemosphere 76(1):48–54. https://doi.org/10.1016/j.chemosphere.2009.02.030

Riolo M, Luz C, Santilli E, Meca G, Cacciola SO (2023) Antifungal activity of selected lactic acid bacteria from olive drupes. Food Biosci 52:102422. https://doi.org/10.1016/j.fbio.2023.102422

Rovetto EI, Luz C, La Spada F, Meca G, Riolo M, Cacciola SO (2023) Diversity of mycotoxins and other secondary metabolites recovered from blood oranges infected by *Colletotrichum*, *Alternaria*, and *Penicillium* species. Toxins (Basel) 15:407. https://doi.org/10.3390/toxins15070407

Saiano F, Ciofalo M, Cacciola SO, Ramirez S (2005) Metal ion adsorption by *Phomopsis* sp. biomaterial in laboratory experiments and real wastewater treatments. Water Res 39:2273–2280. https://doi.org/10.1016/j.watres.2005.04.022

Saldaña-Mendoza SA, Pacios-Michelena S, Palacios-Ponce AS, Chávez-González ML, Aguilar CN (2023) *Trichoderma* as a biological control agent: mechanisms of action, benefits for crops and development of formulations. World J Microbiol Biotechnol 39(10):269. https://doi.org/10.1007/s11274-023-03695-0

Sharma I, Raina A, Choudhary M, Kaul S, Dhar MK (2023) Fungal endophyte bioinoculants as a green alternative towards sustainable agriculture. Heliyon 9(9):e19487. https://doi.org/10.1016/j.heliyon.2023.e19487

Sood M, Kapoor D, Kumar V, Sheteiwy MS, Ramakrishnan M, Landi M, Araniti F, Sharma A (2020) *Trichoderma*: the "secrets" of a multitalented biocontrol agent. Plan Theory 9(6):1–25. https://doi.org/10.3390/plants9060762

Spadaro D, Droby S (2016) Development of biocontrol products for postharvest diseases of fruit: the importance of elucidating the mechanisms of action of yeast antagonists. Trends Food Sci Technol 47:39–49. https://doi.org/10.1016/j.tifs.2015.11.003

Spadaro D, Gullino ML (2004) State of the art and future prospects of the biological control of postharvest fruit diseases. Int J Food Microbiol 91:185–194. https://doi.org/10.1016/S0168-1605(03)00380-5

Sun J, Yuan X, Li Y, Wang X, Chen J (2019) The pathway of 2,2-dichlorovinyl dimethyl phosphate (DDVP) degradation by *Trichoderma atroviride* strain T23 and characterization of a paraoxonase-like enzyme. Appl Microbiol Biotechnol 103(21–22):8947–8962. https://doi.org/10.1007/s00253-019-10136-2

Tian Y, Tan Y, Yan Z, Liao Y, Chen J, De Boevre M, De Saeger S, Wu A (2018) Antagonistic and detoxification potentials of *Trichoderma* isolates for control of zearalenone (ZEN) producing *Fusarium graminearum*. Front Microbiol 8:2710. https://doi.org/10.3389/fmicb.2017.02710

Tijerino A, Cardoza RE, Moraga J, Malmierca MG, Vicente F, Aleu J, Collado IG, Gutiérrez S, Monte E, Hermosa R (2011) Overexpression of the trichodiene synthase gene tri5 increases trichodermin production and antimicrobial activity in *Trichoderma brevicompactum*. Fungal Genet Biol 48:285–296. https://doi.org/10.1016/j.fgb.2010.11.012

Tyśkiewicz R, Nowak A, Ozimek E, Jaroszuk Ściseł J (2022) *Trichoderma*: the current status of its application in agriculture for the biocontrol of fungal phytopathogens and stimulation of plant growth. Int J Mol Sci 23(4). https://doi.org/10.3390/ijms23042329

Umana EJ, Akwaji PI, Markson AA (2016) Bioremediation of spent engine oil contaminated soil by using fungus, *Penicillium* sp. Int Lett Nat Sci 59

Vargas WA, Mandawe JC, Kenerley CM (2009) Plant-derived sucrose is a key element in the symbiotic association between *Trichoderma virens* and maize plants. Plant Physiol 151(2):792–808. https://doi.org/10.1104/pp.109.141291

Ventorino V, Sannino F, Piccolo A, Cafaro V, Carotenuto R, Pepe O (2014) *Methylobacterium populi* VP2: plant growth-promoting bacterium isolated from a highly polluted environment for polycyclic aromatic hydrocarbon (PAH) biodegradation. Sci World J 2014:93179. https://doi.org/10.1155/2014/931793

Vinale F, Sivasithamparam K, Ghisalberti EL, Marra R, Woo SL, Lorito M (2008) *Trichoderma*-plant-pathogen interactions. Soil Biol Biochem 40(1):1–10. https://doi.org/10.1016/j.soilbio.2007.07.002

Vitousek PM, Naylor R, Crews T, David MB, Drinkwater LE, Holland E, Johnes PJ, Katzenberger J, Martinelli LA, Matson PA, Nziguheba G, Ojima D, Palm CA, Robertson GP, Sanchez PA, Townsend AR, Zhang FS (2009) Nutrient imbalances in agricultural development. Science 324(5934):1519–1520. https://doi.org/10.1126/science.1170261

WHO (2018) Arsénico. https://www.who.int/es/news-room/fact-sheets/detail/arsenic

Wightwick A, Walters R, Allinson G, Reichman S, Menzies N (2010) Environmental risks of fungicides used in horticultural production systems. Fungicides

Wilson CL, Chalutz E (1998) Postharvest biological control of *Penicillium* rots of citrus with antagonistic yeasts and bacteria. Sci Hortic 40:105–112. https://doi.org/10.1016/0304-4238(89)90092-7

Wisniewski M, Droby S, Norelli J, Liu J, Schena L (2016) Alternative management technologies for postharvest disease control: the journey from simplicity to complexity. Postharvest Biol Technol 122:3–10. https://doi.org/10.1016/j.postharvbio.2016.05.012

Woo SL, Ruocco M, Vinale F et al (2014) *Trichoderma*-based products and their widespread use in agriculture. Open Mycol J 8(Suppl. 1):71–126

Zehra A, Dubey MK, Meena M, Upadhyay RS (2017) Effect of different environmental conditions on growth and sporulation of some *Trichoderma* species. J Environ Biol 38(2):197–203

Zeilinger S, Gruber S, Bansal R, Mukherjee PK (2016) Secondary metabolism in *Trichoderma* – chemistry meets genomics. Fungal Biol Rev 30(2):74–90

Zhang S, Gan Y, Xu B (2019) Mechanisms of the IAA and ACC-deaminase producing strain of *Trichoderma longibrachiatum* T6 in enhancing wheat seedling tolerance to NaCl stress. BMC Plant Biol 19(1):1–18. https://doi.org/10.1186/s12870-018-1618-5

Zhang JL, Tang WL, Huang QR, Li YZ, Wei ML, Jiang LL, Liu C, Yu X, Zhu HW, Chen GZ, Zhang XX (2021) *Trichoderma*: a treasure house of structurally diverse secondary metabolites with medicinal importance. Front Microbiol 12:723828. https://doi.org/10.3389/fmicb.2021.723828

Zhao L, Zhang YQ (2015) Effects of phosphate solubilization and phytohormone production of *Trichoderma asperellum* Q1 on promoting cucumber growth under salt stress. J Integr Agric 14(8):1588–1597. https://doi.org/10.1016/S2095-3119(14)60966-7

Zin NA, Badaluddin NA (2020) Biological functions of *Trichoderma* spp. for agriculture applications. Ann Agric Sci 65(2):168–178. https://doi.org/10.1016/j.aoas.2020.09.003

Penicillium Genus as a Source of Metabolites for Agricultural Applications

Vishal Gupta, Ayushi Sharma, Gayatri Jamwal, Sushil Kumar Gupta, and Vijay K. Razdan

Abstract Microbial secondary metabolites which are low molecular weight bioactive compounds, are beneficial to agriculture in several ways. However, less attention has been paid to fungal secondary metabolites compared to those from bacteria. An assortment of secondary metabolites produced by the fungi during their growth and development are broadly categorised into antibiotics, alkaloids, mycotoxins, alcohols and fatty acids, which are regulated by various metabolic pathways. These metabolites exhibit marked variations in their chemical structures and biological activities. Though not important for the fungal growth and reproduction, these metabolites determine the interaction between the host and its environment, playing important role in several survival functions. These metabolites are also extensively used in the field of medicine and agriculture as antibiotics, antitumor and antidiarrheal drugs, immunosuppressants, agricultural herbicides, fungicides, bio-insecticides, plant growth stimulators etc. Penicillin, a well-studied secondary metabolite obtained from *Penicillium* sp. was the first antibiotic exhibiting broad spectrum of activities. This species also produces antimalarial (Benzomalvin C), anti-tumour (oxaline), anticholesterolemics (Pravastatin) agents and several other molecules with health benefitting effects. Other fungal species like *Trichoderma viride, Monascus ruber* and *Aspergillus terreus* are also known to produce compounds with antimicrobial and anti-cancer properties. On agriculture front, *Penicillium, Aspergillus* and *Trichoderma* species are known to have biocontrol properties against plant pathogens, with important role in phytoremediation. Using secondary metabolites obtained from fungal organisms for crop improvement can reduce dependence on synthetic substances and impart eco-friendly approach towards sustainable agriculture.

V. Gupta (✉) · A. Sharma · G. Jamwal · V. K. Razdan
Division of Plant Pathology, Faculty of Agriculture, Sher-e-Kashmir University of Agricultural Sciences & Technology of Jammu, Jammu, India

S. K. Gupta
Faculty of Horticulture & Forestry, Sher-e-Kashmir University of Agricultural Sciences & Technology of Jammu, Jammu, India

J. Poveda et al. (eds.), *Fungal Metabolites for Agricultural Applications*, Fungal Biology, https://doi.org/10.1007/978-3-031-76587-2_9

Keywords *Penicillium* · Biocontrol · Biostimulant · Secondary metabolites · Crops

1 Introduction

The genus *Penicillium,* belonging to the family *Trichocomaceae* and having about 350 filamentous fungal species, is one of the most extensively studied and widely distributed genera of fungi. It inhabits diverse environmental conditions worldwide, and produces a wide range of secondary metabolites that have beneficial industrial and agricultural applications. The present chapter discusses the diversity of secondary metabolites produced by various *Penicillium* species and their possible applications as biocontrol agents of plant pathogens, plant growth promoters and biofertilizers for crop improvement in a sustainable way. Colonies of *Penicillium* spp. are usually green in colour, fast growing and exhibiting a characteristic colonial morphology which consists of a dense felt of conidiophores. Chains of single-celled conidia, responsible for the asexual reproduction of the fungus, are produced from a conidiogenous cell called as phialide. Phialides may be produced singly or in groups or branches called as penicillus, giving it a brush-like appearance. Phialides are usually flask-shaped, consisting of a cylindrical basal part and a distinct neck. Conidia, produced in long chains, are globose, ellipsoidal, cylindrical or fusiform, hyaline or greenish, smooth or rough-walled. These fungi are inhibited in an array of environments, including soil, decaying plant material and fruits. Thousands of *Penicillium* isolates have been screened in bioprospecting programmes since the discovery of penicillin and new bioactive metabolites continue to be synthesized (Larsen et al. 2007; Ge et al. 2008; Takahashi and Lucas 2008), indicating their continuous importance as sources of novel bioactive molecules in agricultural industry. *Penicillium* genus has been extensively studied for its diverse range of secondary metabolites, which hold immense potential for agriculture. Various species of *Penicillium viz.*, *P. atrovenetum, P. novaezeelandiae, P. herquei, P. estinogenum, P. paraherquei, P. coralligerum, P. slerotiorum, P. simplicissimum, P. brasilianum, P. janthinellum, P. waksmanii* and *P. asperosporum* have been reported to produce secondary metabolites (Ramirez 1982). These metabolites have significance in enhancing crop productivity, pest and disease management, and overall agricultural sustainability (Fig. 1). From natural fungicides and insecticides to growth-promoting compounds, *Penicillium*-derived secondary metabolites offer promising alternatives to address the challenges faced by modern agriculture under changing climate conditions.

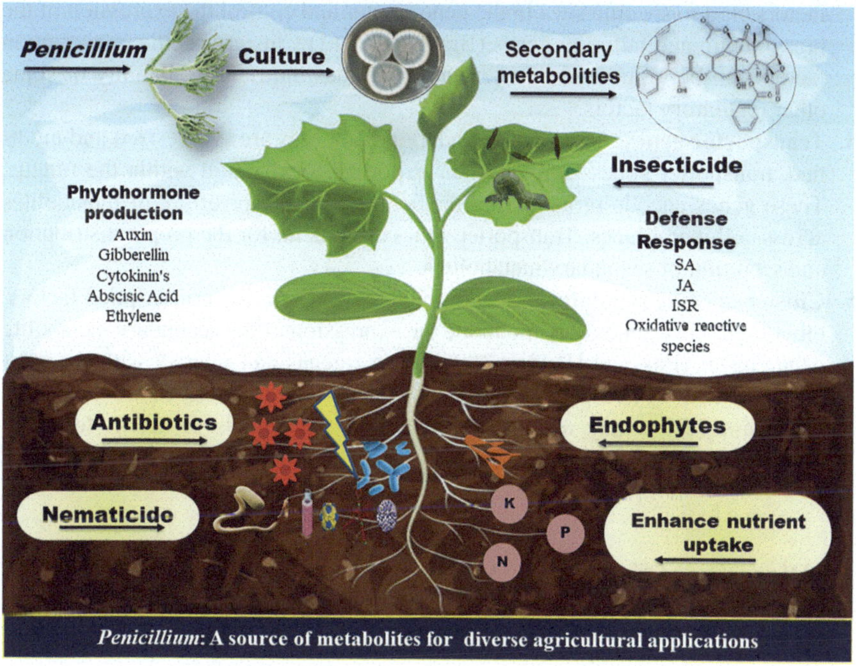

Fig. 1 *Penicillium*: a source of metabolites for diverse agricultural applications

2 Biosynthesis of Secondary Metabolites

Advances in the techniques of molecular biology and genomics have enabled the identification and manipulation of genes encoding enzymes required to produce secondary metabolites, which has led to the discovery of novel compounds and potential therapeutic agents (Martin et al. 2014). These genes have been categorized as:

1. **Biosynthetic genes**: These genes are responsible for the synthesis of the core structures of secondary metabolites. They encode enzymes that catalyze specific chemical reactions, leading to the formation of the metabolite's backbone (Keller 2015). Biosynthetic genes often include key enzymes such as polyketide synthases (PKSs), non-ribosomal peptide synthetases (NRPSs) and terpene synthases.
2. **Tailoring genes:** After the core structure is formed, tailoring genes come into play. These genes encode enzymes that introduce modifications to the core metabolite, such as methylations, oxidations or glycosylation. Tailoring genes can enhance the diversity of secondary metabolites by introducing structural variations.
3. **Cluster-specific transcription factors:** Each secondary metabolite gene cluster is usually regulated by a cluster-specific transcription factor. These transcription

factors bind to specific sites in the gene cluster and control the expression of the biosynthetic and tailoring genes (Brakhage 2013) Cluster-specific transcription factors can be influenced by environmental cues, nutrient availability or some other regulatory factors.

4. **Transporter genes:** Once the secondary metabolites are synthesized and modified, transporter genes facilitate their export or localization within the fungus. These genes encode proteins that mediate the uptake or efflux of metabolites across cell membranes. Transporter genes are crucial for the proper distribution and secretion of secondary metabolites.
5. **Cluster-specific regulators:** In addition to cluster-specific transcription factors, other regulatory genes may modulate the expression of the secondary metabolite gene cluster (Fox and Howlett 2008). These genes can interact with the transcription factors or other regulatory elements within the cluster to fine-tune gene expression. They may respond to environmental signals or act as feedback regulators.

3 Diversity of Secondary Metabolites

These low-molecular-weight metabolites often have potent physiological activities, with penicillin, cephalosporin, ergotrate and the statins being well known fungal secondary metabolites. Advances in molecular biology, bioinformatics and comparative genomics have revealed that though chemically diverse, all secondary metabolites are produced by a few common biosynthetic pathways, and the genes encoding specific fungal secondary metabolites are often clustered near telomeres. Further advances in fungal genomics is bound to accelerate the discovery of new pharmacologically active natural products.

Antibiotics Filamentous fungi produce a wide range of bioactive compounds with important pharmaceutical applications, such as antibiotic penicillins and cholesterol-lowering statins. *Penicillium* is well-known for producing the first antibiotic, penicillin, which revolutionized medicine industry. The genus *Penicillium* is considered as one of the most versatile 'mycofactories', comprising some species able to produce gibberellins and other bioactive compounds that can modulate plant growth and development. Besides penicillin, various *Penicillium* species also synthesize other antibiotics like viridicatin, griseofulvin and patulin. These compounds have potent antimicrobial properties and are used to treat bacterial and fungal infections. *Penicillium*-derived antibiotics hold potential in agricultural applications, offering solutions for disease management and sustainable crop production, thereby reducing the dependence on synthetic pesticides. By inhibiting pathogen growth, *Penicillium* antibiotics help protect plant health and increase production and productivity substantially. Their application effectively contributes to integrated pest management strategies, enhancing soil health by minimizing the impact of soil-borne diseases, and promoting eco-friendly agricultural practices. Production of these antibiotics and their potential applications are listed in the Table 1. However, historically, actinomycetes have been the main source of antibiotics, and the fungi have not been studied to the same extent, despite of their ability to produce diverse arrays of secondary metabolites. Moreover, filamentous fungi have been less studied and characterized at the genomic level though, genomic insight is a key element in efficiently unlocking their potential for new antibiotics and other pharmaceuticals.

Mycotoxins *Penicillium* spp. play important role as decomposers of organic materials, causing destructive rots in the food industry and producing a wide range of mycotoxins. Though they are not considered as human pathogen because they hardly growth at 37 °C, but the ingestion of food or feed contaminated by mycotoxins like ochratoxin A, citrinin, roquefortine C and patulin, produced by *Penicillium* species, have toxic and carcinogenic potential (Table 2).

Enzymes *Penicillium* species are prolific producers of various enzymes, including amylases, cellulases, hemicellulases, proteases and lipases (Brown et al. 1987; Chaabouni et al. 1994). Their remarkable versatility and specificity make them invaluable tools in biotechnology and beyond. These enzymes play a pivotal role in food production, aiding in the fermentation of cheeses, sausages and beverages, enhancing flavours and textures. Lignocellulose, comprising of cellulose, hemicellulose, pectin and aromatic polymer lignin forms the maximum plant biomass on earth. *Penicillium* spp. produce (hemi-) cellulolytic enzymes that have properties for the hydrolysis and utilization of this plant biomass (Mäkelä et al. 2016). It has been reported that *Penicillium*'s cellulase is more efficient for bioconversion, than any other filamentous fungi, as it contains higher levels of β-glucosidase (Vaishnav et al.

Table 1 Antibiotics produced by *Penicillium* species and their applications

Penicillium species	Antibiotic	Application	References
P. chrysogenum, *P. camemberti*, *P. griseoviride*	Penicillin	Antibacterial properties	Fleming (1929)
P. griseofulvum	Griseofulvin	Biocontrol agent and antifungal properties	Kahan et al. (1979)
P. expansum	Diorcinol	Antifungal properties	Camina et al. (2004)
P. crustosum *P. viridicatum*	Penitrem A Viridicatin	Antifungal activities	Jesus et al. (1983), Luckner (1967)
P. purpurogenum	Griseofulvin	Biodegradation of pollutants	Xia et al. (2020)
P. patulum	Patulin	Antifungal properties	Subramanian (1982)
P. notatum	Xanthocillin	Antagonist activities against bacteria	Pfeifer et al. (1972)
P. frequentans	Mycophenolic acid	Antifungal and antiviral properties	Frisvad et al. (2004)
P. verrucosum	Penicillic acid	Antifungal activities	Laich et al. (2002)

Table 2 Mycotoxins produced by *Penicillium* species and their potential applications

Penicillium species	Mycotoxin	Applications	References
P. citrinum	Citrinin, Citreoviridin	Antifungal, antibacterial, antiprotozoal activities	Wakana et al. (2006)
P. roqueforti	Roquefortine C, Citreoviridin	Cheese making industry	Filtonborg et al. (1983)
Penicillium spp.	Gliotoxin, Cytochalasin	Immunosuppressive properties	Aznar et al. (2019)
P. citreonigrum	Citreoviridin	Antimicrobial properties	Suh and Wilcox (1988)
P. verruculosum	Verruculogen, Ochratoxin	Antifeedant activity	Pitt and Leinster (1991)
P. viridicatum	Viridicatin	Antibacterial properties	Cunnigum and Freeman (1953)
P. rubrum	Rubratoxin B	Antifungal properties	Natroi et al. (1970)
P. verucosum	Verrucofortine, Ochratoxin	Biocontrol agent	Larsen et al. (2001)
P. digitatum	Geodin	Toxic to fungus	Koolen et al. (2012)
P. rugulosum	Rugulovasines	Antifungal properties	Abe et al. (1969)
P. albocoremium	Andrastin A	Inhibit protein synthesis	Overy et al. (2005)

Table 3 Some of the important enzymes produced by *Penicillium* species

Penicillium species	Enzymes	References
P. chrysogenum	Penicillin G acylase, Penicillin V acylase	Erickson and Bennett (1965)
P. digitatum	Pectinases, Polygalacturonases, Cellulases	Pereria et al. (2002)
P. verruculosum	Xylanases, Endo-1,4-beta-xylanase	Dashtban et al. (2009)
P. funiculosum	Proteases, Glucanases, Pectinases	Karboune et al. (2008)
P. purpurogenum	Cellulases, Xylanases, Lipases	Sunkar et al. (2020)
P. notatum	Penicillin G acylase, Penicillin V acylase	Mahendiran et al. (2013)
P. griseofulvum	Chitinases, Cellulases, Lipases	Rafiq et al. (2020)
P. citrinum	Lipases, Proteases, Glucanases	Maliszewska and Zboińska (1996)

2018). In agriculture, *Penicillium* enzymes contribute to biofertilizers, enhancing nutrient availability and soil health. Their applications extend to textile and paper industries as well, where they aid in the degradation of organic materials and contribute to sustainable production processes. In environmental bioremediation, these enzymes excel in breaking down pollutants and toxins, offering eco-friendly solutions. Furthermore, they are crucial in molecular biology, facilitating DNA manipulation and protein engineering. Among the potent cellulase producers, *Penicillium* species are in the fore front, specifically regarding biomass conversion for biofuel applications. Some of these enzymes are listed in Table 3.

Nematicidal and Insecticidal Activities *Penicillium* species exhibit remarkable potential as nematicides and insecticides by producing bioactive compounds that target nematode's sensory and physiological systems, thereby disrupting their lifecycle and reducing crop damage. Therefore, these *Penicillium*-based nematicides offer eco-friendly approach in the management of plant-parasitic nematodes and insect pests, thus reducing the dependence on chemical pesticides and minimizing its associated environmental risks. As a non-toxic alternative, *Penicillium*-based nematicides align with the growing demand for safer and more sustainable pest control methods. Their application in integrated pest management can enhance soil health, improve nutrient cycling and promote sustainable agriculture. Their potential impact extends to organic farming systems, ensuring minimal harm to beneficial organisms while effectively managing nematode populations. Ongoing researches are therefore, focused on optimizing the production of *Penicillium* metabolites and their formulation for practical field application. The future of these nematicides holds the key to reducing chemical reliance, promoting biodiversity and fostering resilient agricultural ecosystems (Table 4).

Biocontrol Agent *Penicillium* species are known to produce a series of antibiotics, which potentially inhibit growth of plant pathogens (Yang et al. 2008). Therefore, they exhibit promising potential as biocontrol agents, offering a natural and sustainable approach to manage plant diseases. These fungi possess antagonistic properties, suppressing the growth and activity of various plant pathogens. *Penicillium*-based biocontrol agents can colonize plant surfaces, forming a protective barrier that prevents pathogen invasion. Their ability to produce secondary metabolites, such as antibiotics and enzymes, further contributes to disease suppression. *Penicillium*-based biocontrol strategies reduce the need for chemical fungicides, promoting environment friendly pest management practices. Their application is compatible with integrated pest management systems, ensuring balanced and

Table 4 Important nematicides and their mode of action

Penicillium species	Metabolite	Mode of action	References
P. phoeniceum	Oosporein	Disrupts cellular functions and can lead to paralysis and death in the target organisms.	Christiansen et al. (2022)
P. simplicissimum, *P. paxillin*	Verruculogen	Attacks nervous system of insects and causes muscle spasms.	Zhao et al. (2019)
P. crustosum, *P. melanoconidium*, *P. commune*	Penitrems	Inhibits the growth and development of nematodes and insects.	Li (2010)
P. camemberti, *P. chrysogenum*, *P. commune*, *P. viridicatum*, *P. griseofulvum*	Cyclopiazonic acid	Interferes with calcium regulation in cells, leading to disruptions in cellular functions.	Ostry et al. (2018)
P. bilaiae	Brefeldin	Reduction in gall formation induced by *M. incognita*	Li et al. (2007)

effective disease control. Ongoing research focuses on optimizing formulations and delivery systems for practical field application of these bioagents thereby enhancing their efficacy. As a non-toxic alternative, *Penicillium*-based biocontrol agents uphold food safety standards and safeguard of beneficial micro-organisms. The future holds promise for harnessing the biocontrol potential of *Penicillium*, thus contributing to resilient and sustainable agricultural systems while minimizing adverse ecological impacts. Although *Penicillium* spp. are among the antagonistic microorganisms that have been proved to be active under in vitro or in vivo conditions, yet Boughalleb-M'Hamdi et al. (2018) do not recommend them for biological control assay due to their carcinogenic properties. Below is the list of *Penicillium* species acting as biocontrol agents (Table 5).

Mycorrhizal Symbiosis Certain *Penicillium* species can form beneficial mycorrhizal associations with plant roots. Mycorrhizal fungi facilitate the uptake of water and nutrients, such as nitrogen and phosphorus, in exchange for plant-derived carbohydrates. This mutualistic relationship contributes to improved plant growth and nutrient acquisition. *Penicillium* species in association with Arbuscular mycorrhizal fungi are reported to be beneficial for the wheat growth under saline conditions and are known to solubilize phosphorous in rhizosphere or endosphere (Wakelin et al. 2004; Elgharably an Nafady 2021).

Table 5 *Penicillium* species as biocontrol agents

Penicillium species	Target disease	Target microorganism	Host	References
P. oxalicum	Powdery mildew	*Podosphaera aphanis*	Strawberry	De cal et al. (2008)
P. oxalicum, *P. rubens* strain 212	Fusarium wilt	*Fusarium oxysporum* f. sp. *lycopersici*	Tomato	Larena et al. (2003)
P. oxalicum	Damping-off	*Pythium aphanidermatum*	Cucumber, Tomato	Yang et al. (2014)
P. janthinellum	Rhizoctonia damping-off	*Rhizoctonia* spp.	Cucumber	Siddiqui et al. (2005)
P. bilaii	Fusarium wilt	*Fusarium oxysporum*	Chickpea	Yedidia et al. (2003)
P. commune	Soil-borne pathogens	*Verticillium, Pythium, Rhizoctonia*	Multiple crops	Segone (2021)
P. simplicissimum	Cucumber mosaic disease	*Cucumber mosaic virus*	Cucumber	El-Mougy et al. (2008)
P. citrinum	Sclerotina rot	*Sclerotinia sclerotiorum*	Lettuce, Bean, Sunflower	Shu et al. (2020)
P. citrinum	Gray mould	*Botrytis cinereal*	Chickpea	Sreevidya et al. (2015)
P. freii	Blast	*Pyricularia oryzae*	Rice	Li et al. (2021)

Induced Systemic Resistance Associations between the plants and plant growth-promoting fungi are beneficial to plants, as these determine improved plant growth and induced systemic resistance to invading pathogens. This mutualistic relationship results in improved germination, seedling vigour, biomass production, root hair development, photosynthetic efficiency, flowering and yield. Microbial or fungal metabolites result in enhanced nutrient uptake, phytohormone production and activation of some plant signalling pathways in the plants, thereby triggering plant growth and induced systemic resistance (ISR) responses to pathogen attack. When plants are exposed to *Penicillium* or its metabolites, they activate defence mechanisms, making crops more resistant to diseases and environmental stresses. Application of *Penicillium* conidia or their culture filtrates induce systemic resistance, reduce the disease incidence or may even induce hypersensitive reaction in some crops. The culture filtrate of *Penicillium* sp. has been reported to induce systemic expressions of defence-related genes in plants against virus and its vector (Elsharkawy and Mousa 2015). Increased peroxidase activity, associated with ISR have been reported in the treated plants. Treating leaves with the culture filtrate of *Penicillium* triggers the signal transduction, activating different defence genes in plants, thus protecting them from various diseases (Madi and Katan 1998). *Penicillium*-induced resistance may function independently of the salicylic acid, jasmonic acid and ethylene signalling pathways (Thuerig et al. 2006).

Abiotic Stress Tolerance Some soil microorganisms, including fungi help plants to partially overcome the deleterious effects caused by different environmental stresses, like extremes of temperature, light, flooding, drought, toxic metals, environmental organic contaminants, radiation and high salt. These stresses cause great damage to plants and decrease their productivity. Several fungi known to promote plant growth, facilitate the acquisition of nutritional resources, such as nitrogen, phosphorus or iron, also prevent or decrease the damage to plants due to abiotic stresses. *Penicillium* has various plant-growth-promoting characteristics like production of phytohormones, phosphate solubilization, synthesis of amino acids, secretion of antioxidative enzymes and biologically active secondary metabolites, which favour plant growth and help plants to cope with abiotic stresses. Some *Penicillium* metabolites have been found to enhance plant tolerance to abiotic stresses, such as drought, salinity and extreme temperatures. These compounds can help mitigate the negative effects of adverse environmental conditions on plant growth and productivity. Although plants have the ability to synthesize gibberellins but their levels are lower when plants are under detrimental abiotic conditions, such as salinity stress. *Penicillium* strains and plant host interactions, helps in gibberellins production as the fungal response allows plants to overcome salinity stress. Some of *Penicillium* species are also successfully applied in bioremediation of heavy metal-contaminated soil into soil suitable for agriculture. *Penicillium* as endophyte produce phytohormones like gibberellins, inducing physiological changes in host plants during extreme environmental conditions, thereby improving the plant growth under salinity, drought and heat stresses (Khan et al. 2015). *Penicillium* species stimulant several growth-promoting substances, like auxin,

Table 6 Phytohormones produced by *Penicillium* species

Penicillium species	Phytohormones	References
P. citrinum	Indole-3-acetic acid (IAA)	Khan et al. (2008)
P. chrysogenum	IAA, Gibberellins	Galeano et al. (2023)
P. expansum	Gibberellins, Cytokinins	Hamza et al. (2015)
P. funiculosum	Cytokinins, Abscisic Acid (ABA)	Yurieva et al. (2018)
P. griseofulvum	Gibberellins, Ethylene	Zhang et al. (2017)
P. notatum	IAA, ABA	Tunc et al. (2007)
P. oxalicum	Gibberellins, IAA	Ai et al. (2023)
P. roqueforti	IAA, Cytokinins	Coton et al. (2020)
P. simplicissimum	IAA, ABA, Cytokinins	Chaudhary et al. (2018)
P. verrucosum	IAA, ABA, Gibberellins	Noots et al. (1999)

cytokinin, gibberellin, siderophores which assist in iron uptake, and 1-aminocyclopropane-1-carboxylate (ACC) deaminase, which lowers ethylene levels and reduces stress-induced ethylene-mediated inhibition of growth. *Penicillium* species also produce phytohormones, such as auxins, gibberellins, and cytokinins, which are essential for regulating plant growth and development (Table 6). These hormones influence various processes, including cell division, elongation and differentiation, leading to increased shoot and root growth. Endophytic *Penicillium* species isolated from the coastal sand dune flora have been reported to have plant growth promoting ability and help in the conservation of the sand dune flora. This fungus is known for producing mycotoxin citrinin and cellulose digesting enzymes like cellulase and endoglucanase, as well as xylulase and gibberellins (Khan et al. 2008). *Penicillium* species producing phytohormones, siderophore, ACC deaminase and secondary metabolites play important role in the biosorption of heavy metals like Cd and Pb and decrease their content in soil and plants. Thus, this can be utilized as an excellent technique in the fields contaminated with heavy metals for sustainable agriculture (El-Mahdy et al. 2021).

Root Development *Penicillium* metabolites can stimulate root growth and branching, leading to increased root surface area for nutrient and water uptake. Healthy and well developed root systems are crucial for optimal plant growth and nutrient absorption. *Penicillium* species significantly enhance shoot length, root length, and fresh and dry seedling weight, due to the secretion of various amino acids. They help in plant survival, with increased shoot length, and fresh and dry seedling weights under salt stress conditions. *Penicillium* isolates show antagonistic activity towards *Fusarium* spp., and significantly increase the shoot length in plants infected by the pathogen. Therefore, application of *Penicillium* sp. can act as a biofertilizer and a biocontrol agent to improve plant growth and enhance the plant survival against salt stress and *Fusarium* infection (Radhakrishnan and Kang 2014).

Biofertilizers *Penicillium* species generally occur at greater soil depths than species of other genera, and have low concentrations in rhizosphere soils (Domsch

Table 7 *Penicillium* species as potential biofertilizers

Penicillium species	Enhanced nutrient uptake	References
P. glabrum	Phosphorus, Nitrogen	Smith (2022)
P. bilaiae	Phosphorus, Potassium, Zinc	Lindsay and Schwab (1982)
P. simplicissimum	Nitrogen, Phosphorus, Potassium, Iron	Adam and Brown (2023)
P. rugulosum	Phosphorus, Nitrogen, Potassium	Reyes et al. (1999)
P. funiculosum	Phosphorus, Nitrogen, Potassium, Zinc Calcium	Garcia et al. (2019)

et al. 1993). However, all strains of Penicillium tested so far are known to solubilise metaphosphates and utilize them as phosphorus sources (Phuwiwat and Tong 2001). *Penicillium* metabolites facilitate nutrient solubilization and increase nutrient availability in the rhizosphere (Table 7). Organic acids produced by *Penicillium* can chelate and mobilize essential nutrients like phosphorus, iron, and zinc, making them more accessible to plants. This improved nutrient uptake can result in healthier and more robust plant growth.

Maintaining soil health and increased crop productivity are the two major challenges for sustainable agriculture and food security, and the use of effective microorganisms is considered as the key element to overcome this problem. Soil fungi play a significant role in soil fertility and plant nutrient turnover. *Penicillium* and its secondary metabolite, having the ability of phosphate solubilization, production of siderophore and plant-growth regulators, resulting in the promotion of plant-growth is widely established. Therefore, *Penicillium* as a bioinoculant biofertilizer in enhancing crop productivity and its potential role in sustainable agriculture has a merit. *Penicillium*-based biofertilizers have been developed to enhance soil fertility and plant nutrition. These biofertilizers contain certain beneficial strains that contribute to nutrient cycling and improve nutrient availability for plants. *P. bilaiae* is used as a seed coating in cereals such as wheat and maize (Leggett et al. 2015), and *P. aculeatum* with strong phosphorous solubilization traits has shown to improve the phosphorus fertilizer value of biochar for wheat grown in semi-sterile soil (Efthymiou et al. 2018). *Penicillium* sp. having indole-3-acetic acid and siderophores synthetic capacity, increase total biomass, enhance root vitality, improve Photosynthesis, transpiration, glutamine synthetase, increase leaf photosynthetic pigment content and P and N uptake. Therefore, those *Penicillium* strains which positively improve the plant growth have the potential for use as a biofertilizer (Yuzha et al. 2022).

4 Mechanism of Action

Mode of action of secondary metabolites can vary significantly depending on the specific compound and the target organism. *Penicillium* species are a diverse group of fungi which produce a wide range of these secondary metabolites harbouring various biological activities. These species are well-known for producing antibiotics that inhibit the growth and reproduction of bacteria by targeting essential cellular processes, such as cell wall synthesis or protein synthesis. This mode of action weakens bacterial cells, leading to their eventual death. Many secondary metabolites from *Penicillium* species possess antifungal properties. These metabolites can disrupt the cell membranes of fungal pathogens, inhibit fungal enzyme activity, or interfere with essential cellular processes, ultimately leading to the death of the fungi. Some secondary metabolites produced by *Penicillium* species have nematicidal properties. These compounds target nematode nervous system, leading to its paralysis or death. *Penicillium* species exhibit insecticidal properties by either directly affecting insects or indirectly disrupting their life cycle.

1. **Activation or inhibition of gene expression**: *Penicillium* secondary metabolites directly interact with regulatory proteins or transcription factors, leading to the activation or inhibition of specific genes. This results in the upregulation or downregulation of various pathways involved in plant growth, defense mechanisms, or stress response.
2. **Modulation of signal transduction pathways:** Some *Penicillium* secondary metabolites interfere with signaling pathways within plant cells. They either activate or inhibit specific signaling cascades, thereby modifying the expression of genes associated with these pathways. This influences the process of plant development, immunity, or responses to abiotic or biotic stresses.
3. **Epigenetic modifications:** *Penicillium* secondary metabolites also induce changes in the epigenetic landscape of plant cells. These changes affect gene expression by altering the accessibility of DNA to transcription factors and other regulatory proteins. Epigenetic modifications can be stable and heritable, thus having long-lasting effects on gene regulation and plant phenotype.
4. **Indirect effects on gene expression:** *Penicillium* secondary metabolites indirectly impact gene expression by influencing the activity of enzymes and proteins involved in DNA replication, transcription, or translation. By targeting these molecular processes, these metabolites alter the overall dynamics of gene expression networks in plants.

5 Future Prostects

The *Penicillium* genus is a prolific source of metabolites for revolutionary agricultural applications. Through advanced biotechnology and genomic research, *Penicillium* species are poised to yield bioactive compounds with multifaceted benefits. These metabolites demonstrate potential as eco-friendly biopesticides, targeting pests while minimizing environmental impact. Moreover, *Penicillium*-derived biofertilizers could enhance crop yields by promoting nutrient uptake and soil health. The genus offers solutions to combat emerging plant pathogens, contributing to sustainable crop production. Harnessing *Penicillium* secondary metabolites can lead to innovative plant growth regulators, optimizing growth, flowering and fruiting. These compounds might also serve as natural alternatives to synthetic growth enhancers. Furthermore, *Penicillium*-derived enzymes could revolutionize agricultural processes, from waste management to biofuel production. Collaborative efforts in bioprospecting and synthetic biology can unlock a plethora of novel applications, revolutionizing agriculture and promoting a greener future.

6 Conclusion

The *Penicillium* genus presents a rich reservoir of metabolites with substantial potential for transformative agricultural applications. Its diverse array of bioactive compounds offers innovative solutions for sustainable crop production, disease management and reduced chemical reliance. While challenges exist in extraction and application, interdisciplinary efforts hold the key to unlocking the full benefits of *Penicillium*-derived metabolites.

References

Abe M, Ohmomo S, Ohashi T, Tabuchi T (1969) Isolation of chanoclavine-(I) and two new interconvertible alkaloids, rugulovasine a and B, from the cultures of *Penicillium concavo-rugulosum*. Agric Biol Chem 33:469–471. https://doi.org/10.1271/bbb1961.33.469

Adams R, Brown K (2023) Enhanced nitrogen, phosphorus, potassium, and iron uptake by *Penicillium simplicissimum* in soil. Plant Soil 18:300–315. https://doi.org/10.3389/fpls.2023.1190768

Ai M, Han F, Yang X, Chu H, Luo C, Tan S, Lv S, Qin M, Xie G (2023) Endophytic *Penicillium oxalicum* CX-1 prevented *Phytophthora cactorum* blight on *salvia miltiorrhiza* and promoted plant growth. J Appl Microbiol 134:10. https://doi.org/10.1093/jambio/lxad010

Aznar FT, Cimmino A, Masi M, Rubiales D, Evidente A (2019) Antifeedant activity of long-chain alcohols, and fungal and plant metabolites against pea aphid (*Acyrthosiphon pisum*) as potential biocontrol strategy. Nat Prod Res 33:2471–2479. https://doi.org/10.1080/14786419.2018.1452013

Boughalleb-M'Hamdi N, Salem IB, M'Hamdi M (2018) Evaluation of the efficiency of *Trichoderma, Penicillium,* and *Aspergillus* species as biological control agents against four soil-borne fungi of melon and watermelon. Egypt J Biol Pest Control 28:25. https://doi.org/10.1186/s41938-017-0010-3

Brakhage AA (2013) Regulation of fungal secondary metabolism. Nat Rev Microbiol 11:21–32. https://doi.org/10.1038/nrmicro2916

Brown JA, Sylvia AC, Thomas MW (1987) Enhanced enzyme production by the cellulolytic fungus *Penicillium pinophilum*, mutant strain NTG III/6. Enzyme Microb Technol 9(3):176–180

Camiña M, Franco A, González A, Paz JC (2004) New diorcinol derivatives from *Penicillium paxilli*. J Nat Prod 67:1092–1094. https://doi.org/10.1139/m74-179

Chaabouni SE, Hadj-Taieb N, Mosrati R, Ellouz R (1994) Preliminary assessment of *Penicillium occitanis* cellulase: a further useful system. Enzym Microb Technol 16(6):538–542

Chaudhary S, Shankar A, Singh A, Prasad V (2018) Usefulness of *Penicillium* in enhancing plants resistance to abiotic stresses: an overview. In: Gupta VK, Rodriguez-Couto S (eds) New and future developments in microbial biotechnology and bioengineering, vol 12. Elsevier, pp 277–284. https://doi.org/10.1016/B978-0-444-63501-3.00017-X

Christiansen JV, Isbrandt T, Asferg R, Jarmusch SA, Larsen TO, Frisvad JC (2022) Phoenicin switch: discovering the trigger for radical phoenicin production in multiple wild-type *Penicillium* species. App Env Micro 88:302–322. https://doi.org/10.1128/aem.00302-22

Coton E, Coton M, Hymery N, Mounier J, Jany JL (2020) *Penicillium roqueforti*: an overview of its genetics, physiology, metabolism and biotechnological applications. Fungal Biol Rev 34:59–73. https://doi.org/10.1016/j.fbr.2020.03.001

Cunningham KG, Freeman GG (1953) The isolation and some chemical properties of viridicatin, a metabolic product of *Penicillium viridicatum* Westling. Biochem J 53:328. https://doi.org/10.1042/bj0530328. PMID: 13032073

Dashtban M, Schraft H, Qin W (2009) Fungal bioconversion of lignocellulosic residues; opportunities and perspectives. Int J Biol Sci 5:578–595

De Cal A, Redondo C, Sztejnberg A, Melgarejo P (2008) Biocontrol of powdery mildew by *Penicillium oxalicum* in open-field nurseries of strawberries. Biol Control 47:103–107. https://doi.org/10.1016/j.biocontrol.2008.07.010

Domsch KH, Gams W, Anderson TW (1993) Compendium of soil fungi. IHW-Verlag 1:630–643

Efthymiou A, Jensen B, Jakobsen I (2018) The roles of mycorrhiza and *Penicillium* inoculants in phosphorus uptake by biochar-amended wheat. Soil Biol Biochem 127:168–177. https://doi.org/10.1016/j.soilbio.2018.09.027

Elgharably A, Nafady NA (2021) Inoculation with *Arbuscular mycorrhizae*, *Penicillium funiculosum* and *Fusarium oxysporum* enhanced wheat growth and nutrient uptake in the saline soil. Rhizosphere 18:100345

El-Mahdy OM, Mohamed HI, Mogazy AM (2021) Biosorption effect of *aspergillus Niger* and *Penicillium chrysosporium* for Cd- and Pb-contaminated soil and their physiological effects on *Vicia faba* L. Environ Sci Pollut Res 28:67608–67631. https://doi.org/10.1007/s11356-021-15382-4

El-Mougy NS, El-Gamal NG, Abdalla MA (2008) The use of fungicide alternatives for controlling postharvest decay of strawberry and orange fruits. J Plant Prot Res 48:385–395. https://doi.org/10.2478/v10045-008-0048-z

Elsharkawy MM, Mousa KM (2015) Induction of systemic resistance against papaya ring spot virus (PRSV) and its vector *Myzus persicae* by *Penicillium simplicissimum* GP17-2 and silica (Sio2) nanopowder. Int J Pest Mana 61(4):353–358

Erickson RC, Bennett RE (1965) Penicillin acylase activity of *Penicillium chrysogenum*. Appl Microbiol 13:738–742. https://doi.org/10.1128/am.13.5.738-742.1965

Filtenborg O, Frisvad JC, Svendsen JA (1983) Simple screening method for molds producing intracellular mycotoxins in pure cultures. AEM 45:581–585. https://doi.org/10.1128/aem.45.2.581-585.1983

Fleming A (1929) On the antibacterial action of cultures of a *Penicillium*, with special reference to their use in the isolation of *Bacillus influenza*. Br J Exp Pathol 10:226–236

Fox EM, Howlett BJ (2008) Secondary metabolism: regulation and role in fungal biology. Curr Opin Microbiol 11:481–487. https://doi.org/10.1016/j.mib.2008.10.007

Frisvad JC, Smedsgaard J, Larsen TO, Samson RA (2004) Mycotoxins, drugs and other extrolites produced by species in *Penicillium* subgenus *Penicillium*. Stud Mycol 49:201–241

Galeano RMS, Silva SM, Yonekawa MKA, de Alencar GNC, Giannesi GC, Masui DC, Corrêa BO, da Silva BM, Zanoelo FF (2023) *Penicillium chrysogenum* strain 34-P promotes plant growth and improves initial development of maize under saline conditions. Rhizosphere 26:1007–1010. https://doi.org/10.1016/j.rhisph.2023.100710

Garcia MV, Da Pia AKR, Freire L, Copetti MV, Sant AAS (2019) Effect of temperature on inactivation kinetics of three strains of *Penicillium paneum* and *P. roqueforti* during bread baking. Food Control 96:456–462. https://doi.org/10.1016/j.foodcont.2018.10.002

Ge HM, Shen Y, Zhu CH, Tan SH, Ding HS, Yong C, Tan RX (2008) Penicidones a-c, three cytotoxic alkaloidal metabolites of an endophytic *Penicillium* sp. Phytochemistry 69:571–576. https://doi.org/10.1016/j.phytochem.2007.07.014

Hamza LF, Kamal SA, Hameed IH (2015) Determination of metabolites products by *Penicillium expansum* and evaluating antimicrobial activity. JPP 7:194–220. https://doi.org/10.1016/B978-0-444-63501-3.00017-X

Jesus AE, Steyn PS, van Heerden FR, Vleggaar R, Wessels PL, Hull WE (1983) Tremorgenic mycotoxins from *Penicillium crustosum:* isolation of penitrems A–F and the structure elucidation and absolute configuration of penitrem a. J Chem Soc Perkin Trans 1(1):1847–1856. https://doi.org/10.1039/P19830001847

Kahan JS, Kahan FM, Goegelman R, Currie SA, Jackson M, Stapley EO, Miller TW (1979) Thienamycin: development of imipenem-cilastatin. J Antimicrob Chemother 4:1–5. https://doi.org/10.1093/jac/12.suppl_D.1

Karboune S, Geraert PA, Kermasha S (2008) Characterization of selected cellulolytic activities of multi-enzymatic complex system from *Penicillium funiculosum*. J Agric Food Chem 56:903–909. https://doi.org/10.1021/jf072847l

Keller NP (2015) Fungal secondary metabolism: regulation, function, and drug discovery. Nat Rev Microbiol 17:167–180. https://doi.org/10.1128/am.13.5.738-742.1965

Khan SA, Hamayun M, Yoon M, Kim HY, Suh SY, Hwang SK, Kim JM, Lee I, Choo YS, Yoon UH (2008) Plant growth promotion and *Penicillium citrinum*. BMC Microbiol 8:231–234. https://doi.org/10.1186/1471-2180-8-231

Khan AL, Waqas M, Lee IJ (2015) Resilience of *Penicillium resedanum* LK6 and exogenous gibberellin in improving *Capsicum annuum* growth under abiotic stresses. J Plant Res 128:259–268. https://doi.org/10.1007/s10265-014-0688-1

Koolen HHF, Soares ER, Silva FMAD, Almeida RAD, Souza ADLD, Medeiros LSD, Rodrigues FE, Souza AQLD (2012) An antimicrobial alkaloid and other metabolites produced by *Penicillium* sp. an endophytic fungus isolated from *Mauritia flexuosa*. Quim Nova 35:771–774. https://doi.org/10.1590/S0100-40422012000400022

Laich F, Fierro F, Martín JF (2002) Production of penicillin by fungi growing on food products: identification of a complete penicillin gene cluster in *Penicillium griseofulvum* and a truncated cluster in *Penicillium verrucosum*. AEM 68:1211–1219. https://doi.org/10.1128/AEM.68.3.1211-1219.2002

Larena I, Melgarejo P, De Cal A (2003) Drying of conidia of *Penicillium oxalicum*, a biological control agent against Fusarium wilt of tomato. J Phytopathol 151:600–606. https://doi.org/10.1046/j.0931-1785.2003.00772.x

Larsen TO, Svendsen A, Smedsgaard J (2001) Biochemical characterization of ochratoxin A-producing strains of the genus *Penicillium*. AEM 67:3630–3635. https://doi.org/10.1128/AEM.67.8.3630-3635.2001

Larsen TO, Lange L, Schnorr K, Stender S, Frisvad JC (2007) A novel phenolic compactin analogue from *Penicillium solitum*. Tetrahedron Lett 48:1261–1264. https://doi.org/10.1016/j.tetlet.2006.12.038

Leggett M, Newlands NK, Greenshields D, West L, Inman S, Koivunen ME (2015) Maize yield response to a phosphorus-solubilizing microbial inoculant in field trials. J Agric Sci 153:1464–1478. https://doi.org/10.1017/S0021859614001166

Li SM (2010) Prenylated indole derivatives from fungi: structure diversity, biological activities, biosynthesis and chemoenzymatic synthesis. Nat Prod Rep 27:57–78. https://doi.org/10.1039/B909987P

Li G, Zhang K, Xu J, Dong J, Liu Y (2007) Nematicidal substances from fungi. Recent Pat Biotechnol 1:212–233. https://doi.org/10.2174/187220807782330165

Li Y, Guo L, Zhou Z (2021) Exploring the antifungal mechanism of limonin-loaded eugenol emulsion against *Penicillium italicum:* from the perspective of microbial metabolism. Postharvest Biol Technol 182:111704. https://doi.org/10.1016/j.postharvbio.2021.111704

Lindsay WL, Schwab AP (1982) The chemistry of iron in soils and its availability to plants. J Plant Nutr 5:821–840. https://doi.org/10.1080/01904168209363012

Luckner M (1967) Zur Bildung von Chinolinalkaloiden in Pflanzen: 2. Die fermentative Umwandlung der Penicillium-Alkaloide Cyclopenin und Cyclopenol in Viridicatin und Viridicatol. European J Mol Biol Biochem 2:74–78. https://doi.org/10.1111/j.1432-1033.1967.tb00108.x

Madi L, Katan J (1998) Penicillium *janczewskii* and its metabolites, applied to leaves, elicit systemic acquired resistance to stem rot caused by *Rhizoctonia solani*. Physiol Mol Plant Path 53(3):163–175

Mahendiran B, Duraisamy V, Malarselvi AM (2013) Production, optimization, purification, testing antimicrobial and dye decolorizing properties of penicillin G acylase from *Penicillium chrysogenum*. Euro J Biotechnol Biosci 1:29–36

Mäkelä MR, Mansouri S, Wiebenga A, Rytioja J, de Vries RP, Hildén KS (2016) *Penicillium subrubescens* is a promising alternative for *Aspergillus Niger* in enzymatic plant biomass saccharification. New Biotechnol 33(6):834–841

Maliszewska IH, Zboińska EM (1996) *Penicillium citrinum* protoplasts: preparation, regeneration and lipolytic activity of regenerants. Acta Biotechnol 16:193–198. https://doi.org/10.1002/abio.370160217

Martín JF, García-Estrada C, Zeilinger S (2014) Biosynthesis and molecular genetics of fungal secondary metabolites. Springer. https://doi.org/10.1007/978-1-4939-1191-2

Natori S, Sakaki S, Kurata H, Udagawa SI, Ichinoe M, Saito M, Umeda M, Ohtsubo K (1970) Production of rubratoxin B by *Penicillium purpurogenum*. Appl Microbiol 19:613–617. https://doi.org/10.1128/am.19.4.613-617.1970

Noots I, Delcour JA, Michiels CW (1999) From field barley to malt: detection and specification of microbial activity for quality aspects. Crit Rev Microbiol 25:121–153. https://doi.org/10.1080/10408419991299257

Ostry V, Toman J, Grosse Y, Malir F (2018) Cyclopiazonic acid: 50th anniversary of its discovery. World Mycotoxin J 11:135–148. https://doi.org/10.3920/WMJ2017.2243

Overy DP, Larsen TO, Dalsgaard PW, Frydenvang K, Phipps R, Munro MH, Christophersen C (2005) Andrastin a and barceloneic acid metabolites, protein farnesyl transferase inhibitors from *Penicillium albocoremium*: chemotaxonomic significance and pathological implications. Mycol Res 109:1243–1249. https://doi.org/10.1017/S0953756205003734

Pereira JF, Vieira DQM, Aparecida GE, Muro-Abad JI, Fernandes DAE (2002) Molecular characterization and evaluation of pectinase and cellulase production of *Penicillium* spp. Biotechnol Lett 24:831–838. https://doi.org/10.1023/A:1015502721909

Pfeifer S, Bär H, Zarnack J (1972) Metabolites of the xanthocillin (Brevicid)-producing mutant of *Penicillium notatum*. Pharmazie 27:536–542

Phuwiwat W, Tong K (2001) The effect of Penicillium notatum on plant growth. Fungal Divers 8:143–148

Pitt JI, Leistner L (1991) Toxigenic *Penicillium* species. Mycotox Anim Foods 12:81–99

Radhakrishnan R, Kang S (2014) Characterization of plant growth-promoting traits of *Penicillium* species against the effects of high soil salinity and root disease. J Plant Int 9(1):754–762. https://doi.org/10.1080/17429145.2014.930524

Rafiq M, Nadeem S, Hassan N, Hayat M, Sajjad W, Zada S, Sajjad W, Hasan F (2020) Fungal recovery and characterization from Hindu Kush Mountain range, Tirich Mir glacier, and their potential for biotechnological applications. J Basic Microbiol 60:444–457. https://doi.org/10.1002/jobm.201900608

Ramirez C (1982) Manual and atlas of the Penicillia. Elsevier Biomedical Press

Reyes I, Bernier L, Simard RR, Antoun H (1999) Effect of nitrogen source on the solubilization of different inorganic phosphates by an isolate of *Penicillium rugulosum* and two UV-induced mutants. FEMS Microbiol Ecol 28:281–290. https://doi.org/10.1007/s00248-002-1001-8

Segone GP (2021) Analysis of efficacy of different Penicillium fungal species as bio-control agents against *Fusarium oxysporum*. Doctoral dissertation, North-West University (South Africa). http://hdl.handle.net/10394/38253

Shu Z, Liu Q, Xing C, Zhang Y, Zhou Y, Zhang J, Liu H, Cao M, Yang X, Liu G (2020) Viridicatol isolated from deep-sea *Penicillium griseofulvum* alleviates anaphylaxis and repairs the intestinal barrier in mice by suppressing mast cell activation. Mar Drugs 18:517. https://doi.org/10.3390/md18100517

Siddiqui IR, Singh PK, Singh J, Singh J (2005) Facile synthesis and fungicidal activity of novel 4, 4′-bis [2″-(5‴-substituted rhodanin-3‴-yl) thiazol-4″-yl] bibenzyls. http://nopr.niscpr.res.in/handle/123456789/9197

Smith J (2022) Enhanced phosphorus and nitrogen uptake by *Penicillium glabrum*. J Plant Nutr 25:150–165. https://doi.org/10.1002/bit.260351009

Sreevidya M, Gopalakrishnan S, Melø TM, Simic N, Bruheim P, Sharma M, Srinivas V, Alekhya G (2015) Biological control of *Botrytis cinerea* and plant growth promotion potential by *Penicillium citrinum* in chickpea (*Cicer arietinum* L.). Biocontrol Sci Tech 25:739–755. https://doi.org/10.1080/09583157.2015.1010483

Subramanian T (1982) Colorimetric determination of patulin produced by *Penicillium patulum*. JAOAC 65:5–7. https://doi.org/10.1093/jaoac/65.1.5

Suh H, Wilcox CS (1988) Chemistry of F1, F0-ATPase inhibitors. Stereoselective total syntheses of (+)-citreoviral and (-)-citreoviridin. J Am Chem Soc 110: 470–481. https://doi.org/10.1021/ja00210a026

Sunkar B, Kannoju B, Bhukya B (2020) Optimized production of xylanase by *Penicillium purpurogenum* and ultrasound impact on enzyme kinetics for the production of monomeric sugars from pre treated corn cobs. Front Microbio 11:772. https://doi.org/10.3389/fmicb.2020.00772

Takahashi A, Lucas EMF (2008) Occurrence and structural diversity of fungal metabolites with antibiotic activity. Quim Nova 31:1807–1813. https://doi.org/10.1590/S0100-40422008000700036

Thuerig B, Binder A, Boller T, Guyer U (2006) An aqueous extract of the dry mycelium of *Penicillium chrysogenum* induces resistance in several crops under controlled and field conditions. Eur J Plant Path 114(2):185–197. https://doi.org/10.1007/s10658-005-4512-6

Tunc S, Chollet E, Chalier P, Preziosi-Belloy L, Gontard N (2007) Combined effect of volatile antimicrobial agents on the growth of *Penicillium notatum*. Int J Food Microbiol 113:263–270. https://doi.org/10.1016/j.ijfoodmicro.2006.07.004

Vaishnav N, Singh MA, Adsul DP, Sandhu SK, Mathur A, Puri SK, Singhania RR (2018) *Penicillium*: the next emerging champion for cellulase production. Bioresour Technol Rep 2:131–140

Wakana D, Hosoe T, Itabashi T, Okada K, de Campos TGM, Yaguchi T, Fukushima K, Kawai KI (2006) New citrinin derivatives isolated from *Penicillium citrinum*. J Nat Med 60:279–284. https://doi.org/10.1007/s11418-006-0001-2

Wakelin SA, Warren RA, Harvey PR, Ryder MH (2004) Phosphate solubilization by Penicillium spp. closely associated with wheat roots. Biol Fertil Soils 40:36–43. https://doi.org/10.1007/s00374-004-0750-6

Xia GY, Wang LY, Xia H, Wu YZ, Wang YN, Lin PC, Lin S (2020) Three new polyketides from the endophytic fungus *Penicillium purpurogenum*. J Asian Nat Prod Res 22:233–240. https://doi.org/10.1080/10286020.2019.1699535

Yang XJ, Jiang D, Fu Y, Li G, Lin F (2008) Antifungal substances produced by *Penicillium oxalicum* strain PY-1 – potential antibiotics against plant pathogenic fungi. World J Microbiol Biotechnol 24:909–915. https://doi.org/10.1007/s11274-007-9626-x

Yang Y, Zhao H, Barrero RA, Zhang B, Sun G, Wilson IW, Xi F, Walker KD, Parks JW, Bruce R, Guo G (2014) Genome sequencing and analysis of the paclitaxel-producing endophytic fungus *Penicillium aurantiogriseum* NRRL 62431. BMC Genomics 15:1–14. https://doi.org/10.1186/1471-2164-15-69

Yedidia I, Shoresh M, Kerem Z, Benhamou N, Kapulnik Y, Chet I (2003) Concomitant induction of systemic resistance to *Pseudomonas syringae* pv. *lachrymans* in cucumber by *Trichoderma asperellum* (T-203) and accumulation of phytoalexins. Appl Environ Microbiol 69:7343–7353. https://doi.org/10.1128/AEM.69.12.7343-7353.2003

Yurieva OM, Dragovoz IV, Leonova NO, Biliavska LO, Syrchin SO, Kurchenko IM (2018) Biosynthesis of phytohormones by *Penicillium funiculosum* strains from different ecological niches. J Microbiol 80:44–55

Yuzha L, Jingchun L, Haoliang L, Youfang D, Yanling Y (2022) Screening of phosphorus-solubilizing fungi in the mangrove rhizosphere and their effects on the growth of Tung tree. J Agro-Environ Sci 41:950–958

Zhang D, Zhao L, Wang L, Fang X, Zhao J, Wang X, Li L, Liu H, Wei Y, You X, Cen S (2017) Griseofulvin derivative and indole alkaloids from *Penicillium griseofulvum* CPCC 400528. J Nat Prod 80:371–376. https://doi.org/10.1021/acs.jnatprod.6b00829

Zhao P, Xue Y, Li J, Li X, Zu X, Zhao Z, Quan C, Gao W, Feng S (2019) Non-lipopeptide fungi-derived peptide antibiotics developed since 2000. Biotechnol Lett 41:651–673. https://doi.org/10.1007/s10529-019-02677-3

Aspergillus Genus as a Prolific Source of Bioactive Metabolites for Agricultural Applications

Yunyun Lei, Nan Pu, and Huawei Zhang

Abstract *Aspergillus*, one of the most important filamentous fungi, is widely distributed in nature environments and has been shown to be a rich source of secondary metabolites (SMs) with diverse chemical structures and various biological properties, such as antimicrobial, anticancer, insecticidal, and so on. Using an extensive literature search, this chapter first provides a comprehensive overview of these genus-derived SMs with potential applications in agriculture, including insecticidal agents, fungicides, virucidal chemicals, herbicides, plant growth promoters and other bioactive natural products.

Graphical Abstract

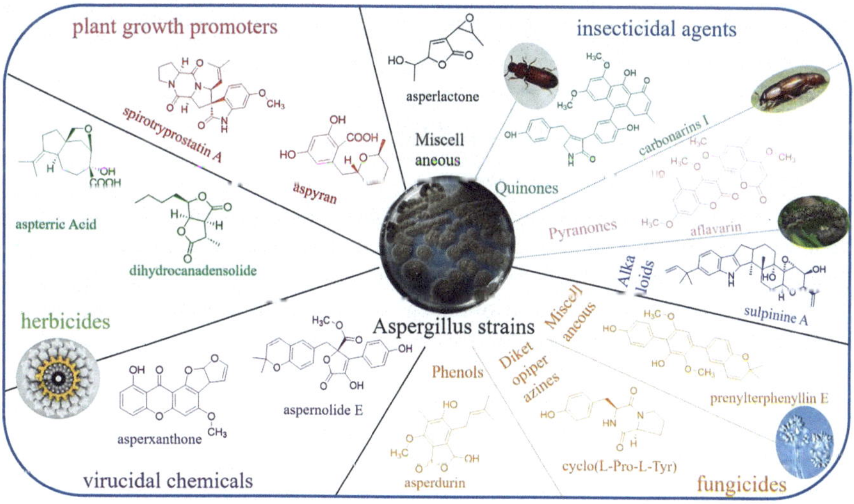

Y. Lei · N. Pu · H. Zhang (✉)
School of Pharmaceutical Sciences, Zhejiang University of Technology, Hangzhou, China
e-mail: hwzhang@zjut.edu.cn

J. Poveda et al. (eds.), *Fungal Metabolites for Agricultural Applications*, Fungal Biology, https://doi.org/10.1007/978-3-031-76587-2_10

Keywords *Aspergillus* · Secondary metabolite · Agricultural applications · Insecticidal effect · Herbicidal activity

1 Introduction

Pesticides are chemical substances used in agriculture to protect crops from insects, fungi, weeds and other pests, and consist of insecticides, antimicrobials, virucides, herbicides and plant growth promoters (Akashe et al. 2018). It was estimated that about 20–30% of agricultural products is lost annually due to insect pests and pathogenic diseases (Rahaman et al. 2018). In the past century, pesticide application had substantially helped humankind in improving crop production (Gilden et al. 2010). Therefore, these chemicals are indispensable in agricultural production nowadays. *Aspergillus*, one of the most commonly discovered fungi in nature, has been shown to an important source of secondary metabolites (SMs) with extraordinary chemical diversity and potent biological activities, including anti-cancer, anti-fungal, anti-bacterial, anti-viral and anti-insect activities, and so on (Chen et al. 2022; De Guzman et al. 1992; Krijgsheld et al. 2013; Sun et al. 2022; Zhang et al. 2020). By extensive literature search using DNP (Dictionary of Natural Products) and SciFinder databases, a total of **108** *Aspergillus*-derived SMs (**1**–**108**, Table 1) with potential of agricultural application had been comprehensively summarized and respectively introduced in this chapter according to their bioactivities.

2 *Aspergillus*-Derived Insecticidal Agents

Insects constitute about 60% of species on earth (Wielkopolan and Obrępalska-Stęplowska 2016), but some of them pose a major threat to agricultural crops, resulting in substantial losses. Among herbivorous insects, *Coleoptera* and *Lepidoptera* account for two-thirds of the known species and often cause significant damage to staple crops such as potato, corn, rice, and cereals.

2.1 Alkaloids

Indole-based compounds are commonly biosynthesized by *Aspergillus* spp. and display insecticidal function presenting intermediate compounds with paralyzing and lethal effects on insects in the kynurenine pathway (Ahuja et al. 2015; Cerstiaens et al. 2003; Kaushik et al. 2013). Compound **1**, a meroterpenoid indole isolated from *A. flavus* sclerotia through solid-substrate fermentation on corn kernels, demonstrated an anti-insecticidal effect against *Helicoverpa zea*, resulting in a 57.2% reduction in weight gain after one week at 200 ppm (Fig. 1) (Laakso et al. 1992).

Table 1 *Aspergillus*-derived secondary metabolites with agricultural applications

No.	Name	Strain	Structure type	Bioactivity
1	Aflatrem B	*A. flavus*	Alkaloid	Insecticidal
2	Aflavazole	*A. flavus*	Alkaloid	Insecticidal
3	Aflavinine:20β,25-dihydroxy	*A. flavus*	Alkaloid	Insecticidal
4	Aflavinine: 20β-hydroxy			
5	Aflavinine: 20β-hydroxy,10β,11α-dihydro11,12-didehydro			
6	Amauromine:2,3-diepimer	*A. ochraceus*	Alkaloid	Insecticidal
7	Secopenitrem B	*A. sulphureus*	Alkaloid	Insecticidal
8	Sulpinine A			
9	Sulpinine B			
10	Sulpinine C			
11	Emindole SC	*A. sclerotiicarbonarius* IBT 28362	Alkaloid	Insecticidal
12	14-hydroxypaspalinine	*A. nomius*	Alkaloid	Insecticidal
13	14-(N,N-dimethyl-l-valyloxy) paspalinine			
14	Tubingensin A	*A. tubingensis*	Alkaloid	Insecticidal
15	Tris(deacetyl)pyripyropene A: 3,7,19-Tri-Ac:Pyripyropene A	*A. fumigatus* FO-1289-2501	Alkaloid	Insecticidal
16	Tris(deacetyl)pyripyropene A: 19-Propanoyl,3,7-di-Ac (pyripyropene B)			
17	Tris(deacetyl)pyripyropene A: 7-Propanoyl, 3,19-di-Ac (pyripyropene C)			
18	Tris(deacetyl)pyripyropene A: 3-Propanoyl, 7,19-di-Ac (pyripyropene D)			
19	Tris(deacetyl)pyripyropene A: 7,19-Dideoxy,3-Ac (pyripyropene G)			
20	Tris(deacetyl)pyripyropene A: 3,17,19-Tripropanoyl Pyripyropene I (1f)			
21	Tris(deacetyl)pyripyropene A: 3,19-Dipropanoyl, 7-Ac Pyripyropene K			
22	16-keto aspergillimide	*Aspergillus* sp. IMI337664	Alkaloid	Insecticidal
23	Aspochracin	*A. ochraceus*	Alkaloid	Insecticidal

(continued)

Table 1 (continued)

No.	Name	Strain	Structure type	Bioactivity
24	Destruxin B	*Aspergillus* sp.	Alkaloid	Insecticidal
25	Sclerolizine	*A. sclerotiicarbonarius* IBT 28362	Alkaloid	Insecticidal
26	Sclerotiamide	*Aspergillus* sp.	Alkaloid	Insecticidal
27	Carbonarin E	*A. carbonarius* NRRL 369	Pyranone	Insecticidal
28	Carbonarin F	*A. carbonarius* NRRL 369	Pyranones	Insecticidal
29	Carbonarin H			
30	Carbonarin J	*A. sclerotiicarbonarius* IBT 28362	Pyranone	Insecticidal
31	Aflavarin	*A. flavus* NRRL 13462	Pyranone	Insecticidal
32	Aspyrone	*A. melleus*	Pyranone	Insecticidal
33	Chrodrimanin B	*Aspergillus* sp.	Pyranone	Insecticidal
34	Territrem A	*A. terreus* 23–1	Pyranone	Insecticidal
35	Arenarin B	*A. arenarius* NRRL 5012	Quinones	Insecticidal
36	Arenarin C			
37	Carbonarin A	*A. carbonarius* NRRL 369	Quinones	Insecticidal
38	Carbonarin B			
39	Carbonarin C	*A. carbonarius* NRRL 369	Quinones	
40	Carbonarin D	*A. carbonarius* NRRL369	Quinones	
41	Carbonarin I	*A. sclerotiicarbonarius* IBT 28362	Quinones	
42	Arenarin A	*A. arenarius* NRRL 5012	Miscellaneous	Insecticidal
43	Carbonarin G	*A. carbonarius* NRRL 369	Miscellaneous	Insecticidal
44	Asperlactone	*A. ochraceus*	Miscellaneous	Insecticidal
45	Asperdurin	*A. donovani* Tu 679	Phenol	Antimicrobial
46	Chlorflavonin	*A. candidus* BRL 274	Phenol	Antimicrobial
47	Asperentin	*Aspergillus* sp. F00785	Phenol	Antimicrobial
48	Clavatol	*A. clavatonanicus*	Phenol	Antimicrobial

(continued)

Table 1 (continued)

No.	Name	Strain	Structure type	Bioactivity
49	Aspewentin D	*A. wentii* SD-310	Phenols	Antimicrobial
50	Aspewentin E			
51	Aspewentin F			
52	Aspewentin G			
53	Aspewentin H			
54	Aspewentin A			
55	7,11-epoxy-1-hydroxy-1,3,5-bisabolatrien-15-oic acid	*A. flavipes* 297	Phenol	Antimicrobial
56	Asperazine	*Aspergillus niger*	Alkaloid	Antimicrobial
57	Fumitremorgin C	*A. fumigatus* M580	Alkaloid	Antimicrobial
58	Cyclo(L-Pro-L-Tyr)			
59	Diorcinol L	*A. tenseensis*	Miscellaneous	Antimicrobial
60	(R)-diorcinol B	*A. tenseensis*	Miscellaneous	Antimicrobial
61	(S)-diorcinol B			
62	Prenylterphenyllin B	*Aspergillus* sp. YXf3	Miscellaneous	Antimicrobial
63	Prenylterphenyllin E			
64	2'-O-methylprenylterphenyllin			
65	Spiculisporic acid F	*A. seaurchin* HDf2	Miscellaneous	Antimicrobial
66	Spiculisporic acid G			
67	Kojic acid	*Aspergillu* sp.	Miscellaneous	Antimicrobial
68	Aspochalasins F	*Aspergillus* sp. FO-4282	Miscellaneous	Antimicrobial
69	Aspochalasins G			
70	Versicoloids A	*A. versicolor* SCSIO 05879	Miscellaneous	Antimicrobial
71	Versicoloids B			
72	Cottoquinazoline A			
73	Aspewentin I	*A. wentii* SD-310	Miscellaneous	Antimicrobial
74	Aspewentin J	*A. wentii* SD-310	Miscellaneous	
75	Aspewentin K	*A. wentii* SD-310	Miscellaneous	Antimicrobial
76	Aspewentin L	*A. wentii* SD-310	Miscellaneous	Antimicrobial
77	Aspewentin M	*A. wentii* SD-310	Miscellaneous	Antimicrobial
78	Wentinoid A	*A. wentii* SD-310	Miscellaneous	Antimicrobial
79	Wentinoid B	*A. wentii* SD-310	Miscellaneous	Antimicrobial
80	Wentinoid C	*A. wentii* SD-310	Miscellaneous	
81	Wentinoid D	*A. wentii* SD-310	Miscellaneous	
82	Wentinoid E	*A. wentii* SD-310	Miscellaneous	
83	Wentinoid F	*A. wentii* SD-310	Miscellaneous	

(continued)

Table 1 (continued)

No.	Name	Strain	Structure type	Bioactivity
84	Butyrolactone IV	*A. versicolor*	Butyrolactones	Antiviral
85	Aspernolide B			
86	Aspernolide A			
87	Aspernolide E			
88	Butyrolactone V			
89	Versicolactone A			
90	Asperphenol C			
91	Aspernolide C			
92	Versicolactone B			
93	Versicolactone C			
94	Versicolactone D			
95	Versicolactone E			
96	Aspernolide D			
97	Versicolactone F	*A. versicolor*	Butyrolactone	Antiviral
98	Terrephenol A	*A. terreus*	Phenyl butyrolactones	Antiviral
99	Terrephenol B			
100	Terrephenol C			
101	Butyrolactones III			
102	Asperxanthone	*Aspergillus* sp.	Phenolic polyketides	Antiviral
103	Asperbiphenyl			
104	Aspterric acid	*A. terreus*	Carotane-typesesquiterpene	Herbicidal
105	Dihydrocanadensolide	*Aspergillus* sp.	Furanone	Herbicidal
106	6,14,15-trihydroxy-8-drimen-12,11-olide: 6β-form, 6-benzoyl, 14,15-di-ac	*Aspergillus* sp.	Penylated diketopiperazine alkaloid	
107	Aspyran	*Aspergillus* sp.	Phenolic acid	Plant growth promoters
108	Spirotryprostatin A	*A. fumigatus*	Alkaloid	

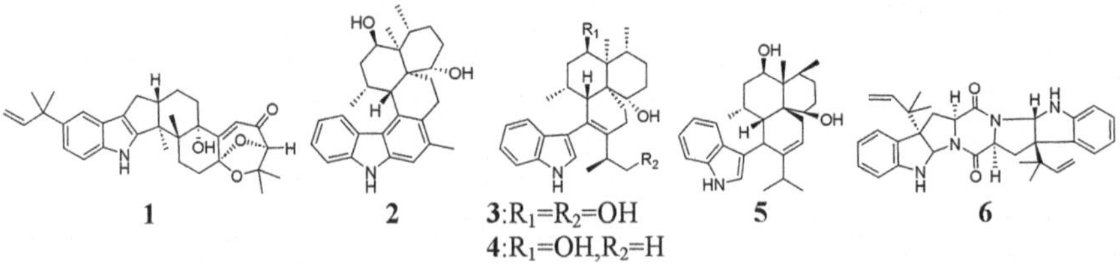

Fig. 1 *Aspergillus*-derived insecticidal agents (**1**–**6**)

Aflavazole (**2**), a novel carbazole metabolite isolated from *A. flavus* sclerotia, showed insecticidal activity against *Carpophilus hemipterus* at 100 ppm (TePaske et al. 1990). Three aflavinine derivatives (**3**–**5**) from *A. flavus* deterred feeding by *C. hemipterus*, leading to mortality (Gloer et al. 1988). Amauromine-2,3-diepimer

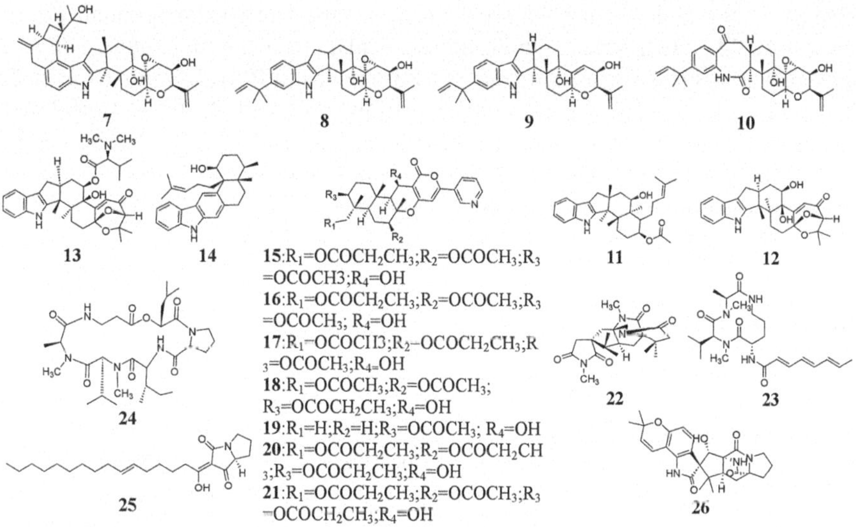

Fig. 2 *Aspergillus*-derived insecticidal agents (**7–26**)

(**6**), one of diketopiperazine alkaloids from *A. ochraceus* sclerotia, reduced *H. zea* weight gain by 30% at 370 ppm (Florecita et al. 1992). Four new anti-insect indole analogs (**7–10**, Fig. 2) were detected in the sclerotia of *A. sulphureus* using solid substrate fermentation on corn kernels and compound **8** exhibited the highest level of effectiveness against *H. zea* with a 10% mortality rate and an impressive 96.0% reduction in weight gain at 100 ppm (Laakso et al. 1991).

Emindole SC (**11**) produced by strain *A. sclerotiicarbonarius* IBT 28362 was shown to have insecticidal property against *Drosophila* larvae with a LD_{50} value of 1.83 ± 0.08 μM (Petersen et al. 2015). Two new paspalinine derivatives (**12–13**) with anti-insectan activity from the sclerotia of *A. nomius* possessed the capacity to achieve a substantial 90% reduction in *H. zea* weight gain at 100 ppm (Staub et al. 1993). Tubingensin A (**14**), a new carbazole alkaloid biogenetically related to aflavinines, was characterized from the sclerotia of *A. tubingensis* and displayed moderate activity against *H. zea* with an 11% mortality at 125 ppm (TePaske et al. 1989). Chemical analysis of *A. fumigatus* FO-1289-2501 afforded seven polyoxygenated metabolites (**15–21**), which exhibited insecticidal properties against a range of pests, including *H. zea, Diamondback moth, Tenebrio molitor, Aphids, Myzus persicae*, and so on (Goto et al. 2018; Horikoshi et al. 2017, 2018; Tomoda et al. 1996). Particularly, compound **15** demonstrated a broad spectrum of insecticidal properties and **20** had better insecticidal activity against *M. persicae* with a LC_{90} value of 0.043 ppm. 16-Keto aspergillimide (**22**), a novel anthelmintic compound, has been isolated from the strain *Aspergillus* sp. IMI 337664 and tested against *Trichostrongylus colubriformis* (Banks et al. 1997). Aspochracin (**23**), a new insecticidal substance isolated from the culture filtrate of *A. ochraceus*, exhibited insecticidal activity against fall webworm and last instar silkworm larvae, resulting in mortality rates of

33% and 100% at doses of 85 γ/g and 170 γ/g, respectively (Myokei et al. 1969). Compound **24** demonstrated significant insecticidal effects on *Anastrepha obliqua*, achieving an impressive mortality rate of 81.3% (Kodaira 1961; Lozano-Tovar et al. 2023). Sclerolizine (**25**) was also obtained from strain IBT 28362 and displayed insecticidal effect on *D. melanogaster* larvae at 1.83 ± 0.08 μM (Petersen et al. 2015). The prenylated alkaloid (**26**) demonstrated remarkable larvicidal activity against *H. zea* with 83.2% of mortality at 500 ppm (Zhang et al. 2019).

2.2 Pyranones

Carbonarins E, F, and H (**27–29**, Fig. 3) were isolated from *A. carbonarius* NRRL 369. Compounds **27** and **29** reduced the growth rate of *H. zea* by 26% at 100 ppm and *C. hemipterus* by 20% at 45 ppm, respectively. Carbonarin J (30), also from strain IBT 28362, showed insecticidal efficacy against *D. melanogaster* larvae with an LD_{50} of 1.83 ± 0.08 μM (Alfatafta et al. 1997; Petersen et al. 2015). Aflavarin (**31**), extracted from *A. flavus* NRRL 13462 using solid substrate fermentation on corn kernels, showed anti-insect activity against *C. hemipterus* adults and larvae, reducing feeding by 66% and 53% at 100 ppm, respectively (Tepaske et al. 1992). Aspyrone (**32**), a nematicidal compound produced by *A. melleus*, exhibited strong nematicidal effects against *Pratylenchus penetrans* with 39.0% and 80.8% reduction at 100 mg/L and 300 mg/L, respectively (Kimura et al. 1996). Chrodrimanin B (**33**) were isolated from a marine-derived strain *Aspergillus* collected in the South China Sea and found to have potent insecticidal activity against the third instar larva of the silkworm with a LD_{50} value of 10 μg/g of diet (Hayashi et al. 2012; Wei et al. 2011). Territrem A (**34**), extracted from *A. terreus* 23–1, resulted in a 96.0% reduction in *H. zea* weight gain at 250 ppm (Dowd et al. 1992).

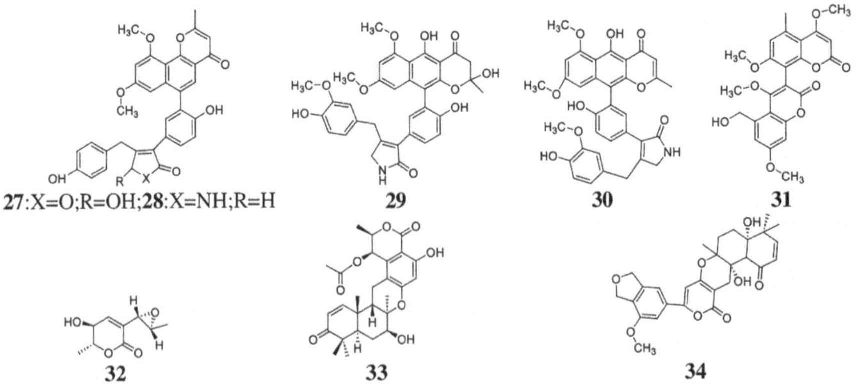

Fig. 3 *Aspergillus*-derived insecticidal agents (**27–34**)

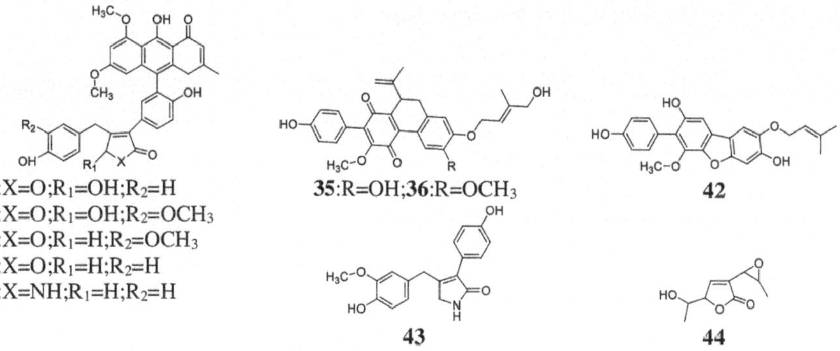

Fig. 4 *Aspergillus*-derived insecticidal agents (**35–44**)

2.3 Quinones

Two terphenyl-type metabolites (**35** and **36**, Fig. 4) were purified from the sclerotia of *A. arenarius* NRRL 5012 using solid substrate fermentation on corn kernels and showed a modest impact against the *C. hemipterus* with 20% and 13% feeding reduction at 100 ppm, respectively (Oh et al. 1998). Carbonarins A-D (**37–40**) and I (**41**) displayed various insecticidal property (Petersen et al. 2015), in which compounds **37** and **38** respectively caused 31%, 33% feeding reduction against *C. hemipterus* adults at 100 ppm (Alfatafta et al. 1997) and **41** showed inhibitory effect on *D. melanogaster* larvae with a LD_{50} value of 1.83 ± 0.08 μM (Petersen et al. 2015).

2.4 Miscellaneous

Arenarin A (**42**), also produced by strain NRRL 5012, had a 13% reduction in feeding activity against *C. hemipterus* adults and larvae at 100 ppm (Oh et al. 1998). Compound **43**, isolated from strain NRRL 369, caused a 33% reduction in *C. hemipterus* feeding at 105 ppm (Alfatafta et al. 1997). Asperlactone (**44**), a furanone substance produced by *A. ochraceus*, exhibited a remarkable growth-enhancing effect on *Tribolium castaneum.* The values for insect growth regulators in *T. castaneum* were respectively determined as 2.9, 2.6, and 2.8 at 10, 1, and 0.1 μg/μL (Balcells et al. 1995).

3 *Aspergillus*-Derived Fungicides

Fusarium and *Colletotrichum* spp. are two of the most common plant pathogens, usually causing anthracnose diseases and great losses to cucurbit crop production (Costa et al. 2019). A growing number of evidence indicates that several *Aspergillus* species including *A. niger, A. flavus,* and *A. terreus* have hyperparasitism on *F. oxysporum* f. sp. *melonis* (Pfliegler et al. 2020).

3.1 Phenols

A new type of bentiloxin (**45**, Fig. 5) produced by strain *A. donovani* Tu 679 possessed significant antifungal activity against *Botrytis cinerea* and *Rhizoctonia solani* with MIC values of 30 and 10 pg/mL, respectively (Mühlenfeld and Achenbach 1988). Compound **46**, a chloroflavone isolated from *A. candidus* BRL 274, exhibited strong inhibitory activity (MIC <1 μg/mL) against *A. amstelodami*, *A. fumigatus*, and *A. ochraceous* (Richards et al. 1969).

A new asperentin-type compound **47** was isolated from a halotolerant *Aspergillus* sp. strain F00785 endotrophic on marine alga and displayed strong inhibitory activity against *Colletotrichum* spp. and *C. sac* at 5 mg/mL (Tang et al. 2014). *A. clavatonanicus* isolated from southern yew was found to produce clavatol (**48**), which had mild inhibitory effects on several plant pathogenic fungi, including *B. cinerea, Didymella bryoniae*, *F. oxysporum* f. sp. *cucuminum*, *R. solani* and *Pyium ultimum* (Zhang et al. 2008). Five new phenolic derivatives, aspewentins D–H (**49–53**) and aspewentin A (**54**), were obtained from the marine-derived strain *A. wentii* SD-310. Compounds 49 and 53 had MIC values of 2 μg/mL and 4 μg/mL against *F. graminearum*, respectively (Li et al. 2016). Compound **55** was purified from the marine strain *A. flavipes* 297 and exhibited noticeable anti-fungal activity against three plant pathogens *B. cinerea*, *F. oxysporum* and *Valsa mali* with the same MIC value of 8 μg/mL (Chen et al. 2021; Guo et al. 2018).

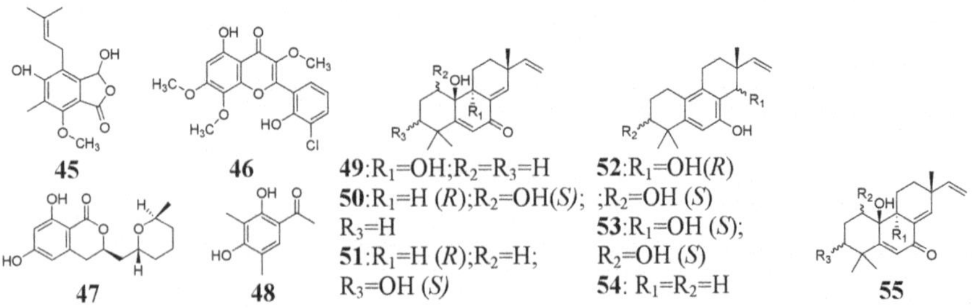

Fig. 5 *Aspergillus*-derived fungicides (**45–55**)

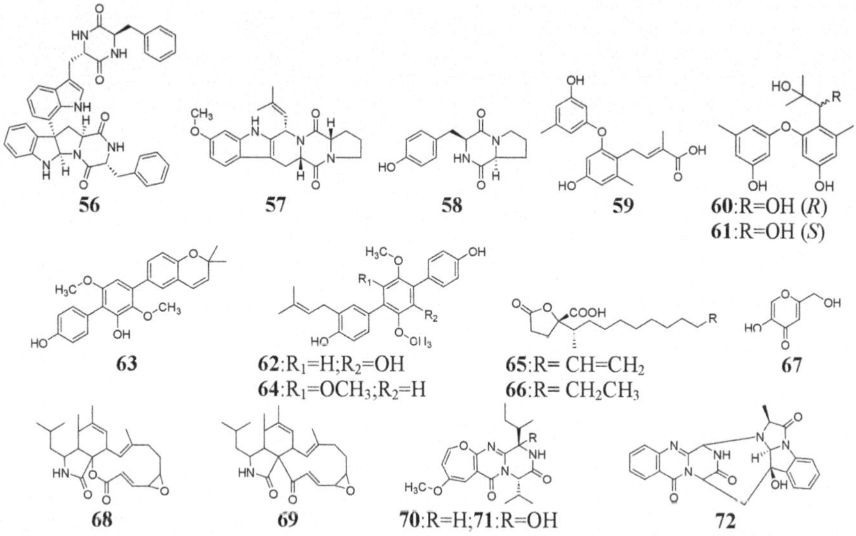

Fig. 6 *Aspergillus*-derived fungicides (**56–72**)

3.2 Diketopiperazines

Asperazine (**56**, Fig. 6), a dimeric diketopiperazine consists of a tryptophan and a phenylalanine, displayed a MIC value of 60 mg/mL to inhibit *Fol spores* (Zulqarnain et al. 2020). Compounds **57** and **58** were produced by the marine strain *A. fumigatus* M580, and the later substance showed anti-fungal effect on *Pseudomonas syringae* pv. *syringae* with a MIC value of 15 μg/mL (Barrow and Sun 1994; Castaldi et al. 2022; Kumar et al. 2012; Liao et al. 2022).

3.3 Miscellaneous

Two new diphenyl ethers, diorcinol L (**59**) and (*R*)-diorcinol B (**60**) together with (*S*)-diorcinol B (**61**), were separated from the seaweed-derived endophytic fungus *A. tenseensis* and displayed certain antibacterial activity against *Cochliobolus heterostrophus* and *Gaeumannomyces graminis*. Compound **60** showed considerable inhibitory effect against *C. heterostrophus* with MIC values of 4 μg/mL. (Li et al. 2018). Compounds **62–64** were new parabiphenyls produced by strain *Aspergillus* sp. YXf3 and showed certain antibacterial activity against rice blast fungus *E. amylofora* with the same MIC value of 10 μg/mL (Yan et al. 2017). Two new antibiotics, spiculisporic acids F (**65**) and G (**66**), were obtained from the fermentation broth of strain *A. seaurchin* HDf2 and exhibited certain antibacterial effect on *Ralstonia solanacearum* (Wang et al. 2015). Kojic acid (**67**), one natural pyrone produced by

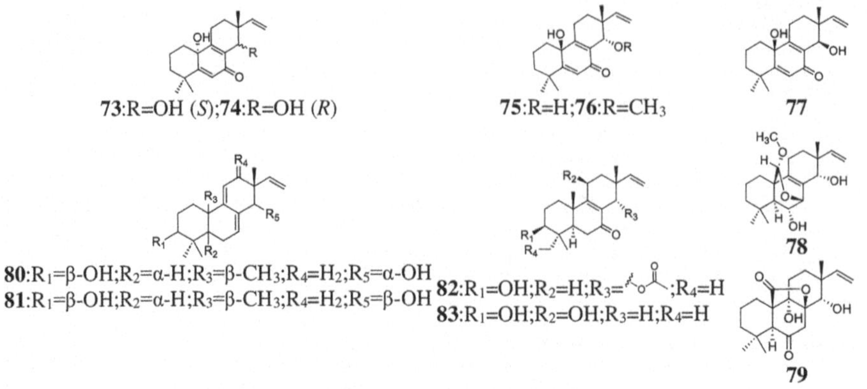

Fig. 7 *Aspergillus*-derived fungicides (**73**–**83**)

Aspergillus and *Penicillium*, usually acted as a chemical sensitizer to enhance the activity of antifungal agents against pathogenic *Aspergillus* and yeast strains (Kim et al. 2012). Two new aspochalasins, aspochalasins F (**68**) and G (**69**), were derived from soil strain *Aspergillus* sp. FO-4282, and compound **69** demonstrated strong *in vitro* inhibitory effect on *Pyricularia oryzae* KF180 (Fang et al. 1997).

Versicoids A (**70**) and B (**71**) together with cottoquinazoline A (**72**) were novel oxepine-containing diketopiperazine alkaloids isolated from the culture extract of marine strain *A. versicolor* SCSIO 05879, and demonstrated potent anti-fungal activities against phytopathogens *C. acutatum, Magnaporthe oryzae* and *F. oxysporum* with the same MIC value of 1.6 μg/mL (Wang et al. 2016). Four new unusual 20-nor-isopimarane diterpenoid epimers aspewentin I-L (**73–76**, Fig. 7), a new methylated derivative aspewentin M (**77**) and six new isohymarane-type diterpene derivatives wentinoids A-F (**78–83**) were isolated and characterized from the marine fungus *A. wentii* SD-310. Bioassay results suggested that compound **76** had strong inhibitory activity against *F. graminearum* head blight with a MIC value of 4 μg/mL and **78** showed remarkable antimicrobial effect on *F. oxysporum* f. sp. *lycopersici, F. graminearum* and *Botryosphaeria dothidea* with MIC values of 8, 4, 1 and 4 mg/mL, respectively (Li et al. 2017) (73–83, Fig. 7).

4 *Aspergillus*-Derived Virucidal Chemicals

Tobacco mosaic virus was one of the relatively common viruses in agriculture. It can cause poor plant growth and even affect overall yield and quality, having a significant impact on crops. Therefore, it is necessary to take preventive measures to control its spread (Abdelkhalek et al. 2019). Five new butyrolactones (**84–88**, Fig. 8) together with nine known butyrolactones (**89–97**) were obtained and characterized from an endophytic fungus *A. versicolor* (Ye et al. 2014a). Among these

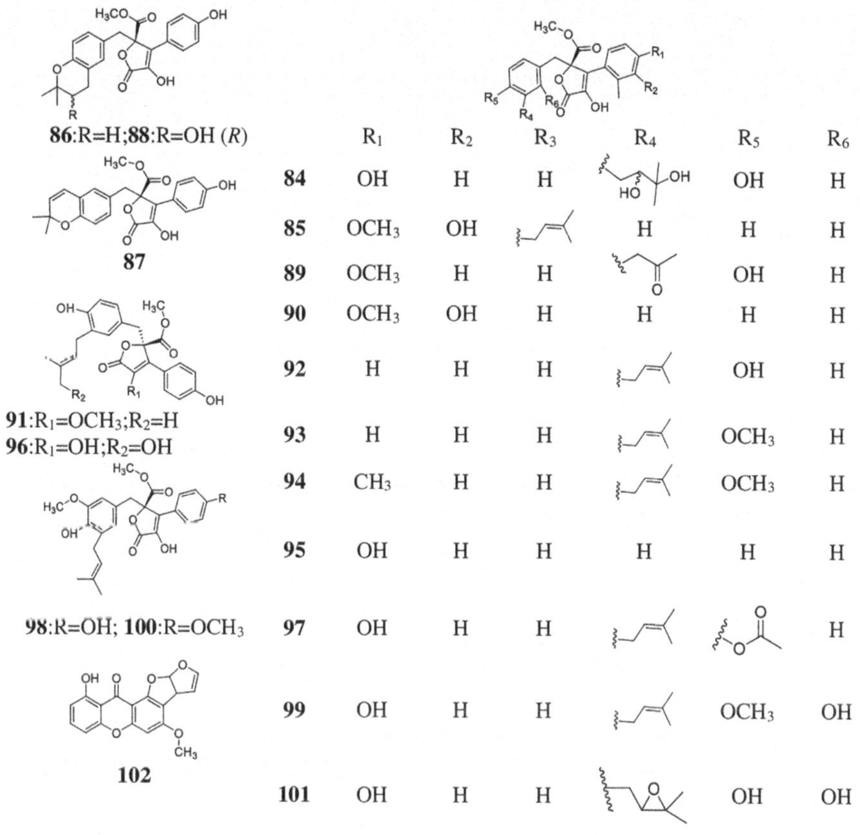

Fig. 8 *Aspergillus*-derived virucidal chemicals (**84**–**102**)

compounds, **89** was shown to have the highest virus inhibition rate at 46.4% (Li et al. 2015; Zhou et al. 2015a, b, 2016). Four new phenyl butyrolactones (**98**–**101**) were isolated from the fermentation products of *A. terreus*, and compound **98** displayed the highest virus inhibition rate at 35.2% (Ye et al. 2014b). Two new phenolic polyketides, asperxanthone (**102**) and asperbiphene (**103**, Fig. 9), were produced by an unidentified *Aspergillus* strain from seawater and displayed inhibitory activities against TMV with rates of 62.9%, 35.5%, respectively (Wu et al. 2009).

5 *Aspergillus*-Derived Herbicides

Arabidopsis thaliana is an annual weed belonging to Brassicaceae family. (Gepstein and Horwitz 1995). Compound **104** (Fig. 9), a carotane-type sesquiterpene from *A. terreus*, was shown to possess the potent ability to hinder pollen development of *A. thaliana* at 38 μM (Shimada et al. 2002).

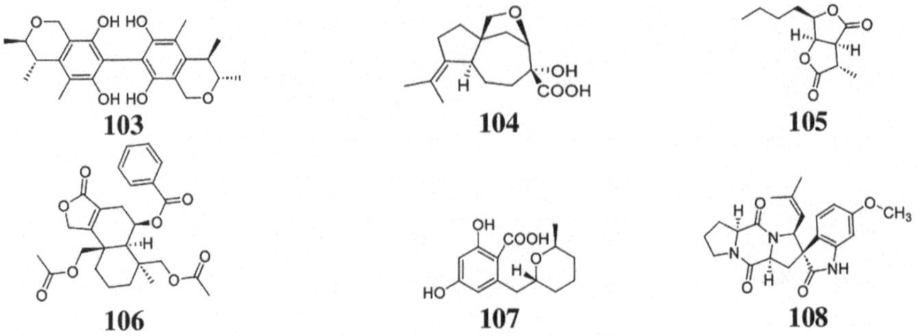

Fig. 9 *Aspergillus*-derived virucidal chemical (**103**), herbicides (**104**–**106**) and plant growth promoters (**107** and **108**)

Chemical investigation of an unknown *Aspergillus* strain resulted in the isolation of a furanone derivative (**105**) and a prenylated diketopiperazine alkaloid (**106**), in which compound **106** had significant inhibitory activity against the seed germination of *A. thaliana* with a MIC value of 200 μg/mL (Ma et al. 2022) (Fig. 9).

6 *Aspergillus*-Derived Plant Growth Promoters

Agricultural biostimulants include diverse substances and microorganisms that enhance plant growth (Calvo et al. 2014). Aspyran (**107**, Fig. 9) was a phenolic acid derivative and displayed activity for enhancing the root lettuce elongation in the concentration range of 0.01–1 mM and promoted the root of rice seedling growth to 132% at a concentration of 1 mM compared with 61% of control at 1 mM (Kimura et al. 2012). Spirotryprostatin A (**108**) from a marine-derived strain of *A. fumigatus*, it provided a stimulating effect on the growth of soybean seedlings at 10^{-16}, 10^{-13}, and 10^{-10} M, the root length reached 108%, 109%, and 114% respectively (Anisimov et al. 2013).

7 Conclusions

In summary, secondary metabolites of *Aspergillus* strains are one of important reservoirs of natural products with insecticidal, herbicidal and fungicidal properties and have potential application in agriculture, such as the insecticide sulpinine A (**8**), the antiviral agent asperxanthone (**102**), the antibacterial chemical chlorflavonin (**46**), the herbicide aspterric acid (**104**) and the plant growth promoter spirotryprostatin A (**108**). Pesticides as toxic substances have played a tremendous role in controlling weeds, insect infestation and plant diseases. However, the resistance of pests, weeds and phytopathogens in agriculture continues to rise (Tabashnik et al.

2013). It is urgent for global scientists to develop more effective and environmentally friendly pesticides. Therefore, more efforts should be made to develop novel strategies for efficient discovery of new *Aspergillus*-derived SMs, such as OSMAC (one strain many compounds) (Pan et al. 2019), genome mining (Romsdahl and Wang 2019), molecular networking (Nothias et al. 2020), heterogeneous expression of SM biosynthetic gene cluster (Wösten et al. 2013).

References

Abdelkhalek A, Ismail IA, Dessoky ES, El-Hallous EI, Hafez E (2019) A tomato kinesin-like protein is associated with tobacco mosaic virus infection. Biotechnol Biotec Eq 33:1424–1433

Ahuja I, Van Dam NM, Winge P, Trælnes M, Heydarova A, Rohloff J, Langaas M, Bones AM (2015) Plant defence responses in oilseed rape mineless plants after attack by the cabbage moth *Mamestra brassicae*. J Exp Bot 66:579–592

Akashe MM, Pawade UV, Nikam AV (2018) Classification of pesticides: a review. Int J Res Ayurveda Pharm 9:144–150

Ali Alfatafta A, Dowd PF, Wicklow DT 1997. Carbonarin antiinsectan metabolites. US Patent No. 5672621

Anisimov MM, Chaikina EL, Afiyatullov SS, Vashchenko AP (2013) Influence of alkaloids from the marine-derived strain of *Aspergillus fumigatus* on the root growth in soybean (*Glycine max* (L.) Merr.) seedlings. Russ Agr Sci 39:102–104

Balcells M, Canela R, Coll J, Sanchís V, Torres M (1995) Effect of fungal metabolites and some derivatives against Tribolium castaneum (Herbst) and Nezara viridula (L.). J Pestic Sci 45:319–323

Banks RM, Blanchflower SE, Everett JR, Manger BR, Reading C (1997) Novel anthelmintic metabolites from an Aspergillus species; the Aspergillimides. J Antibiot 50:840–846

Barrow CJ, Sun HH (1994) Spiroquinazoline, a novel substance p inhibitor with a new carbon skeleton, isolated from *Aspergillus flavipes*. J Nat Prod 57:471–476

Calvo P, Nelson L, Kloepper JW (2014) Agricultural uses of plant biostimulants. Plant Soil 383:3–41

Castaldi S, Cimmino A, Masi M, Evidente A (2022) Bacterial lipodepsipeptides and some of their derivatives and cyclic dipeptides as potential agents for biocontrol of pathogenic bacteria and fungi of agrarian plants. J Agri Food Chem 70:4591–4598

Cerstiaens A, Huybrechts J, Kotanen S, Lebeau I, Meylaers K, De Loof A, Schoofs L (2003) Neurotoxic and neurobehavioral effects of kynurenines in adult insects. Biochem Biophys Res Commun 312:1171–1177

Chen Y, Zhu HY, Xu LC, Wang SP, Liu S, Liu GD, Luo WH, Cao GY, Zhang ZX (2021) Antimicrobial and cytotoxic phenolic bisabolane sesquiterpenoids from the fungus *Aspergillus flavipes* 297. Fitoterapia 155:105038

Chen YH, Chen PH, Yang CH, Wu XN, Luo LZ, Zhang WY (2022) Comprehensive study on the chemical profile and anti-tumor activity of secondary metabolites produced by *Aspergillus Niger*. Curr Pharm Anal 18:141–160

Costa JFO, Kamei SH, Silva JRA, Miranda ARGDS, Netto MB, Da Silva SJC, Correia KC, De Andrade Lima GS, Assunção IP (2019) Species diversity of colletotrichum infecting annona spp. in Brazil. Eur J Plant Pathol 153:1119–1130

De Guzman FS, Gloer JB, Wicklow DT, Dowd PF (1992) New diketopiperazine metabolites from the sclerotia of *Aspergillus ochraceus*. J Nat Prod 55:931–939

Dowd PF, Peng FC, Chen JW, Ling KH (1992) Toxicity and anticholinesterase activity of the fungal metabolites territrems to the corn earworm, Helicoverpa zea. Entomol Exp Appl 65:57–64

Fang F, Ui H, Shiomi K, Masuma R, Yamaguchi Y, Zhang Y, Zhang CG, Zhang XW, Tanaka Y, Omura S (1997) Two new components of the aspochalasins produced by *Aspergillus* sp. J Antibiot 50:919–925

Florecita S, Patrick F, James B, Donald T (1992) N-methylepiamauromine, epiamauromine and cycloechinulin antiinsectan metabolites. World Intellectual Property Organization (WO 93/01718)

Gepstein S, Horwitz BA (1995) The impact of arabidopsis research on plant biotechnology. Biotechnol Adv 13:403–414

Gilden RC, Huffling K, Sattler B (2010) Pesticides and health risks. Jognn J Obst Gyn Neo 39:103–110

Gloer JB, TePaske MR, Sima JS, Wicklow DT, Dowd PF (1988) Antiinsectan aflavinine derivatives from the sclerotia of Aspergillus flavus. J Org Chem 53:5457–5460

Goto K, Horikoshi R, Mitomi M, Oyama K, Hirose T, Sunazuka T, Ōmura S (2018) Synthesis and insecticidal efficacy of pyripyropene derivatives focusing on the C-1, C-7, and C-11 positions' substituent groups. J Antibiot 71:785–797

Guo ZY, Tan MH, Liu CX, Lv MM, Deng ZS, Cao F, Zou K, Proksch P (2018) Aspergoterpenins A–D: four new antimicrobial bisabolane sesquiterpenoid derivatives from an endophytic fungus *Aspergillus versicolor*. Molecules 23:1291

Hayashi H, Oka Y, Kai K, Akiyama K (2012) A new meroterpenoid, Chrodrimanin C, from YO-2 of Talaromyces sp. Biosci Biotech Bioch 76:745–748

Horikoshi R, Goto K, Mitomi M, Oyama K, Sunazuka T, Ōmura S (2017) Identification of pyripyropene A as a promising insecticidal compound in a microbial metabolite screening. J Antibiot 70:272–276

Horikoshi R, Goto K, Mitomi M, Oyama K, Sunazuka T, Ōmura S (2018) Insecticidal properties of pyripyropene A, a microbial secondary metabolite, against agricultural pests. J Pestic Sci 43:266–271

Kaushik N, Kaushik N, Attri P, Kumar N, Kim C, Verma A, Choi E (2013) Biomedical importance of indoles. Molecules 18:6620–6662

Kim JH, Chang PK, Chan KL, Faria N, Mahoney NCG, Kim Y, Martins N, Campbell B, Kim YK, De Martins ML, Campbell BC (2012) Enhancement of commercial antifungal agents by kojic acid. Int J Mol Sci 13:13867–13880

Kimura Y, Nakahara S, Fujioka S (1996) Aspyrone, a nematicidal compound isolated from the fungus, Aspergillus melleus. Biosci Biotech Bioch 60:1375–1376

Kimura Y, Shimomura N, Tanigawa F, Fujioka S, Shimada A (2012) Plant growth activities of aspyran, asperentin, and its analogues produced by the fungus *Aspergillus* sp. Z Naturforsch C 67:587–593

Kodaira Y (1961) Toxic substances to insects, produced by Aspergillus ochraceus and oopsra destructor. Agric Biol Chem 25:261–262

Krijgsheld P, Bleichrodt R, Van Veluw GJ, Wang F, Müller WH, Dijksterhuis J, Wösten HAB (2013) Development in *Aspergillus*. Stud Mycol 74:1–29

Laakso JA, Gloer JB, Wicklow DT, Dowd PF (1991) Sulpinines A-C and secopenitrem B: new anti-insectan metabolites from the sclerotia of Aspergillus sulphureus. J Org Chem 57:2066–2071

Laakso JA, TePaske MR, Dowd PF (1992) Sulpinine, secopemtrem B and aflatrem B anti-insectan metabolites. US Patent No. 5130326

Li Y, Zhou K, Wang Y, Dong W, Ji BK, Du G, Yang HY, Hu Q, Ye Y (2015) A new butyrolactone from the fermentation products of endophytic fungus *Aspergillus versicolor*. Asian J Chem 27:3529–3531

Li XD, Li X, Li XM, Xu GM, Liu Y, Wang BG (2016) Aspewentins D–H, 20-Nor-isopimarane derivatives from the deep sea sediment-derived fungus *Aspergillus wentii* SD-310. J Nat Prod 79:1347–1353

Li X, Li XD, Li XM, Xu GM, Liu Y, Wang BG (2017) Wentinoids A–F, six new isopimarane diterpenoids from *Aspergillus wentii* SD-310, a deep-sea sediment derived fungus. RSC Adv 7:4387–4394

Li XD, Li X, Li XM, Xu GM, Liu Y, Wang BG (2018) 20-Nor-isopimarane epimers produced by *Aspergillus wentii* SD-310, a fungal strain obtained from deep sea sediment. Mar Drugs 16:440

Liao Y, Hong X, Yang J, Qin JJ, Zhang B, Lin J, Shao Z, Wang W (2022) Structure elucidation of a novel cyclic tripeptide from the marine-derived fungus *Aspergillus ochraceopetaliformis* DSW-2. Nat Prod Res 36:3572–3578

Lozano-Tovar MD, Ballestas Álvarez KL, Sandoval-Lozano LA, Palma Mendez GM, Barrera-Cubillos GP (2023) Study on the insecticidal activity of entomopathogenic fungi for the control of the fruit fly (anastrepha obliqua), the main pest in mango crop in Colombia. Arch Microbiol 205:83

Ma JJ, Lu CH, Tang YJ, Shen YM (2022) Phytotoxic metabolites isolated from *Aspergillus* sp. an endophytic fungus of crassula arborescens. Molecules 27:7710

Mühlenfeld A, Achenbach H (1988) Asperdurin, ein neues antimykotisches phthalid aus *Aspergillus duricaulis*. Arch Pharm 321:803–805

Myokei R, Sakurai A, Chang CF, Kodaira Y, Takahashi N, Tamura S (1969) Aspochracin, a new insecticidal metabolite of Aspergillus ochraceus: Part I. Isolation, structure and biological activities Part II. Synthesis of hexahydroaspochracin. Agric Biol Chem 33:1491–1500

Nishanth Kumar S, Mohandas C, Siji JV, Rajasekharan KN, Nambisan B (2012) Identification of antimicrobial compound, diketopiperazines, from a bacillus sp. N strain associated with a rhabditid entomopathogenic nematode against major plant pathogenic fungi. J Appl Microbiol 113:914–924

Nothias LF, Petras D, Schmid R, Dührkop K, Rainer J, Sarvepalli A, Protsyuk I, Ernst M, Tsugawa H, Fleischauer M, Aicheler F, Aksenov A, Alka O, Allard PM… Dorrestein PC. (2020) Feature-based molecular networking in the GNPS analysis environment. Nat Methods 17:905–908

Oh H, Gloer JB, Wicklow DT, Dowd PF (1998) Arenarins A–C: new cytotoxic fungal metabolites from the sclerotia of *Aspergillus arenarius*. J Nat Prod 61:702–705

Pan R, Bai XL, Chen JW, Zhang HW, Wang H (2019) Exploring structural diversity of microbe secondary metabolites using OSMAC strategy: a literature review. Front Microbiol 10:294

Petersen LM, Frisvad JC, Knudsen PB, Rohlfs M, Gotfredsen CH, Larsen TO (2015) Induced sclerotium formation exposes new bioactive metabolites from *Aspergillus sclerotiicarbonarius*. J Antibiot 68:603–608

Pfliegler WP, Pócsi I, Győri Z, Pusztahelyi T (2020) The aspergilli and their mycotoxins: metabolic interactions with plants and the soil biota. Front Microbiol 10:2921

Rahaman MM, Islam KS, Jahan M (2018) Rice farmers' knowledge of the risks of pesticide use in Bangladesh. J Health Pollut 8:181203

Richards M, Bird AE, Munden JE (1969) Chlorflavonin: a new antifungal antibiotic. J Antibiot 22:388–389

Romsdahl J, Wang CCC (2019) Recent advances in the genome mining of *Aspergillus* secondary metabolites (covering 2012–2018). Med Chem Comm 10:840–866

Shimada A, Kusano M, Takeuchi S, Fujioka S, Inokuchi T, Kimura Y (2002) Aspterric acid and 6-hydroxymellein, inhibitors of pollen development in arabidopsis thaliana, produced by *Aspergillus terreus*. Z Naturforsch C 57:459–464

Staub GM, Gloer KB, Gloer JB, Wicklow DT, Dowd PF (1993) New paspalinine derivatives with antiinsectan activity from the sclerotia of Aspergillus nomius. Tetrahedron Lett 34:2569–2572

Sun LX, Wang HN, Yan MC, Sai CM, Zhang Z (2022) Research advances of bioactive sesquiterpenoids isolated from marine-derived *Aspergillus* sp. Molecules 27:7376

Tabashnik BE, Brévault T, Carrière Y (2013) Insect resistance to bt crops: lessons from the first billion acres. Nat Biotechnol 31:510–521

Tang Q, Guo K, Li XY, Zheng XY, Kong XJ, Zheng ZH, Xu QY, Deng XM (2014) Three new asperentin derivatives from the algicolous fungus *Aspergillus* sp. f00785. Mar Drugs 12:5993–6002

TePaske MR, Gloer JB, Wicklow DT, Dowd PF (1989) Tubingensin A: an antiviral carbazole alkaloid from the sclerotia of Aspergillus tubingensis. J Org Chem 54:4743–4746

TePaske MR, Gloer JB, Wicklow DT, Dowd PF (1990) Aflavazole: a new antiinsectan carbazole metabolite from the sclerotia of Aspergillus flavus. J Org Chem 55:5299–5301

Tepaske MR, Gloer JB, Wicklow DT, Dowd PF (1992) Aflavarin and β-aflatrem: new anti-insectan metabolites from the sclerotia of Aspergillus flavus. J Nat Prod 55:1080–1086

Tomoda H, Tabata N, Yang DJ, Namatame I, Tanaka H, Omura S, Kaneko T (1996) Pyripyropenes, novel ACAT inhibitors produced by Aspergillus fumigatus. iv. structure elucidation of pyripyropenes M to R. J Antibiot 49:292–298

Wang R, Guo ZK, Li XM, Chen FX, Zhan XF, Shen MH (2015) Spiculisporic acid analogues of the marine-derived fungus, *Aspergillus candidus* strain HDf2, and their antibacterial activity. Antonie Van Leeuwenhoek 108:215–219

Wang JF, He WJ, Huang XL, Tian XP, Liao SR, Yang B, Wang FZ, Zhou XJ, Liu YH (2016) Antifungal new oxepine-containing alkaloids and xanthones from the deep-sea-derived fungus *Aspergillus versicolor* SCSIO 05879. J Agric Food Chem 64:2910–2916

Wei MY, Chen GY, Wang Y, Zhang XL, Wang CY, Shao CL (2011) Isolation, 1H, 13C NMR assignments, and crystal structure of Chrodrimanin B from a marine fungus Aspergillus sp. Chem Nat Compd 47:571–573

Wielkopolan B, Obrępalska-Stęplowska A (2016) Three-way interaction among plants, bacteria, and coleopteran insects. Planta 244:313–332

Wösten H, Van Veluw GJ, De Bekker C, Krijgsheld P (2013) Heterogeneity in the mycelium: implications for the use of fungi as cell factories. Biotechnol Lett 35:1155–1164

Wu ZJ, Ouyang MA, Tan QW (2009) New asperxanthone and asperbiphenyl from the marine fungus *Aspergillus* sp. Pest Manag Sci 65:60–65

Yan W, Wuringege LSJ, Guo ZK, Zhang WJ, Wei W, Tan RX, Jiao RH (2017) New p-terphenyls from the endophytic fungus *Aspergillus* sp. YXf3. Bioorg Med Chem Lett 27:51–54

Ye YQ, Xia CF, Yang JX, Yang YC, Qin Y, Gao XM, Gang D, Li XM, Hu QF (2014a) Butyrolactones derivatives from the fermentation products of an endophytic fungus *Aspergillus versicolor*. Bull Korean Chem Soc 35:3059–3062

Ye YQ, Xia CF, Yang JX, Yang YC, Gao XM, Du G, Yang HY, Li XM, Hu QF (2014b) Phenyl derivated butyrolactones from the fermentation products of an endophytic fungus Aspergillus terreus. Heterocycles 89:2177–2183

Zhang CL, Zheng BQ, Lao JP, Mao LJ, Chen SY, Kubicek CP, Lin FC (2008) Clavatol and patulin formation as the antagonistic principle of *Aspergillus clavatonanicus*, an endophytic fungus of taxus mairei. Appl Microbiol Biotechnol 78:833–840

Zhang P, Yuan XL, Du YM, Zhang HB, Shen GM, Zhang ZF, Liang YJ, Zhao DL, Xu K (2019) Angularly prenylated indole alkaloids with antimicrobial and insecticidal activities from an endophytic fungus Fusarium sambucinum TE-6L. J Agric Food Chem 67:11994–12001

Zhang SS, Zhu A, Bai X, Zhu HJ, Cao F (2020) Alkaloids and polyketides from the marine-derived fungus *Aspergillus versicolor*. Chem Nat Compd 56:964–967

Zhou M, Du G, Yang HY, Xia CF, Yang JX, Ye YQ, Gao XM, Li XN, Hu QF (2015a) Antiviral butyrolactones from the endophytic fungus *Aspergillus versicolor*. Planta Med 81:235–240

Zhou M, Lou J, Li YK, Wang YD, Zhou K, Ji BK, Dong W, Gao XM, Du G, Hu QF (2015b) Butyrolactones from the endophytic fungus *Aspergillus versicolor* and their anti-tobacco mosaic virus activity. J Braz Chem Soc 26:545–549

Zhou K, Zhu L, Wang XL, Zhang TD, Wang YD, Dong W, Ji BK, Yang H, Du G, Hu Q, Zhou M (2016) Butyrolactones from the fermentation products of the endophytic fungus *Aspergillus versicolor*. Chem Nat Compd 52:591–594

Zulqarnain, Iqbal Z, Cox R, Anwar J, Ahmad N, Khan K, Iqbal M, Manzoor N, Khattak SU (2020) Antifungal activity of compounds isolated from *Aspergillus Niger* and their molecular docking studies with tomatinase. Nat Prod Res 34:2642–2646

Fusarium Genus as a Source of Metabolites for Agricultural Applications

Diane Yimta Youmbi

Abstract *Fusarium* is a multifaced genus composed of both devastating plant pathogens and beneficial fungi species. When *Fusarium* is not causing disease in plants, it is producing a wide array of structurally diverse and unique metabolites with potential application in various industries including agriculture. Metabolites produced by *Fusarium* have multidimensional bioactivities. In the agricultural system, they have been reported to affect plant health both directly and indirectly. *Fusarium* can produce molecules such as phytohormones for development, siderophores for iron acquisition, enzymes for nutrient mobilization and volatile metabolites that can function as signal molecules and play important roles in plant growth and development. Fusarium also produces non-volatiles and volatiles metabolites with ability to inhibit various phytopathogens and nematodes. Metabolites produced by *Fusarium* can also impact plant physiology and induce plant defense systems to resist against biotic and abiotic stresses. This chapter provides a brief overview of the beneficial and potential application of *Fusarium* metabolites for crops improvement.

Keywords *Fusarium* spp. · Metabolites · Potential · Agriculture

1 Introduction

Agriculture is the backbone of the economic system of many developing countries, accounting for 29% of GDP and 65% of jobs. Globally, agriculture produces every day an average of 23.7 million tons of food, provides livelihoods for 2.5 billion people, and is the largest source of income and jobs for poor and rural households (FAO 2023). However, the agriculture industry faces multiple challenges including

D. Y. Youmbi (✉)
Antimicrobial and Biocontrol Agents Unit, Laboratory for Phytobiochemistry and Medicinal Plants Studies, Department of Biochemistry, Faculty of Science, University of Yaounde 1, Yaounde, Cameroon

J. Poveda et al. (eds.), *Fungal Metabolites for Agricultural Applications*, Fungal Biology, https://doi.org/10.1007/978-3-031-76587-2_11

the rapid soil degradation, fast depletion of groundwater, excessive use of pesticides and fertilizers, extreme weather events, and the relentless pests and pathogens attack. Collectively these problems are putting immense stress on farming and threatening productivity.

Plant diseases caused by phytopathogens, such as fungi, bacteria, viruses, protozoa, as well as insects and parasites, are among the most devastating problems in agriculture (Lebeda 2002), affecting the productivity of major agricultural crops including the world's most important food species such as cereals, vegetables, and industrial crops with an estimated annual losses of about 13% to 22% (Wang et al. 2021). Regardless of the pathogenic strains involved, the management of plant diseases hugely rely on the application of agrochemicals (Borrelli et al. 2018; Burdon and Zhan 2020). Agrochemicals application is often the major control option used to immediately mitigate the effects of pathogens attack either by killing them or constraining their growth and reproduction (Nazarov et al. 2020). This approach of fighting pathogens has shown efficiency and their potential benefits in improving economic efficiency of the agricultural production systems and addressing food security concerns cannot be underrated. In fact, the application of agrochemicals is mostly responsible for the significant increase in food production witnessed since the 1930s. To date, there are more than 600 different chemical control agents on the market (fungicides, pesticides, herbicides, nematicides, molluscicides, rodenticides, and antibiotics), and its economic sector is now valued at more than USD 50 billion (Devi et al. 2022; Nazarov et al. 2020).

Although efficient, this method of managing phytopathogens has multiple drawbacks including their adverse effects on the environment and human health (Devi et al. 2022). Therefore, more ecofriendly methods are now gaining popularity, particularly biological methods that are increasingly been used as remedy to control various plant pathogens (He et al. 2016). Examples of biological control include the use of parasitic insects for controlling insect pests, bacteriophages, insects for weed control, and antagonist strains and their bioactive metabolites to control bacteria, fungi, and viruses. Microbial biocontrol agents can protect plants by various mechanisms including nutrient competition, modulating the growth conditions for the pathogen, inducing resistance or priming plants without any direct interaction with the pathogen. The direct antagonism can be achieved via mechanisms such as hyperparasitism and the production of bioactive metabolites (Köhl et al. 2019). These microbial biocontrol agents belong to several groups of microbes including bacteria and fungi. Biocontrol fungi are among the naturally occurring microorganisms. They can be found in the plant rhizosphere, competing with pathogens for nutrients and space, or can be found colonizing healthy plant tissues without causing harm (endophytes).

Metabolites produced by biocontrol fungi have broad range of applications for different industries, including the wood and food industry and agronomy. They can be used to improve the general performance of crops or to protect them against biotic or abiotic stresses which results in increase crop production (Rodrigo et al.

Table 1 A summary of a few plant growth-promoting activity of *Fusarium spp*

Strains	Metabolites	Implicate plant	References
F. proliferatum KGL0401	Gibberellins acid	Waito-c rice seedlings	Rim (2013)
Fusarium spp	Gibberellin acid, indole acetic acid and auxins	*Euphorbia pekinensis*	Ben Rhouma et al. (2020)
Fusarium spp	Indole acetic acid and gibberellin acid		Leblanc (2013)
Fusarium spp	Siderophore	Banana	Al-Ani (2019)
F. oxysporum	Siderophore		Wijesundera et al. (1995)
F. oxysporum	VOCs	*A. thaliana*	Bitas et al. (2015)
F. oxysporum NRRL26379 and NRRL38335	VOCs	*A. thaliana*	Li and Kang (2018)
F. oxysporum MSA 35	VOCs	Lettuce	Minerdi et al. (2011)

2021; Xia et al. 2022). The *Fusarium* genus composed of 300 known species is well known for their capacity to produce diverse chemically unique metabolites exhibiting diverse biological activities with potential application in agriculture (Ahmed et al. 2023; Toghueo 2020). Some metabolites are produced by many species, while others are limited to a few or only a single species (Frisvad et al. 2008). They vary in structure and purpose. While some metabolites can work as toxins against animals and/or plants (Divekar et al. 2022), some are small volatiles molecules with ability to promote plant growth, stimulate plant defense systems, act as antimicrobial against phytopathogens or as deterrent against pests and herbivores (Bitas et al. 2015). The present chapter aims to provide a brief overview of the potential agricultural application of metabolites produced by *Fusarium* species (Tables 1, 2, 3, and 4)

2 Fusarium Is Not Only a Pathogen

Fusarium species are among the soilborne fungi that fascinate microbial ecologists, plant pathologists, mycologists, medical doctors, evolutionists, geneticists, and taxonomists (Steinberg et al. 2016). At first glance, the name *Fusarium* almost always referee to devastating pathogen, agents of chaos and threat for agricultural productivity (Ejaz et al. 2023). To be fair *Fusarium* species are very serious plant pathogens, causing wide range of diseases including rots, blights, cankers, and wilts to over hundred plant species (horticultural and ornamental plants and forest trees) in both agricultural and natural setting (Ma et al. 2013). However, *Fusarium* species are not only pathogenic, because some nonpathogenic strains have beneficial effects on plants. Nonpathogenic *Fusarium* from the rhizosphere, epysphere and endosphere of plants play important beneficial roles, protecting the plant against biotic

Table 2 Activity of *Fusarium spp* metabolites in inhibition of phytopathogenes

Strains	Metabolites	activities	References
F. oxysporum EF119	Fusaric acid	Inhibit in vitro and in vivo *P. capsici and P. infestans with* IC_{50} values <1 μg mL^{-1}	Son et al. (2008)
F. verticillioides strain WF18	Isocoumarin A, emodin and sesquiterpenoid phytohormone (+)-abscisic acid	Inhibit *Aspergillus austroafricanus* Eis.1, *A. versicolor* EIS2, *P. fungicola* AT100, and *A. tubingensis* AN103 MIC values ranging from 0.3 to 1.3 μg mL^{-1}	Ebrahim et al. (2020)
Fusarium spp	Bikaverin	*Antifungal activity against P. notatum, Alternaria humicola*, and *A. Flavus*	Limón et al. (2010)
Fusarium sp	Helvolic acid methyl ester, helvolic acid, and hydrohelvolic acid	Inhibitory activity against *B. cinerea, C. gloeosporioides, F. oxysporum f. sp. niveum, F. graminearum*, and *P. capsic* with MIC values ranging from 12.5 to 25 μg/mL	Liang et al. (2016)
Fusarium sp. K432	Fusarielins A-D	Fusarielin A exhibited antifungal potency, of *Pyricularia oryzae*	Kobayashi et al. (1995)
F. equiseti D39	Equisentin	Anti-phytopathogenic activity against *Pseudomonas syringae*	Zhao et al. (2019)
F. tricinctum CECT 20150	Compounds eniatins A, A(1), B, B(1)	Inhibition of a wide range of plant pathogens including *Fusarium verticilloides, F. sporotrichioides, F. tricinctum, F. poae, F. oxysporum, F. proliferatum, Beauveria bassiana, Aspergillus flavus, A.parasiticus, A.fumigatus, A. ochraceus*, and *P. expansum*	Meca et al. (2010)
F. oxysporum MSA35	VOCs	Inhibit the mycelial growth of *F. oxysporum*	
F. oxysporum MSA35	α-Humulene	*Inhibit the growth of F. oxysporum f. sp. Lactucae* Fuslat10	Yoshida et al. (1976)
F. oxysporum CanR-46	5-hexenoic acid, (±)-limonene, 3,4-2H-dihydropyran, and octanoic acid	Inhibitory activity against *V. dahlia*	Zhang et al. (2015)

and abiotic stresses, pest, and herbivore attacks. Many strains produce important bioactive metabolites that can be applied for agricultural improvement (Steinberg et al. 2016; Toghueo 2020). This review will focus on the potential application of metabolites produced by *Fusarium* species in agricultural systems to improve plant growth and protection against various stresses.

Table 3 Activity of *Fusarium spp* metabolites in inhibition against nematodes

Strain	Metabolites	Activities	References
Fusarium spp	Culture filtrates	Inhibition of *Meloidogyne* spp. larvae	Ramadhan et al. (2021)
F. oxysporum and *F. solani*	Culture filtrates	*Inhibition of Meloidogyne javanica*	Ali et al. (2023)
F. oxysporum 162	4-hydroxybenzoic acid, indole-3-acetic acid (IAA), and gibepyrone D	Inhibition of nematodes with LC $_{50}$ values of 104, 117 and 134 μg mL^{-1}, respectively for each metabolite	Bogner et al. (2017)
F. oxysporum	Bikaverin, 3-*O*-methyl-8-*O*--methyl fusarubin, 8-*O*-methyl fusarubin, anhydrofusarubin, and fusarubin	Inhibition of nematodes	Kundu et al. (2016)
F. chlamydosporum	Chlamydosporin	Exhibit significant phytotoxic activity against the radicle growth of *Echinochloa crusgalli*	Wang et al. (2020)
F. oxysporum	VOCs	Nematicidal potential towards *M. incognita*	Freire et al. (2012)

Table 4 Activity of *Fusarium spp* metabolites in induction of resistance

Strains	Metabolites	Activities	References
Fusarium spp	Benzyl isothiocyanate	Induce the resistance of *Lepidium sativum* seedlings against *Pythium ultimum*	Ishimoto et al. (2004)
F. oxysporum Fo47	PR-1 protein and chitinases	Induced resistance on tomato against fusarium wilt	Duijff et al. (1998)
Fusarium sp	Phenolic metabolites	Disease resistance against virulent plant pathogens	Chaĭkovskaia et al. (1975)
F. oxysporum	VOCs	Increase salt tolerance in *A. thaliana* and confer resistance to disease caused by *P. syringae*	Li and Kang (2018)

3 Application of *Fusarium* Metabolites in Agriculture

Fusarium is a rich source of hundreds of compounds belonging to various chemical classes such as butenoides, alkaloids, terpenoids, cytochalasins, phenalenones, xanthones, sterols, diphenyl ether, and anthraquinone derivatives, along with over hundred compounds with unique chemical structures (Ahmed et al. 2023). These metabolites have multidimensional bioactivities and great potential for agriculture.

4 Fusarium Metabolites Can Promote Plants Growth

Microbes synthetize various molecules that affect plant health both directly and indirectly, such as phytohormones for development, siderophores for iron acquisition, enzymes for nutrient mobilization and antibiotics. Volatile metabolites also produced by microbes travel through air and porous soils and function as signal molecules, and play important roles in plant growth and development.

5 Phytohormones and Siderophores Produced by *Fusarium*

Plant hormones are essential regulatory molecules that influence the development and production yield of plants in both normal and stress like conditions. Phytohormones are critical for growth and development, helping the plant control and regulate several processes such as seeds germination, flowering, and fruit ripening (Pal et al. 2023). *Fusarium* species have been found to produce such metabolites (Bogner et al. 2017). For instance, KGL0401, a strain of *F. proliferatum* isolated from roots of *Physalis alkekengi var. francheti* plant was reported to produce several new gibberellins acid (GAs) and significantly improved the growth of waito-c rice seedlings. Following their investigation, the authors reported the characterization of the gibberellin biosynthetic gene cluster from *Fusarium proliferatum* KGL0401 (Rim 2013). Another *Fusarium* isolate was found to promote the growth of the medicinal plant *Euphorbia pekinensis*. Subsequent analysis revealed that the isolate could produce several hormones including gibberellin acid (GA), indole acetic acid (IAA) and auxins (Ben Rhouma et al. 2020). In another investigation, the screening for genes underlying the production of secondary metabolites in *Fusarium* isolates obtained from the rhizosphere of native perennial legumes and grasses growing in monoculture and polyculture show the presence of genes underlying the production of plant hormones such as indoleacetic acid and gibberellic acid (Leblanc 2013). *Fusarium* can therefore promote the growth of plants via the production of phytohormones.

Besides phytohormones, iron is a vital mineral that plays a very important functions in plants. To acquire iron, studies have shown that *Fusarium* species can produce siderophores molecules. These relatively low molecular weight compounds act as iron chelator to increase the availability of this mineral under limited conditions which in consequence promote growth of plants. In planta studies reveal that isolates of nonpathogenic Fusarium could produce siderophores to sequester environmental iron which reduce the *Fusarium* wilt disease of banana caused by pathogenic *F. oxysporum* f. sp. *cubense* tropical race 4 (*FocTR4)* LJ27 strain (Al-Ani 2019). Another investigation also reveal the ability of non-pathogenic *Fusarium ozysporum* isolates to produce wide range of siderophores to impact plant growth and disease suppression (Wijesundera et al. 1995).

6 *Fusarium* Volatiles Organic Compounds (VOCs) Can Promote Plant Growth

In recent years, small gaseous signaling molecules emitted by beneficial microbes named volatile organic compounds (VOCs) have emerged as a promising biotechnological tool to promote the germination, growth, and development of crops (Almeida et al. 2023; Naznin et al. 2014). Several *Fusarium* strains have been reported to produce bioactive volatiles organic compounds with the ability to induce significant morphological changes in plants. The investigation of the effect of VOCs produced by 46 strains of *Fusarium* on *A. thaliana* ecotype Col-0 revealed that 39 strains significantly enhanced plant growth. Moreover, *F. oxysporum* volatiles caused multiple changes in *A. thaliana*. For instance, compared to control plants, cocultivation with strains NRRL 26379 and NRRL 38335 increased the total leaf area by 2.7- and four-fold, respectively, and the chlorophyll content by three-fold, increased the mass (4.8- and four-fold, respectively) and total length (3.6- and 5.2-fold, respectively) of roots, and more than doubled the lateral root density. Further analysis revealed the involvement of the auxin signaling in mediating growth enhancement by *F. oxysporum* volatiles (Bitas et al. 2015). Following this study, the effect of volatiles from *F. oxysporum* (NRRL26379 and NRRL38335) on *A. thaliana* under salinity was also investigated. The results revealed that in the presence of 100 mM NaCl, *A. thaliana* seedlings exposed to *Fusarium* VOCs displayed much more robust foliar growth than control seedlings and had increased leaf chlorophyll content and lateral root density (Li and Kang 2018). In another investigation, volatiles produced by *F. oxysporum* MSA 35 increase in the root length (95.6%), shoot length (75.0%), and fresh weight (85.8%) of lettuce. Moreover, leaf chlorophyll content was also increased (68%) with volatile exposure. Further analysis using semi-quantitative and quantitative reverse transcription-PCR revealed a significant change in expansin gene expression level between leaf (6.7-fold) and roots (4.4-fold) after exposure to *Fusarium* volatiles. β-Caryophyllene was found to be one of the volatiles responsible for the plant growth promotion effect (Minerdi et al. 2011). Previously, this β-Caryophyllene metabolite was also reported to enhance lettuce growth (Minerdi et al. 2011).

7 Metabolites from *Fusarium* Can Inhibit Phytopathogens

Members of the genus *Fusarium* are well-known producers of secondary metabolites with structural and chemical diversity. These metabolites belong to several chemical classes and can exhibit diverse biological activities (Fig. 1). In this section, we will focus on the potential application of these secondary metabolites as antimicrobial agents. Several metabolites from *Fusarium* species have demonstrated very good activity against bacterial and fungal pathogens of agricultural

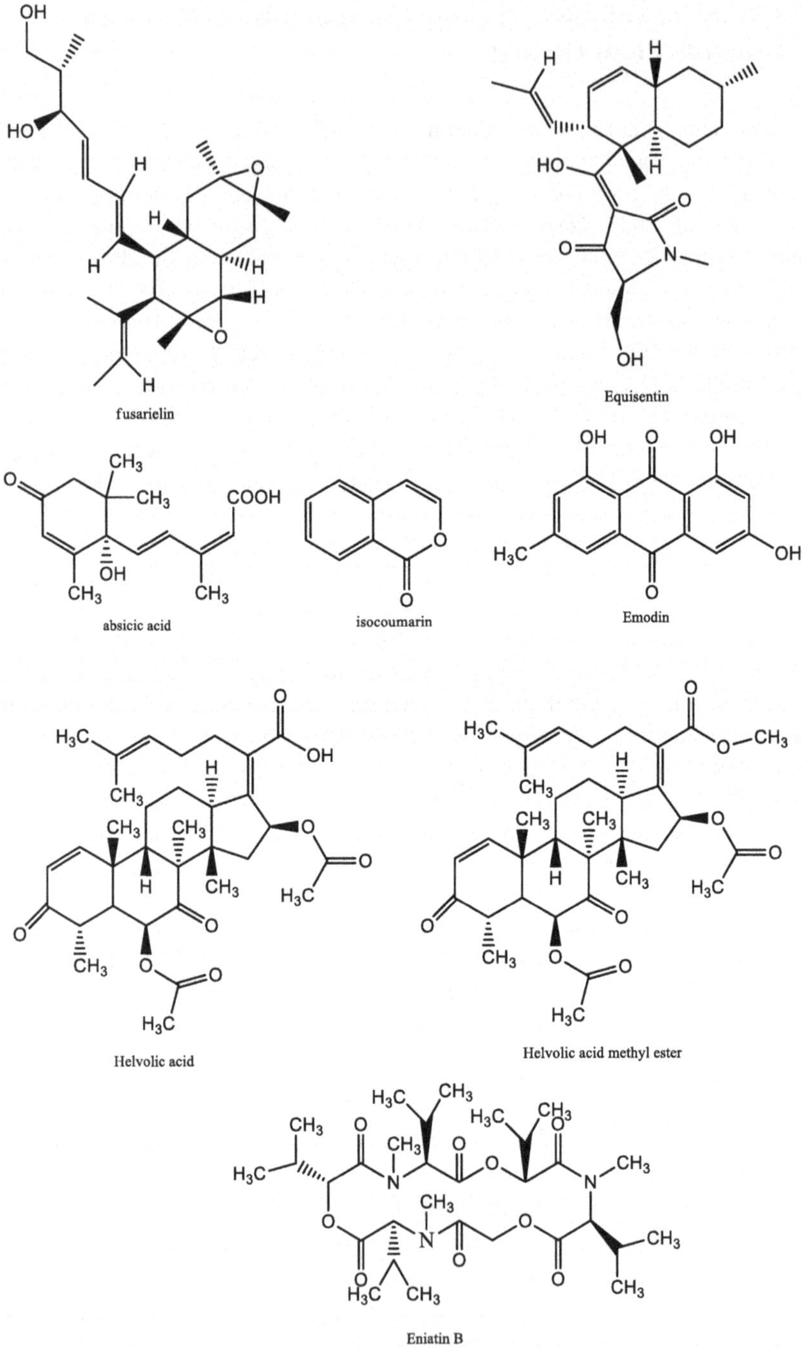

Fig. 1 Selected Fusarium metabolites with activity against plant pathogens

importance. Indeed, using both MS and NMR analyses, an antinomycete metabolite, Fusaric acid was identified from *F. oxysporum* EF119. This compound shows potent inhibition *in vitro* and *in vivo* against *P. capsici* (causative agent of wheat leaf rust) and *P. infestans* (causative agent of tomato late blight) with IC_{50} values <1 μg mL^{-1}. It also completely suppressed the growth of various bacteria (IC_{50} values ranged from 0.2 to 12 μg mL^{-1}, Conc. < 100 μg mL^{-1}). Thus, it could be used as a biocontrol agent towards tomato late blight produced by *P. infestans* (Son et al. 2008). The investigation of crude ethyl acetate (EtOAc) extract from endophytic fungus *Fusarium verticillioides* strain WF18 afforded three compounds including one new polyketide metabolite, isocoumarin A, and two known compounds, emodin and sesquiterpenoid phytohormone (+)-abscisic acid (ABA). All three compounds inhibited fungal pathogens such as *Aspergillus austroafricanus* Eis.1, *A. versicolor* EIS2, *P. fungicola* AT100, and *A. tubingensis* AN103 with MIC values ranging from 0.3 to 1.3 μg mL^{-1} (Ebrahim et al. 2020). Bikaverin is a reddish pigment produced *Fusarium* species was found to have antifungal activity against multiple fungal pathogens including *P. notatum*, *Alternaria humicola*, and *A. flavus* (Limón et al. 2010). Three compounds including a new helvolic acid derivative named helvolic acid methyl ester, and two known helvolic acid compounds, helvolic acid and hydrohelvolic acid, were isolated from the endophytic fungus *Fusarium* sp. obtained from *Ficus carica* leaves. All three compounds exhibited inhibitory activity against plant pathogens *B. cinerea*, *C. gloeosporioides*, *F. oxysporum f. sp. niveum*, *F. graminearum*, and *P. capsic* with MIC values ranging from 12.5 to 25 μg/mL (Liang et al. 2016). From the investigation of another strain of *Fusarium* sp. K432, Fusarielins A-D were isolated. Only Fusarielin A exhibited antifungal potency, inducing a significant deformation of *Pyricularia oryzae* (Kobayashi et al. 1995). Equisentin, a compound isolated from the marine-derived fungus *Fusarium equiseti* D39 was found to displayed a remarkable anti-phytopathogenic activity against *Pseudomonas syringae* by damaging cell membrane and ensuing leakage of the intracellular components (Zhao et al. 2019). this compound was previously reported for its antibiotic activity against a wide range of bacteria (Vesonder et al. 1979). Compounds eniatins A, A(1), B, B(1) isolated from *Fusarium tricinctum* CECT 20150 and grown in a liquid medium of potato were shown to display potent inhibition of a wide range of plant pathogens including *Fusarium verticilloides*, *F. sporotrichioides*, *F. tricinctum*, *F. poae*, *F. oxysporum*, *F. proliferatum*, *Beauveria bassiana*, *Aspergillus flavus*, *A. parasiticus*, *A. fumigatus*, *A. ochraceus*, and *P. expansum* (Meca et al. 2010).

Volatiles organic compounds emitted by *Fusarium* species have also been reported for their inhibitory activity against various plant pathogens (Fig. 2). In fact, volatile organic compounds (VOCs) emitted by *Fusarium oxysporum* MSA35 was found to inhibit the mycelial growth of different formae speciales of *F. oxysporum*. Further analysis revealed that these VOCs could also repress the expression of two putative virulence genes in *F. oxysporum lactucae* strain Fuslat10. The analysis of VOCs emitted by MSA35 revealed the presence of eight compounds including β-caryophyllene; iso-caryophyllene; α-humulene; 2-epi-trans-caryophyllene; isolongifolene; phenolic compound; longipinalolo; 5,8-cyclocaryophyllan-4-olo.

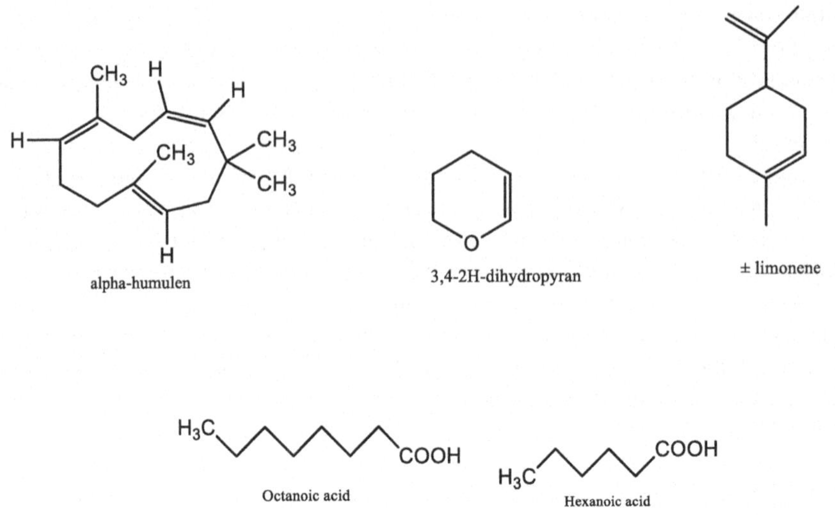

Fig. 2 VOCs with activity against plant pathogens

Off the two compounds α-humulene and β-caryophyllene tested for their effects of on the growth of *F. oxysporum f. sp. Lactucae* Fuslat10, only α-humulene was active at various concentrations (Yoshida et al. 1976). In another investigation, VOCs produced by *F. oxysporum* CanR-46 were found to inhibit the mycelia growth of 14 fungal pathogens including *V. dahliae*. The compositional analysis shows that CanR-46 VOCs were composed of 19 compounds belonging to classes of alkenes, esters, alkanes, organic acids, and alcohols. Compounds, 5-hexenoic acid, (±)-limonene, 3,4-2H-dihydropyran, and octanoic acid exhibited inhibitory activity against *V. dahlia*. These metabolites could delay *V. dahliae* conidial germination, suppress *V. dahliae* germ-tube elongation, and cause *V.dahliae* hyphal shriveling and collapse (Zhang et al. 2015).

8 *Fusarium* Metabolites Can Inhibit Nematodes

Bioactive metabolites from *Fusarium* species can also inhibit the proliferation of plant parasitic nematodes (Fig. 3). For instance, culture filtrates of several *Fusarium* species were found to inhibit *Meloidogyne* spp. larvae (Ramadhan et al. 2021). Culture filtrate from several other fungi including *F.oxysporum* and *F.solani* were also found to inhibit the root knot nematode *Meloidogyne javanica* (Ali et al. 2023). Among metabolites produced by the fungal endophyte *Fusarium oxysporum* 162, three potent nematode-antagonistic compounds including 4-hydroxybenzoic acid,

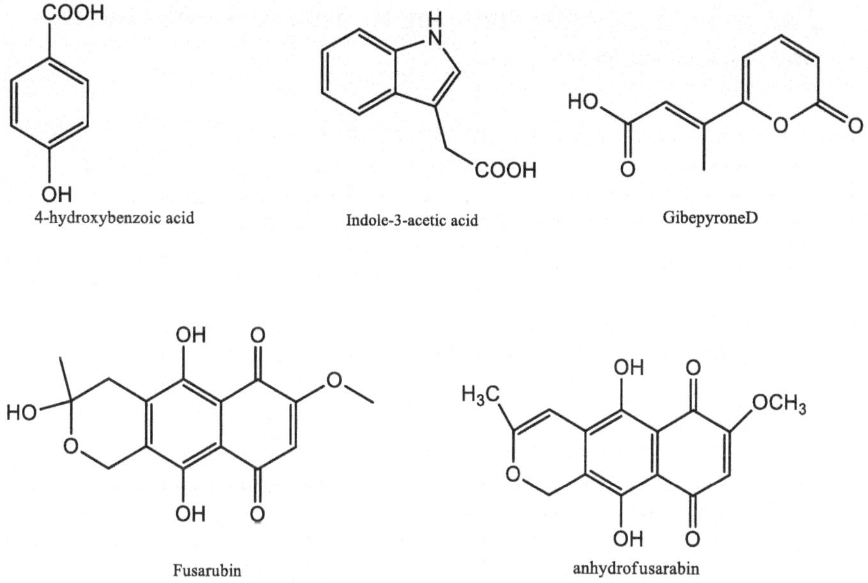

Fig. 3 Selected compounds with nematicidal activity

indole-3-acetic acid (IAA), and gibepyrone D with LC $_{50}$ values of 104, 117 and 134 μg mL^{-1}, respectively, were identified. These compounds may serve as important leads for the development of novel and environmentally friendly nematicide chemicals (Bogner et al. 2017). In another study, ethyl acetate (EtOAc) extract of several nonpathogenic strains of *Fusarium oxysporum* were found to inhibit nematodes. The chromatographic separation of EtOAc extract afforded five compounds, bikaverin, 3-*O*-methyl-8-*O*-methyl fusarubin, 8-*O*-methyl fusarubin, anhydrofusarubin, and fusarubin. These compounds were tested against *M. incognita* and Fusarubin (LC_{50} 248.9 μg mL^{-1}) was found to be highly active followed by anhydrofusarubin (LC_{50} 257.6 μg mL^{-1}). Therefore, Fusarubin derivatives have the potential to be further develop into nematicidal formulation to combat *M. incognita* (Kundu et al. 2016). Chlamydosporin, a new metabolite from the endophytic fungus *Fusarium chlamydosporum* was also reported to exhibit significant phytotoxic activity against the radicle growth of *Echinochloa crusgalli* (Wang et al. 2020). VOCs emitted by *F. oxysporum* isolate 21 obtained from coffee plant rhizosphere displayed nematicidal potential towards *M. incognita*. The compositional analysis reveals the presence of 38 metabolites among which five including dioctyl disulfide, 1-(2-hydroxyethoxy) tridecane, 4-methyl-2,6-di-tert-butylphenol, caryophyllene, and acoradiene were above 1%. These VOCs could be useful for the development of bio-control agent for *Meloidogyne* spp. in coffee fields (Freire et al. 2012).

9 *Fusarium* Metabolites Inducing Resistance Against Biotic and Abiotic Stresses

In addition to their activity against phytopathogens and their ability to promote the growth of various crops, metabolites produced by *Fusarium* can also induce the resistance of plants against stresses. The ability of *Fusarium* metabolites to impact plant physiology and improve plant defense systems has been investigated. For instance, two *Fusarium* strains, Ls-F-in-4-1 and Rs-F-in-11, isolated from roots of Brassicaceae plants, were able to induce the resistance of *Lepidium sativum* seedlings against *Pythium ultimum* via increasing the synthesis of benzyl isothiocyanate and its precursor glucotropaeolin, in the roots of the host plants (Ishimoto et al. 2004). In another investigation, *F. oxysporum* strain Fo47 was found to induced resistance and suppress fusarium wilt of tomato via the accumulation of PR-1 protein and chitinases (Duijff et al. 1998). In *Hordeum vulgare*, an endophytic *Fusarium* sp. could confer disease resistance against virulent plant pathogens via the accumulation of phenolic metabolites (Chaĭkovskaia et al. 1975). VOCs emitted by *F. oxysporum* were also reported to increase salt tolerance in *A. thaliana* and confer resistance to disease caused by *P. syringae* (Li and Kang 2018).

10 Conclusion

Fusarium species are well-known pathogens causing serious problems in agricultural systems. However, nonpathogenic species from this genus can also have significant agricultural benefits. This genus produces a variety of chemically diverse secondary metabolites among which only a few have been characterized. These secondary metabolites have been shown to have diverse functions such as antimicrobial to fight against phytopathogens, improve the growth of plants (Phytohormones), or to induce host defense response against various biotic and abiotic stresses. The overall findings presented in the present chapter demonstrate the potential agricultural application of *Fusarium* metabolites and support the continue investigation of these fungi as sources of potential plant growth promotion and biocontrol agents for agricultural benefits.

Acknowledgments I thank Dr. Toghueo Kuipou Marie Rufin for the critical reading and advice provided during the writing of this chapter.

References

Ahmed AM, Mahmoud BK, Millán-Aguiñaga N, Abdelmohsen UR, Fouad MA (2023) The endophytic *Fusarium* strains: a treasure trove of natural products. RSC Adv 13:1339–1369. https://doi.org/10.1039/D2RA04126J

Al-Ani LKT (2019) Secondary metabolites of non-pathogenic Fusarium: scope in agriculture. In: Singh HB, Keswani C, Reddy MS, Sansinenea E, García-Estrada C (eds) Secondary metabolites of plant growth promoting Rhizomicroorganisms. Springer Singapore, Singapore, pp 59–76. https://doi.org/10.1007/978-981-13-5862-3_3

Ali WM, Abdel-Mageed MA, Hegazy MGA, Abou-Shlell MK, Sultan SME, Salama EAA, Yousef AF (2023) Biocontrol agent of root-knot nematode Meloidogyne javanica and root-rot fungi, Fusarium solani in okra morphological, anatomical characteristics and productivity under greenhouse conditions. Sci Rep 13:11103. https://doi.org/10.1038/s41598-023-37837-z

Almeida OAC, De Araujo NO, Mulato ATN, Persinoti GF, Sforça ML, Calderan-Rodrigues MJ, Oliveira JVDC (2023) Bacterial volatile organic compounds (VOCs) promote growth and induce metabolic changes in rice. Front Plant Sci 13:1056082. https://doi.org/10.3389/fpls.2022.1056082

Ben Rhouma M, Kriaa M, Ben Nasr Y, Mellouli L, Kammoun R (2020) A new Endophytic Fusarium Oxysporum gibberellic acid: optimization of production using combined strategies of experimental designs and potency on tomato growth under stress condition. Biomed Res Int 2020:1–14. https://doi.org/10.1155/2020/4587148

Bitas V, McCartney N, Li N, Demers J, Kim J-E, Kim H-S, Brown KM, Kang S (2015) Fusarium Oxysporum volatiles enhance plant growth via affecting auxin transport and signaling. Front Microbiol 6. https://doi.org/10.3389/fmicb.2015.01248

Bogner CW, Kamdem RST, Sichtermann G, Matthäus C, Hölscher D, Popp J, Proksch P, Grundler FMW, Schouten A (2017) Bioactive secondary metabolites with multiple activities from a fungal endophyte. Microb Biotechnol 10:175–188. https://doi.org/10.1111/1751-7915.12467

Borrelli VMG, Brambilla V, Rogowsky P, Marocco A, Lanubile A (2018) The enhancement of plant disease resistance using CRISPR/Cas9 technology. Front Plant Sci 9. https://doi.org/10.3389/fpls.2018.01245

Burdon JJ, Zhan J (2020) Climate change and disease in plant communities. PLoS Biol 18:e3000949. https://doi.org/10.1371/journal.pbio.3000949

Chaĭkovskaia SM, Vakulenko NA, Braginskaia PS, Makarova RA (1975) Microbiological study of gentamiycin. Antibiotiki 20:903–907

Devi PI, Manjula M, Bhavani RV (2022) Agrochemicals, environment, and human health. Annu Rev Environ Resour 47:399–421. https://doi.org/10.1146/annurev-environ-120920-111015

Divekar PA, Narayana S, Divekar BA, Kumar R, Gadratagi BG, Ray A, Singh AK, Rani V, Singh V, Singh AK, Kumar A, Singh RP, Meena RS, Behera TK (2022) Plant secondary metabolites as defense tools against herbivores for sustainable crop protection. Int J Mol Sci 23:2690. https://doi.org/10.3390/ijms23052690

Duijff BJ, Pouhair D, Olivain C, Alabouvette C, Lemanceau P (1998) Implication of systemic induced resistance in the suppression of Fusarium wilt of tomato by Pseudomonas fluorescens WCS417r and by nonpathogenic Fusarium oxysporum Fo47. Eur J Plant Pathol 104:903–910. https://doi.org/10.1023/A:1008626212305

Ebrahim W, Özkaya FC, Ebada SS (2020) Antifungal metabolites from endophytic fungus Fusarium verticillioides strain WF18. South Afr J Bot 133:40–44. https://doi.org/10.1016/j.sajb.2020.06.029

Ejaz MR, Jaoua S, Ahmadi M, Shabani F (2023) An examination of how climate change could affect the future spread of Fusarium spp. around the world, using correlative models to model the changes. Environ Technol Innov 31:103177. https://doi.org/10.1016/j.eti.2023.103177

FAO (2023) State of food security and nutrition in the world 2023: urbanization, Agrifood system transformation. Food & Agriculture org, S.l

Freire ES, Campos VP, Pinho RSC, Oliveira DF, Faria MR, Pohlit AM, Noberto NP, Rezende EL, Pfenning LH, Silva JRC (2012) Volatile substances produced by Fusarium oxysporum from coffee rhizosphere and other microbes affect Meloidogyne incognita and Arthrobotrys conoides. J Nematol 44:321–328

Frisvad JC, Andersen B, Thrane U (2008) The use of secondary metabolite profiling in chemotaxonomy of filamentous fungi. Mycol Res 112:231–240. https://doi.org/10.1016/j.mycres.2007.08.018

He D, Zhan J, Xie L (2016) Problems, challenges and future of plant disease management: from an ecological point of view. J Integr Agric 15:705–715. https://doi.org/10.1016/S2095-3119(15)61300-4

Ishimoto H, Fukushi Y, Tahara S (2004) Non-pathogenic Fusarium strains protect the seedlings of Lepidium sativum from Pythium ultimum. Soil Biol Biochem 36:409–414. https://doi.org/10.1016/j.soilbio.2003.10.016

Kobayashi H, Sunaga R, Furihata K, Morisaki N, Iwasaki S (1995) Isolation and structures of an antifungal antibiotic, Fusarielin a, and related compounds produced by a Flusarium sp. J Antibiot (Tokyo) 48:42–52. https://doi.org/10.7164/antibiotics.48.42

Köhl J, Kolnaar R, Ravensberg WJ (2019) Mode of action of microbial biological control agents against plant diseases: relevance beyond efficacy. Front Plant Sci 10:845. https://doi.org/10.3389/fpls.2019.00845

Kundu A, Saha S, Walia S, Dutta TK (2016) Anti-nemic secondary metabolites produced by Fusarium oxysporum f. sp. ciceris. J Asia Pac Entomol 19:631–636. https://doi.org/10.1016/j.aspen.2016.06.003

Lebeda A (2002) Characterization and identification of entomopathogenic and mycoparasitic fungi using RAPD-PCR technique. Plant Prot Sci 38:36–36. https://doi.org/10.17221/4819-PPS

LeBlanc N (2013) Influence of plant diversity and perennial plant identity on *Fusarium* communities in soil. Retrieved from the University of Minnesota Digital Conservancy, https://hdl.handle.net/11299/175529

Li N, Kang S (2018) Do volatile compounds produced by *Fusarium oxysporum* and *Verticillium dahliae* affect stress tolerance in plants? Mycology 9:166–175. https://doi.org/10.1080/21501203.2018.1448009

Liang X-A, Ma Y-M, Zhang H-C, Liu R (2016) A new helvolic acid derivative from an endophytic *Fusarium* sp. of *Ficus carica*. Nat Prod Res 30:2407–2412. https://doi.org/10.1080/14786419.2016.1190722

Limón MC, Rodríguez-Ortiz R, Avalos J (2010) Bikaverin production and applications. Appl Microbiol Biotechnol 87:21–29. https://doi.org/10.1007/s00253-010-2551-1

Ma L-J, Geiser DM, Proctor RH, Rooney AP, O'Donnell K, Trail F, Gardiner DM, Manners JM, Kazan K (2013) *Fusarium* Pathogenomics. Ann Rev Microbiol 67:399–416. https://doi.org/10.1146/annurev-micro-092412-155650

Meca G, Soriano JM, Gaspari A, Ritieni A, Moretti A, Mañes J (2010) Antifungal effects of the bioactive compounds enniatins A, A1, B, B1. Toxicon 56:480–485. https://doi.org/10.1016/j.toxicon.2010.04.013

Minerdi D, Bossi S, Maffei ME, Gullino ML, Garibaldi A (2011) Fusarium oxysporum and its bacterial consortium promote lettuce growth and expansin A5 gene expression through microbial volatile organic compound (MVOC) emission: plant growth promotion by antagonistic Fusarium oxysporum. FEMS Microbiol Ecol 76:342–351. https://doi.org/10.1111/j.1574-6941.2011.01051.x

Nazarov PA, Baleev DN, Ivanova MI, Sokolova LM, Karakozova MV (2020) Infectious plant diseases: etiology, current status, problems and prospects in plant protection. Acta Nat 12:46–59. https://doi.org/10.32607/actanaturae.11026

Naznin HA, Kiyohara D, Kimura M, Miyazawa M, Shimizu M, Hyakumachi M (2014) Systemic resistance induced by volatile organic compounds emitted by plant growth-promoting fungi in Arabidopsis thaliana. PLoS One 9:e86882. https://doi.org/10.1371/journal.pone.0086882

Pal P, Ansari SA, Jalil SU, Ansari MI (2023) Regulatory role of phytohormones in plant growth and development. In: Plant hormones in crop improvement. Elsevier, pp 1–13. https://doi.org/10.1016/B978-0-323-91886-2.00016-1

Ramadhan MN, Amin N, Nasruddin A (2021) The effectiveness of filtrate culture with Fusarium spp. fungi and against Meloidogyne spp. nematodes by in-vitro tomato plants. IOP Conf Ser Earth Environ Sci 807:022083. https://doi.org/10.1088/1755-1315/807/2/022083
Rim S-O (2013) Characterization of gibberellin biosynthetic gene cluster from Fusarium proliferatum. J Microbiol Biotechnol 23:623–629. https://doi.org/10.4014/jmb.1212.12029
Rodrigo S, García-Latorre C, Santamaria O (2021) Metabolites produced by fungi against fungal Phytopathogens: review. Implement Perspect Plants 11:81. https://doi.org/10.3390/plants11010081
Son SW, Kim HY, Choi GJ, Lim HK, Jang KS, Lee SO, Lee S, Sung ND, Kim J-C (2008) Bikaverin and fusaric acid from Fusarium oxysporum show antioomycete activity against Phytophthora infestans. J Appl Microbiol 104:692–698. https://doi.org/10.1111/j.1365-2672.2007.03581.x
Steinberg C, Lecomte C, Alabouvette C, Edel-Hermann V (2016) Root interactions with nonpathogenic Fusarium oxysporum. In: Vos CMF, Kazan K (eds) Belowground Defence strategies in plants, signaling and communication in plants. Springer International Publishing, Cham, pp 281–299. https://doi.org/10.1007/978-3-319-42319-7_12
Toghueo RMK (2020) Bioprospecting endophytic fungi from *Fusarium* genus as sources of bioactive metabolites. Mycology 11:1–21. https://doi.org/10.1080/21501203.2019.1645053
Vesonder RF, Tjarks LW, Rohwedder WK, Burmeister HR, Laugal JA (1979) Equisetin, an antibiotic from Fusarium equiseti NRRL 5537, identified as a derivative of N-methyl-2,4-pyrollidone. J Antibiot (Tokyo) 32:759–761. https://doi.org/10.7164/antibiotics.32.759
Wang Z-F, Zhang W, Xiao L, Zhou Y-M, Du F-Y (2020) Characterization and bioactive potentials of secondary metabolites from *Fusarium chlamydosporum*. Nat Prod Res 34:889–892. https://doi.org/10.1080/14786419.2018.1508142
Wang Y-P, Pan Z-C, Yang L-N, Burdon JJ, Friberg H, Sui Q, Zhan J (2021) Optimizing plant disease management in agricultural ecosystems through rational in-crop diversification. Front Plant Sci 12. https://doi.org/10.3389/fpls.2021.767209
Wijesundera R, Amarasekera A, Yapa D (1995) Production of a siderophore by the fungus fusarium oxysporum. J Natl Sci Found 23:97. https://doi.org/10.4038/jnsfsr.v23i3.5846
Xia Y, Liu J, Chen C, Mo X, Tan Q, He Y, Wang Z, Yin J, Zhou G (2022) The multifunctions and future prospects of endophytes and their metabolites in plant disease management. Microorganisms 10:1072. https://doi.org/10.3390/microorganisms10051072
Yoshida K, Ichiman Y, Ohtomo T (1976) Mouse virulent strain of Staphylococcus epidermidis. Relation of antiphagocytic activity to the protection-inducing antigen. Jpn J Microbiol 20:209–217
Zhang Q, Yang L, Zhang J, Wu M, Chen W, Jiang D, Li G (2015) Production of anti-fungal volatiles by non-pathogenic Fusarium oxysporum and its efficacy in suppression of Verticillium wilt of cotton. Plant Soil 392:101–114. https://doi.org/10.1007/s11104-015-2448-y
Zhao D, Han X, Wang D, Liu M, Gou J, Peng Y, Liu J, Li Y, Cao F, Zhang C (2019) Bioactive 3-Decalinoyltetramic acids derivatives from a marine-derived strain of the fungus Fusarium equiseti D39. Front Microbiol 10:1285. https://doi.org/10.3389/fmicb.2019.01285

Clonostachys Genus as a Source of Metabolites for Agricultural Applications

Meena Venkatesan Ruppavalli, Iruthayasamy Johnson, Sambasivam Periyannan, and Karthikeyan Muthusamy

Abstract The book chapter delves into the multifaceted world of *Clonostachys*, a group of free-living saprophytic ascomycete fungi that pervade diverse natural habitats, from soil to decomposing organic matter. With their remarkable potential as biocontrol agents against seed and soil-borne pathogens, *Clonostachys* species has reaped significant attention. These fungi exhibit fascinating parasitic and symbiotic relationships with both plants and animals, making them ubiquitous inhabitants of terrestrial and aquatic environments, while also offering a wide array of applications in agriculture. A distinguishing feature of *Clonostachys* species is their capacity to produce a diverse range of secondary metabolites, boasting potent antifungal, antibacterial, entomopathogenic, and nematicidal properties that can combat a multitude of plant pathogens and insects. In this chapter, we provide a comprehensive overview of the biology, phylogeny, diversity, biocontrol activities, and metabolites of *Clonostachys* species, shedding light on their significant contributions to the fields of agriculture.

M. V. Ruppavalli
Department of Plant Pathology, SRS Institute of Agriculture and Technology, Dindigul, Tamil Nadu, India

Department of Plant Pathology, Tamil Nadu Agricultural University, Coimbatore, Tamil Nadu, India

I. Johnson
Department of Plant Pathology, Tamil Nadu Agricultural University, Coimbatore, Tamil Nadu, India

S. Periyannan
Centre for Crop Health, School of Agriculture and Environmental Science, University of Southern Queensland, Toowoomba, QLD, Australia

K. Muthusamy (✉)
Department of Plant Pathology, Tamil Nadu Agricultural University, Coimbatore, Tamil Nadu, India

Centre for Crop Health, School of Agriculture and Environmental Science, University of Southern Queensland, Toowoomba, QLD, Australia
e-mail: karthikeyan.m@tnau.ac.in

J. Poveda et al. (eds.), *Fungal Metabolites for Agricultural Applications*, Fungal Biology, https://doi.org/10.1007/978-3-031-76587-2_12

Keywords Antagonist · Biocontrol agent · *Clonostachys* · Secondary metabolites

1 Introduction

The fungal genus *Clonostachys*, formerly known as *Gliocladium*; teleomorph of *Bionectria*, belongs to the class Sordariomycetes, order Hypocreales, and the family Bionectriaceae (Han et al. 2020; Rossman et al. 2013). These fungi are renowned for their versatility as destructive saprophytic mycoparasites (Rodriguez et al. 2011) and are widely distributed across the globe. They thrive in diverse environments, including lichens, various decaying organic materials, and have also been found as potential endophytes and epiphytes on living plants (Torcato et al. 2020).

Several *Clonostachys* species are mycoparasites that live off other fungi, absorbing nutrients to sustain their growth. They achieve mycoparasitic activity by stimulating high molecular weight proteins, ultimately breaking down their host fungi (Yang and Chen 2004). Mycoparasitism often involves the binding of mono or disaccharides, such as galactose or trehalose, to lectins on the host fungus's surface, leading to agglutination (Sun et al. 2020). Once in contact with the host's surface, mycoparasites penetrate host cells using specialized infection structures, intertwining with the host's mycelium (Zin and Badaluddin 2020). Through the production of cell-wall-degrading enzymes (CWDEs), they degrade the host cell surface (Qualhato et al. 2013). Ultimately, mycoparasites obtain nutrients from the host fungus by releasing toxins or antibiotics that deteriorate the invading host, often leading to its death (Ajith and Lakshmidevi 2010). *Clonostachys* species can also parasitize myxomycetes, nematodes, thrips, ticks, spiders, and leafhoppers (Toledo et al. 2006; Wang et al. 2023).

Currently, the genus *Clonostachys* comprises approximately 44 species, with nearly 18 of them extensively studied for their secondary metabolites against various fungal and bacterial plant pathogens. Notable examples include *Gliocladium roseum*, *C. candelabrum*, *C. rosea*, *C. rogersoniana*, *Nectria coccinea*, *N. inventa*, and *N. lucida* (Han et al. 2020). *C. rosea*, in particular, is considered a promising biocontrol agent against several economically significant plant pathogens, including Sclerotia of *S. sclerotiorum*, *Phymatotrichum omnivorum* Duggar, *Pythium tracheiphilum*, *Helminthosporium solani*, *Verticillium dahlia*, *Rhizoctonia solani*, *Fusarium culmorum*, and *Botrytis cinerea* (Sutton et al. 1997; Lysøe et al. 2017). The modes of action employed by *Clonostachys* species include competition, lysis, mycoparasitism, and systemic resistance (Roberti et al. 2008). Additionally, they produce various antifungal secondary metabolites, such as polyketides and nitrogenous compounds, which exhibit antimicrobial, antifungal, insecticidal, antibacterial, and nematocidal properties (Rodriguez et al. 2011).

Similar to *Trichoderma* species, *Clonostachys* are the effective biocontrol agents that are involved in the production of cell wall-degrading and cellulolytic enzymes (Bischof et al. 2016). They can directly antagonize other plant pathogenic fungi by secreting enzymes like endochitinases, cellulases, pectate lyases, β-1,3 and β-1,6-glucanases, proteases, and antibiotics such as gliotoxin, viridin, and peptaibols (Karlsson et al. 2015; Mukherjee et al. 2013; Atanasova et al. 2018). Furthermore, *Clonostachys* can promote plant growth and induce systemic resistance against pathogen attacks. They also indirectly induce host defense responses, resulting in root surface colonization (Roberti et al. 2008; Dubey et al. 2016). The ability to tolerate the production of secondary metabolites is a notable feature in the biology of *C. rosea* (Karlsson et al. 2015). This chapter aims to provide a comprehensive understanding of the biological mechanisms underpinning *Clonostachys* species' interactions with plant pathogens, ultimately contributing to the enhancement of agricultural production systems.

2 Biology and Diversity of *Clonostachys* Fungi

Clonostachys rosea, initially described as *Gliocladium roseum* by Bainier in 1907, has earned recognition in the scientific community. Advances in morphology and DNA sequencing, as demonstrated by Schroers et al. in 1999, prompted the reclassification of *G. roseum* to *C. rosea*. This remarkable soil fungus exhibits a global presence, demonstrating a remarkable capacity for decomposing decaying plant material and thriving as a competitive saprophyte. Beyond its saprotrophic ability, *C. rosea* is renowned as a soilborne ascomycete with well-established mycoparasitic and antagonistic abilities against various economically significant plant pathogens and nematodes (Rodriguez et al. 2011; Sutton et al. 1997; Luongo et al. 2005; Zhang et al. 2008). It showcases an impressive adaptability, inhabiting a wide array of environmental conditions and displaying a diverse host range, from tropical to arid deserts (Toledo et al. 2006).

Clonostachys rosea has been documented in an array of environments, including grasslands, woodlands, freshwater ecosystems, forests, wildlands, and coastal soils with neutral to alkaline pH (Zhai et al. 2016). Soil serves as a prominent habitat, and this fungus has been recurrently associated with cysts of *Globodera* and *Heterodera* nematodes, as well as deceased insects and other living nematodes (Atanasova et al. 2018). In addition to its competence as a mycoparasite, it has revealed an endophytic capability, colonizing various plant tissues. Throughout its life cycle, the fungus produces two distinct types of spores: conidia and chlamydospores (Sun et al. 2019). Conidia may vary in size, with dimensions ranging from 2.9 to 1.6 μm in length and 0.3 to 0.1 μm in width. Chlamydospores are relatively larger, measuring 5.0 to 5.8 μm in length and 0.4 μm in width. Conidiophores may be either verticillate or penicillately branched, conidia originating from verticillate branches generally being more elongated than those from penicillately branched conidiophores (Schroers et al. 1999; Sutton et al. 1997). Chlamydospores are produced under adverse environmental conditions, such as unfavorable pH levels and low temperatures, and they exhibit greater resistance to environmental stressors (Sun et al. 2018). Colony color varies depending on conditions, appearing greyish white to yellowish-orange on potato dextrose agar medium (Schroers et al. 1999; Sutton et al. 1997).

3 Morphology of *Clonostachys*

The sterile mycelium of *C. rosea* is hyaline and subcortical, aggregating locally to form prosenchymatous tissue, and then evolving into erumpent stromata. The stroma consists of angular to subglobose cells that either merge into sporodochia or give rise to perithecia on the surface (Karlsson et al. 2015). Perithecia are typically found in clusters of 5–50, occasionally solitary, and are well-developed, erumpent, and typically associated with the stroma or ascomata. The stroma, often globose to sub-globose, varies in color, from orange to yellowish-orange or brownish-orange when dry, and lacks a noticeable papilla. The perithecial wall ranges in thickness from 25 to 60 μm, with 2 to 4 layers of angular cells merging at the stroma's base, leading to a small ostiole in the outer region. The outer ostiolar apparatus transitions into the oblong inner region, lined with thin-walled periphyses (Schroers et al. 1999). Asci are clavate, narrow, measuring 41–49 μm in length and 4.5–6.3 μm in width. Perithecia may also contain apical paraphyses. Ascospores are ellipsoidal to oblong-ellipsoidal, bi-celled, and vary in size from 7.5 to 14.5 μm in length and 2.3 to 4.7 μm in width, often exhibiting spines but occasionally being smooth-walled (Zhao et al. 2023). *C. rosea* colonies may vary in color, with older colonies displaying a faded center and yellow pigmentation diffusing beyond the colony. The conidial mass may appear as orange to yellow granules due to agar pigmentation. Conidiophores are dimorphic, with primary conidiophores being verticillate and secondary conidiophores penicillate (Bainier 1907). Primary conidiophores, prevalent in earlier colony parts, display aerial mycelium, monoverticillate with short side branches, often bearing deviating branches and phialides. The stipe measures between 25–70 μm in length and 3.5–5.5 μm in width at the base. Phialides are typically formed in whorls of 2–5, tapering towards the end. Each phialide produces small, hyaline, slightly curved conidia with laterally displaced hilae (a scar at the base of a conidium) (Moreira et al. 2016).

4 Biodiversity and Phylogeny of *Clonostachys*

The asexual stage of the genus *Clonostachys* was initially recognized by Corda in 1839, characterized by its penicillate conidiophores and imbricate conidia (Schroers et al. 1999). The teleomorphic stage, known as *Bionectria*, ascomata are typically located on pseudoparenchymatous stromata that can arise directly on substrates, displaying white, pale yellowish, or orange-brown walls that may be either warted or smooth (Wang et al. 2023). These ascomata are mainly subglobose to globose and contain thick-walled ascospores that are often spinulate, striated, or slightly warted in appearance (Moreira et al. 2016). To distinguish among different isolates of *Clonostachys* were resolved into taxonomic complexities, DNA sequencing of the Internal Transcribed Spacer regions and the β-tubulin (*TUB2*) gene (Wang et al. 2023; Zeng and Zhuang 2022; Prasher and Chauhan 2017). Additional DNA sequences, including ATP citrate lyase (*ACL1*), the larger subunit RNA polymerase II (*RPB1*), and Translation elongation factor 1-α (*TEF1*) gene regions, have been used to further differentiate between species (Moreira et al. 2016; Wang et al. 2023; Torcato et al. 2020). Multi-locus phylogenetic approaches provide valuable insights into species identity within *Clonostachys*. Phylogenetic analysis of nuclear ribosomal large subunit (*nrLSU*) sequences also contributes to our understanding of the phylogenomic relationships between *Clonostachys* and other related genera (Prasher and Chauhan 2017; Luo and Zhuang 2010). The genome size of *C. rosea* varies from 55.4 to 58.3 Mbp, with more than 14,000 genes identified, particularly those encoding cell wall-degrading enzymes, such as glucanases, proteinases, and chitinases, as well as polyketide synthases, ABC transporters, and monooxygenases (Karlsson et al. 2015; Dubey et al. 2016).

5 Diverse Applications of *Clonostachys* Fungi

Clonostachys is a versatile genus of fungi that thrives as a saprophyte in soil. This chapter delves into the numerous applications of *Clonostachys*, shedding light on its pivotal role in various fields.

5.1 As a Biocontrol Agent

Clonostachys is a remarkable multifaceted organism. It acts as an endophyte, colonizing living plants, a saprophyte, digesting soil debris, and a parasite of fungi and nematodes (Anwar et al. 2018). The fungus emits a diverse array of volatile chemical molecules that prove detrimental to various organisms, including fungi, bacteria, and insects (Geiger et al. 2022). This exceptional adaptability positions act as an effective biological tool for pest and pathogen management strategy (Pastor et al. 2023; Zhang et al. 2008). Notably, *C. rosea* emerges as a potent biocontrol agent against destructive seed and soil-borne pathogens such as *Sclerotinia sclerotiorum*, *Fusarium culmorum*, and *Fusarium oxysporum* (Rodriguez et al. 2011; Jensen et al. 2007). An endophyte fungus *Clonostaschys rosea* EC 28, isolated from *Ceiba pentandra*, showed the antifungal properties against *A. burnsii* and *S. rolfsii* infecting tomato. Using, Open silica gel column chromatography method, the secondary metabolites were extracted and purified, wherein verticillin D inhibited spore germination of *A. burnsii* and the mycelial growth of *S. rolfsii* (Dominique et al. 2023). *Clonostachys rosea* was also shown to create glisoprenin compounds that particularly suppressed appressorium production by the rice blast fungus *Magnaporthe oryzae* (Thines et al. 1998). An array of isolates also demonstrates their biocontrol prowess against various pathogens, as summarized in Table 1 and Fig. 1.

Table 1 Different species of *Clonostachys* acting against plant pathogens

Isolate name	Pathogen	Crop	References
Clonostachys rosea strain BAFC387	*Sclerotinia sclerotiorum*	Soil suppressive pathogens	Rodriguez et al. (2011)
	Rhizoctonia solani		
	Fusarium solani		
	Colletotrichum truncatum		
	Macrophomina phaseolina		
C. rosea strain IK726	*Fusarium culmorum*	Barley	Jensen et al. (2007)
C. rosea, strain CR47	*Fusarium culmorum*	Wheat	Roberti et al. (2006)
C. rosea f. *catenulata*	*Fusarium oxysporum* f. sp. *radicis-cucumerinum*	Cucumber	Chatterton and Punja (2010)
C. rosea f. *catenulata*	*Botrytis cinerea*	Geranium	Chatterton and Punja (2012)
C. byssicola and Clonostachys rosea	*Phytophthora palmivora*	Cocoa	Krauss et al. (2013)
C. rosea strain ACM941	*Fusarium graminearum*	Wheat	Demissie et al. (2018)
C. byssicola, C. rhizophaga and *C. rosea*	*H. vastatrix*	Coffee	Kapeua-Ndacnou et al. (2023)
Gliocladium roseum	*Botrytis cinerea*	Chickpea	Burgess et al. (1997)
C. rosea strain 67-1	*Botrytis cinerea*	Tomato	Hasan et al. (2022)
	Sclerotinia sclerotiorum	Soybean	Liu et al. (2016), Sun et al. (2017) and Zhang et al. (2004)
	Fusarium oxysporum f. sp. cucumerinum	Cucumber	Tian et al. (2014)
C. rosea isolate IK726	*Bipolaris sorokiniana*	Barley	Jensen et al. (2016)
	Fusarium graminearum	Wheat	Kosawang et al. (2014)
	Helminthosporium solani	Potato	Lysøe et al. (2017)
C. rosea strain WY-1	*Sclerotinia sclerotiorum*	Cucumber	Wu et al. (2018) and Mouekouba et al. (2014)
C. rosea	*Botrytis cinerea*	Tomato	Meng et al. (2022)
C. chloroleuca, C. pseudochroleuca, C. rhizophaga	*Alternaria grandis*	Potato	Da Silva et al. (2021)
C. rosea TNAU-CR 02	*Sclerotinia sclerotiorum*	Cabbage	Venkatesan et al. (2023)

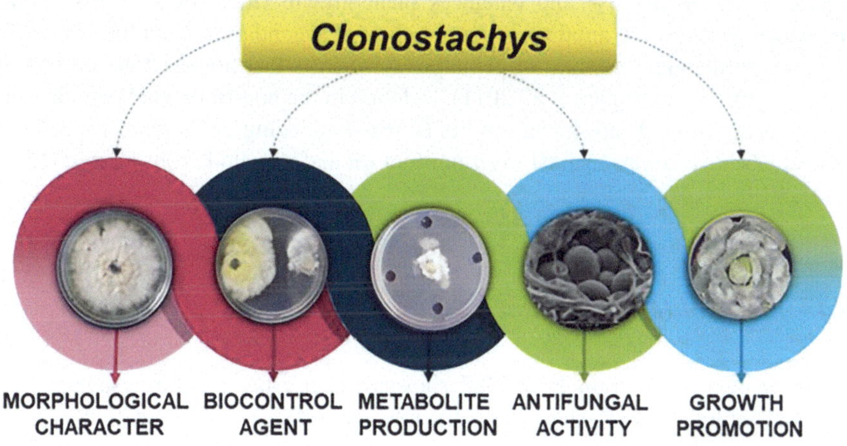

Fig. 1 Schematic representation of *Clonostachys* sp. for agriculture applications

5.2 *As an Entomopathogenic Fungus*

Clonostachys rosea stands out as a potential entomopathogenic fungus, effectively combating leafhopper pests like *Sonesimia grossa* and *Oncometopia tucumana*, whiteflies (*Bemisia tabaci*), and aphids (*Aphis gossypii*) (Toledo et al. 2006; Anwar et al. 2018). Limited studies have explored the pathogenicity of *Clonostachys* against insects, but it has been observed that the mortality of infected insects correlates with the number of spores in contact with them (Mohammed et al. 2022; Al-Zurfi et al. 2023). The use of entomopathogenic fungi as biocontrol agents hinges on the optimal preparation and timely application of inoculum. Notably, *Clonostachys* inoculation exhibits no phytotoxic effects, and the production of enzymes, such as proteases and chitinases, facilitates fungal penetration through the cuticle layer of insect larvae, leading to their mortality (Saranraj and Jayaprakash 2017). *C. rosea* demonstrates substantial mortality rates against various pests, including adult aphids, mango hoppers, Asian citrus psyllids, whiteflies, and other insect stages (Tamta et al. 2022; Yang et al. 2021; Mohammed et al. 2018).

5.3 *Nematicidal Activity*

Clonostachys demonstrates nematophagous capabilities, effectively reducing nematode populations through antagonistic activities. The effectiveness of this fungus is contingent on factors such as temperature and humidity (Cristóbal-Alejo et al. 2021). Notably, *C. rosea* strain TH27 exhibits remarkable nematicidal activity, reducing nematode infections in tomato roots and improving agronomic attributes (Cristóbal-Alejo et al. 2021). In the field, *C. rosea* strain IK726 reduces nematode infections by up to 73% and enhances plant growth (Iqbal et al. 2018). Non-ribosomal peptides are another important class of secondary metabolites in fungi, *C. rosea* producing a mixture of peptaibol substances that hindered the growth of *S. sclerotiorum* (Rodriguez et al. 2011). Deletion in the non-ribosomal peptide synthetase genes of nps1, nps4, and nps5 in *C. rosea*, reducing the mutants capacity to protect wheat seedlings against fusarium foot rot and nematode root disease (Iqbal et al. 2018).

5.4 Metabolite Production

Clonostachys rosea strain BAFC387 showcases an array of abilities, including mycelial inhibition and sclerotia production, attributed to a combination of peptaibiotic metabolite production and mycoparasitic activity against *S. sclerotiorum* (Rodriguez et al. 2011). The peptaibiotic metabolite compounds produced from BAFC387 are frequently involved in antagonistic activity against fungal pathogens. This versatile fungus is also a prolific producer of secondary metabolites that play a pivotal role in antagonizing various pathogens. Some of these compounds, such as bisorbicillinoids and TMC151 C and E, exhibit potent antibacterial activity against six different bacteria (Zhai et al. 2016). Other metabolites, like 11-deoxyverticillin A, verticillin A, Sch52900, and Sch52901, show activity against nematodes and insects (Iqbal et al. 2019; Dong et al. 2005). Using one strain many compounds (OSMAC) methodology, the endophytic fungus *Clonostachys rosea* B5-2 isolated from mangrove plants induced four different new metabolite compounds *viz.*, (−)-dihydrovertinolide, clonostach acids A, B, C added to an unknown compound called (−)-vertinolide (Supratman et al. 2021). Even at a lower concentration of 50 mg l^{-1} the new compound (−)-dihydrovertinolide showed phytotoxic effect against lettuce seedlings. *C. rosea* is also known for its polyketides, including Clonorosein A-D, which display remarkable antifungal activity against economically destructive fungal pathogens (Fatema et al. 2018). During mycoparasitic activity, *C. rosea* involved in induction and biosynthesis of various secondary metabolites and activates different defense related enzyme (Nygren et al. 2018). The mycoparasite *C. rosea* ACM941 act as an efficient biocontrol against fusarium head blight of cereal crop incited by *Fusarium graminearum.* The antagonistic activity of *C. rosea* ACM941 showed upregulation of polyketide synthase gene along with the accumulation of equivalent secondary metabolite products. Correspondingly, fusarielin and fusarin biosynthesis were downregulated (Demissie et al. 2018). These secondary metabolites offer diverse possibilities for enhancing pest and pathogen management in agriculture. For further information, see Table 2 for a list of metabolites produced by *Clonostachys* and their associated biological activities.

Table 2 Secondary metabolites of *Clonostachys* sp. and their biological activities

S. no.	Metabolite class	Name of the metabolite	Name of the organism	Biological activity	References
1	Nitrogen-containing compounds	Cyclosporin A and Cyclosporin C	*Nectria* sp. F-4908	Antimicrobial, antiviral, phytotoxic and enzyme-inhibitory activities	Wang et al. (2019)
		Haematocin	*Nectria haematococca*	Antifungal activity	Suzuki et al. (2000)
		Peptaibiotics	*Clonostachys rosea* BAFC3874	Antifungal activity and antibacterial	Rodriguez et al. (2011)
		Fusarin C	*Nectria coccinea* A56-9	Antifungal activity	Eilbert et al. (1997)
		Gliocladiosin A	*Clonostachys rogersoniana*	Antibacterial activity	Wang et al. (2023)
2	Polyketides	Nectrianolins A-C	*Nectria pseudotrichia* GUHN1	Antifungal activity	Escudero-Leyva et al. (2023)
		Herbarin	*Nectria pseudotrichia* 120-1NP	Antibacterial and phytotoxic activities	Ariefta et al. (2018)
		Dihydrotrichodimer ether A	*Clonostachys rosea* YRS-06	Antibacterial activity	Zhai et al. (2016)
		Usnic acid	*Bionectria ochroleuca* Bo-1	Antibacterial activity	Liu et al. (2014)
		Tetrahydrotrichodimero	*Clonostachys rosea* YRS-06	Antibacterial activity	Zhai et al. (2016)
		Clonoresin A–D	*Clonostachys rosea*	Antifungal activity	Fatema et al. (2018)
3	Terpenoids	Helvolic acid	*Nectria* sp.	Antimicrobial activity	Kong et al. (2018)
		Glisoprenin D	*Gliocladium roseum* HA190-95	Inhibition on appressorium formation of *Magnaporthe grisea*	Thines et al. (1998)
		Ascochlorin	*Nectria* sp.	Antifungal activity	Molnár et al. (2010)
		Glisoprenin A	*Gliocladium* sp. FO 1513	Inhibition on acyl-CoA:cholesterol acyltransferase	Nishida et al. (1992)
		Ilicicolin C = LL-Z 1272δ	*Nectria galligena*	Inhibitory activity on AChE and α-glucuronidase	Gutiérrez et al. (2005)

5.5 *Antagonistic Activity of* Clonostachys *Metabolites*

Antifungal activity experiments based on *C. rosea* growth curves established different set of diffusible, nonvolatile chemical compounds that are directly involved in secondary metabolism. High mycelium growth suppression and sclerotia formation inhibition were reported, showing that metabolites found in *C. rosea* exudates mediated these inhibitory effects. This inhibition may be related to both the nature and concentration of the implicated metabolites (Hadacek and Greger 2000; Singh et al. 2002). Secondary metabolites produced by fungi are highly essential, but they have been shown to influence growth and development as well as interactions with other species. Some of the secondary metabolites released by *C. rosea* may inhibit pathogen development. The volatile as well as non-volatile organic metabolite compounds produced from the *Clonostachys* sp. directly enters into the plasma membrane, triggering modifications in the osmotic balance of the cell (Degenkolb et al. 2006). In case of biocontrol and pathogen interaction, due to parasitic activity it showed changes in pathogen mycelium, including melanization and hyphal cell wall lysis (Tyśkiewicz et al. 2022) The peptaibiotic metabolite compounds directly interrupt fungal growth leading to parasitize the rest of the mycelium (Castro et al. 2023). Metabolites produced from *C. rosea* BAFC3874 triggered optimistic effects on lettuce plants and ultimately results in significant increases in dry shoot weights as well it inhibits several soil-borne pathogens such as *Rhizoconia solani*, *Fusarium solani* and *Macrophomina phaseolina*. Moreover, it has capacity to ameliorate a wide range of abiotic and biotic stresses, particularly, by enhancing nutrient uptake in plants (Shoresh et al. 2010; Jan et al. 2021). Disruption of a polyketide synthase-encoding gene in *C. rosea* resulted in a loss in the ability to produce the previously unknown polyketides clonorosein A and B, which exhibited antifungal activity against *B. cinerea* and *F. graminearum* (Fatema et al. 2018). Besides, deletion of the polyketide synthase pks29 gene resulted in mutation with an reduced ability to control fusarium foot rot on barley (Fatema et al. 2018).

Secondary metabolites produced from *C. rosea*, such as bisorbicillinoids and TMC-151 C and E, demonstrated substantial antibacterial activity against six different bacteria, including *Bacillus megaterium*, *B. subtilis*, *Clostridium perfringens*, *Micrococcus tetragenus*, *Staphylococcus aureus*, and *Escherichia coli* (Zhai et al. 2016). Inoculated mycoparasites enhances plant resistance to pathogens by activating defence enzymes or defense-related genes in plants. The incidence of grey mould disease caused by *B. cinerea* in tomato leaves was reduced when the tomato plant was inoculated with *C. rosea*, wherein the biocontrol agent significantly increases three defence enzymes including glutathione S-transferases, phenyl ammonia lyase and polyphenol oxidase (Mouekouba et al. 2014). A similar behaviour has been observed in potato silver scurf induced by *Helminthosporium solani*. When *C. rosea* was co-inoculated into potato tubers with *H. solani*, the symptoms of silver scurf were reduced compared to *H. solani* infection alone, thereby it increases the genes involved in plant defence systems *viz.*, lipid transporters, glucan endo-1,3-b glucanase, and drug resistance, were upregulated in the potato tubers

(Lysøe et al. 2017). Four compounds obtained from C. rosea includes 11′-deoxyverticillin A, verticillin A, Sch52900 and Sch52901, act against Caenorhabditis elegans and Panagrellus redivivus (Dong et al. 2005).

6 Conclusion

This chapter underscores the paramount role of *Clonostachys* fungi in promoting sustainable agriculture and safeguarding the environment. The genus's versatility, comprising 44 species, has led to its extensive adoption worldwide for a myriad of applications, including biocontrol agents, entomopathogenic fungi, and nematicides. The production of secondary metabolites by *Clonostachys* offers exciting possibilities for pest and pathogen management. Further research into the biological activities, structure-activity relationships, mechanisms of action, and biosynthesis of these metabolites will be instrumental in expanding our knowledge and potential applications. As molecular technology continues to advance, understanding and harnessing the genetic and physiological capabilities of *Clonostachys* for various purposes will be of paramount importance. The upcoming decade holds significant promise for this multifaceted genus and its potential benefits for humanity.

References

Ajith P, Lakshmidevi N (2010) Effect of volatile and non-volatile compounds from *Trichoderma* spp. against *Colletotrichum capsici* incitant of anthracnose on bell peppers. Nat Sci 8(9):265–269

Al-Zurfi S, Kareem AA, Alamry AT, Kadhem ZJ, Port G, Sanderson R (2023) The efficiency of *Beauveria bassiana*, *Metarhizium anisopliae* and *Lecanicillium muscarium* against different stages of the flour beetle, *Tribolium castaneum* (Herbst) (Coleoptera: Tenebrionidae). J Kerbala Agric Sci 10(2):15–32. https://doi.org/10.59658/jkas.v10i2.1182

Anwar W, Ali S, Nawaz K, Iftikhar S, Javed MA, Hashem A, Alqarawi AA, Abd Allah EF, Akhter A (2018) Entomopathogenic fungus *Clonostachys rosea* as a biocontrol agent against whitefly (*Bemisia tabaci*). Biocontrol Sci Tech 28(8):750–760. https://doi.org/10.1080/09583157.2018.1487030

Ariefta NR, Kristiana P, Aboshi T, Murayama T, Tawaraya K, Koseki T, Kurisawa N, Kimura K-I, Shiono Y (2018) New isocoumarins, naphthoquinones, and a cleistanthane-type diterpene from *Nectria pseudotrichia* 120-1NP. Fitoterapia 127:356–361. https://doi.org/10.1016/j.fitote.2018.03.012

Atanasova L, Dubey M, Grujić M, Gudmundsson M, Lorenz C, Sandgren M, Kubicek CP, Jensen DF, Karlsson M (2018) Evolution and functional characterization of pectate lyase PEL12, a member of a highly expanded *Clonostachys rosea* polysaccharide lyase 1 family. BMC Microbiol 18(1):1–19. https://doi.org/10.1186/s12866-018-1310-9

Bainier G (1907) *Gliocladium roseum* sp. nov. et *Cephalosporium acremonium* (Corda). Bull Soc Mycol France 23:111–114. https://doi.org/10.1186/s12934-016-0507-6

Bischof RH, Ramoni J, Seiboth B (2016) Cellulases and beyond: the first 70 years of the enzyme producer *Trichoderma reesei*. Microb Cell Factories 15(1):1–13. https://doi.org/10.1186/s12934-016-0507-6

Burgess D, Bretag T, Keane P (1997) Biocontrol of seedborne *Botrytis cinerea* in chickpea with *Gliocladium roseum*. Plant Pathol 46(3):298–305. https://doi.org/10.1046/j.1365-3059.1997.d01-17.x

Castro GS, Sousa TF, Da Silva GF, Pedroso RC, Menezes KS, Soares MA, Dias GM, Santos AO, Yamagishi ME, Faria JV (2023) Characterization of peptaibols produced by a marine strain of the fungus *Trichoderma endophyticum* via mass spectrometry, genome mining and phylogeny-based prediction. Metabolites 13(2):221. https://doi.org/10.3390/metabo13020221

Chatterton S, Punja Z (2010) Factors influencing colonization of cucumber roots by *Clonostachys rosea* f. *catenulata*, a biological disease control agent. Biocontrol Sci Tech 20(1):37–55. https://doi.org/10.1080/09583150903350253

Chatterton S, Punja Z (2012) Colonization of geranium foliage by *Clonostachys rosea* f. *catenulata*, a biological control agent of botrytis grey mould. Botany 90(1):1–10. https://doi.org/10.1139/b11-076

Corda AKJ (1839) Pracht-flora europaeischer, schimmelbildungen. Bei G. Fleischer, Leipzig

Cristóbal-Alejo J, Diaz-Braga A, Herrera-Parra E, Heredia G, Medina-Baizabal IL, Canto-Canche B, Tzec-Sima M, Gamboa-Angulo M (2021) *Clonostachys rosea* selected by nematicidal screening and its efficacy against *Meloidogyne incognita* in a greenhouse. Biocontrol Sci Tech 31(12):1283–1297. https://doi.org/10.1080/09583157.2021.1944607

Da Silva HA, Teixeira WD, Borges ÁV, Silva Junior AL, Alves KS, Rodrigues Junior OM, De Abreu LM (2021) Biocontrol of potato early blight and suppression of *Alternaria grandis* sporulation by *Clonostachys* spp. Plant Pathol 70(7):1677–1685. https://doi.org/10.1111/ppa.13402

Degenkolb T, Gräfenhan T, Nirenberg HI, Gams W, Brückner H (2006) *Trichoderma brevicompactum* complex: rich source of novel and recurrent plant-protective polypeptide antibiotics (peptaibiotics). J Agric Food Chem 54(19):7047–7061. https://doi.org/10.1021/jf060788q

Demissie ZA, Foote SJ, Tan Y, Loewen MC (2018) Profiling of the transcriptomic responses of *Clonostachys rosea* upon treatment with *Fusarium graminearum* secretome. Front Microbiol 9:1061. https://doi.org/10.3389/fmicb.2018.01061

Dominique S, Alex PG, Dodehe Y, Georges AL (2023) Biological control of *Clonostachys rosea* EC 28 against *Alternaria burnsii* and *Sclerotium rolfsii*. Res Adv Microbiol Biotechnol 3:40–57. https://doi.org/10.9734/bpi/ramb/v3/5301A

Dong JY, He HP, Shen YM, Zhang KQ (2005) Nematicidal epipolysulfanyldioxopiperazines from *Gliocladium roseum*. J Nat Prod 68(10):1510–1513. https://doi.org/10.1021/np0502241

Dubey M, Jensen DF, Karlsson M (2016) The ABC transporter ABCG29 is involved in H2O2 tolerance and biocontrol traits in the fungus *Clonostachys rosea*. Mol Gen Genomics 291:677–686. https://doi.org/10.1007/s00438-015-1139-y

Eilbert F, Thines E, Arendholz W, Sterner O, Anke H (1997) Fusarin C,(7Z)-fusarin C and (5Z)-fusarin C; inhibitors of dihydroxynaphthalene-melanin biosynthesis from *Nectria cocdnea* (*Cylindrocarpon* sp.). J Antibiot 50(5):443–445. https://doi.org/10.7164/antibiotics.50.443

Escudero-Leyva E, Granados-Montero MDM, Orozco-Ortiz C, Araya-Valverde E, Alvarado-Picado E, Chaves-Fallas JM, Aldrich-Wolfe L, Chaverri P (2023) The endophytobiome of wild Rubiaceae as a source of antagonistic fungi against the American leaf spot of coffee (*Mycena citricolor*). J Appl Microbiol 134(5):lxad090. https://doi.org/10.1093/jambio/lxad090

Fatema U, Broberg A, Jensen DF, Karlsson M, Dubey M (2018) Functional analysis of polyketide synthase genes in the biocontrol fungus *Clonostachys rosea*. Sci Rep 8(1):15009. https://doi.org/10.1038/s41598-018-33391-1

Geiger A, Karácsony Z, Geml J, Váczy KZ (2022) Mycoparasitism capability and growth inhibition activity of *Clonostachys rosea* isolates against fungal pathogens of grapevine trunk diseases suggest potential for biocontrol. PLoS One 17(9):e0273985. https://doi.org/10.1371/journal.pone.0273985

Gutiérrez M, Theoduloz C, Rodríguez J, Lolas M, Schmeda-Hirschmann G (2005) Bioactive metabolites from the fungus *Nectria galligena*, the main apple canker agent in Chile. J Agric Food Chem 53(20):7701–7708. https://doi.org/10.1021/jf051021l

Hadacek F, Greger H (2000) Testing of antifungal natural products: methodologies, comparability of results and assay choice. Phytochem Anal 11(3):137–147. https://doi.org/10.1002/(SICI)1099-1565(200005/06)11:3<137::AID-PCA514>3.0.CO;2-I

Han P, Zhang X, Xu D, Zhang B, Lai D, Zhou L (2020) Metabolites from *Clonostachys* fungi and their biological activities. J Fungi 6(4):229. https://doi.org/10.3390/jof6040229

Hasan R, Lv B, Uddin MJ, Chen Y, Fan L, Sun Z, Sun M, Li S (2022) Monitoring mycoparasitism of *Clonostachys rosea* against *Botrytis cinerea* using GFP. J Fungi 8(6):567. https://doi.org/10.3390/jof8060567

Iqbal M, Dubey M, Mcewan K, Menzel U, Franko MA, Viketoft M, Jensen DF, Karlsson M (2018) Evaluation of *Clonostachys rosea* for control of plant-parasitic nematodes in soil and in roots of carrot and wheat. Phytopathology 108(1):52–59. https://doi.org/10.1094/PHYTO-03-17-0091-R

Iqbal M, Dubey M, Broberg A, Viketoft M, Jensen DF, Karlsson M (2019) Deletion of the nonribosomal peptide synthetase gene nps1 in the fungus *Clonostachys rosea* attenuates antagonism and biocontrol of plant pathogenic *Fusarium* and nematodes. Phytopathology 109(10):1698–1709. https://doi.org/10.1094/PHYTO-02-19-0042-R

Jan R, Asaf S, Numan M, Lubna L, Kim KM (2021) Plant secondary metabolite biosynthesis and transcriptional regulation in response to biotic and abiotic stress conditions. Agronomy 11(5):968. https://doi.org/10.3390/agronomy11050968

Jensen DF, Knudsen IM, Lü Beck M, Mamarabadi M, Hockenhull J, Jensen B (2007) Development of a biocontrol agent for plant disease control with special emphasis on the near commercial fungal antagonist *Clonostachys rosea* strain 'IK726'. Australas Plant Pathol 36:95–101. https://doi.org/10.1071/AP07009

Jensen B, Lübeck PS, Jørgensen HJ (2016) *Clonostachys rosea* reduces spot blotch in barley by inhibiting prepenetration growth and sporulation of *Bipolaris sorokiniana* without inducing resistance. Pest Manag Sci 72(12):2231–2239. https://doi.org/10.1002/ps.4260

Kapeua-Ndacnou M, De Abreu LM, De Macedo DM, Da Nóbrega TF, Pereira CM, Evans HC, Barreto RW (2023) Assessing the biocontrol potential of *Clonostachys* species isolated as endophytes from coffea species and as mycoparasites of *Hemileia* rusts of coffee in Africa. J Fungi 9(2):248. https://doi.org/10.3390/jof9020248

Karlsson M, Durling MB, Choi J, Kosawang C, Lackner G, Tzelepis GD, Nygren K, Dubey MK, Kamou N, Levasseur A (2015) Insights on the evolution of mycoparasitism from the genome of *Clonostachys rosea*. Genome Biol Evol 7(2):465–480. https://doi.org/10.1093/gbe/evu292

Kong F-D, Huang XL, Ma QY, Xie QY, Wang P, Chen PW, Zhou LM, Yuan JZ, Dai HF, Luo DQ (2018) Helvolic acid derivatives with antibacterial activities against *Streptococcus agalactiae* from the marine-derived fungus *Aspergillus fumigatus* HNMF0047. J Nat Prod 81(8):1869–1876. https://doi.org/10.1021/acs.jnatprod.8b00382

Kosawang C, Karlsson M, Vélëz H, Rasmussen PH, Collinge DB, Jensen B, Jensen DF (2014) Zearalenone detoxification by zearalenone hydrolase is important for the antagonistic ability of *Clonostachys rosea* against mycotoxigenic *Fusarium graminearum*. Fungal Biol 118(4):364–373. https://doi.org/10.1016/j.funbio.2014.01.005

Krauss U, Ten Hoopen M, Rees R, Stirrup T, Argyle T, George A, Arroyo C, Corrales E, Casanoves F (2013) Mycoparasitism by *Clonostachys byssicola* and *Clonostachys rosea* on *Trichoderma* spp. from cocoa (*Theobroma cacao*) and implication for the design of mixed biocontrol agents. Biol Control 67(3):317–327. https://doi.org/10.1016/j.biocontrol.2013.09.011

Liu Q, Jiang D, Qi Y, Chen C, Xie X, Sun L (2014) Isolation, identification and activity analysis of antimicrobial compound from *Bionectria ochroleuca* strain Bo-1. Acta Phytophylacica Sinica 41(1):41–44

Liu JY, Li SD, Sun MH (2016) Transaldolase gene Tal67 enhances the biocontrol activity of *Clonostachys rosea* 67-1 against *Sclerotinia sclerotiorum*. Biochem Biophys Res Commun 474(3):503–508. https://doi.org/10.1016/j.bbrc.2016.04.133

Luo J, Zhuang WY (2010) New species and new Chinese records of Bionectriaceae (Hypocreales, Ascomycota). Mycol Prog 9:17–25. https://doi.org/10.1007/s11557-009-0613-5

Luongo L, Galli M, Corazza L, Meekes E, Haas LD, Van Der Plas CL, Köhl J (2005) Potential of fungal antagonists for biocontrol of *Fusarium* spp. in wheat and maize through competition in crop debris. Biocontrol Sci Tech 15(3):229–242. https://doi.org/10.1080/09583150400016852

Lysøe E, Dees MW, Brurberg MB (2017) A three-way transcriptomic interaction study of a biocontrol agent (*Clonostachys rosea*), a fungal pathogen (*Helminthosporium solani*), and a potato host (*Solanum tuberosum*). Mol Plant-Microbe Interact 30(8):646–655. https://doi.org/10.1094/MPMI-03-17-0062-R

Meng F, Lv R, Cheng M, Mo F, Zhang N, Qi H, Liu J, Chen X, Liu Y, Ghanizadeh H (2022) Insights into the molecular basis of biocontrol of *Botrytis cinerea* by *Clonostachys rosea* in tomato. Sci Hortic 291:110547. https://doi.org/10.1016/j.scienta.2021.110547

Mohammed AA, Kadhim JH, Kamaluddin ZN (2018) Selection of highly virulent entomopathogenic fungal isolates to control the greenhouse aphid species in Iraq. Egypt J Biol Pest Control 281-7. https://doi.org/10.1186/s41938-018-0079-3

Mohammed AA, Ahmed FA, Younus AS, Kareem AA, Salman AM (2022) Molecular identification of two entomopathogenic fungus *Clonostachys rosea* strains and their efficacy against two aphid species in Iraq. J Genet Eng Biotechnol 20(1):1–8. https://doi.org/10.1186/s43141-022-00347-y

Molnár I, Gibson DM, Krasnoff SB (2010) Secondary metabolites from entomopathogenic Hypocrealean fungi. Nat Prod Rep 27(9):1241–1275. https://doi.org/10.1039/C001459C

Moreira GM, Abreu LM, Carvalho VG, Schroers HJ, Pfenning LH (2016) Multilocus phylogeny of *Clonostachys* subgenus Bionectria from Brazil and description of *Clonostachys chloroleuca* sp. nov. Mycol Prog 15:1031–1039. https://doi.org/10.1007/s11557-016-1224-6

Mouekouba LD, Zhang L, Guan X, Chen X, Chen H, Zhang J, Zhang J, Li J, Yang Y, Wang A (2014) Analysis of *Clonostachys rosea*-induced resistance to tomato gray mold disease in tomato leaves. PLoS One 9(7):e102690. https://doi.org/10.1371/journal.pone.0102690

Mukherjee PK, Horwitz BA, Herrera-Estrella A, Schmoll M, Kenerley CM (2013) *Trichoderma* research in the genome era. Annu Rev Phytopathol 51:105–129. https://doi.org/10.1146/annurev-phyto-082712-102353

Nishida H, Huang XH, Tomoda H, Omura S (1992) Glisoprenins, new inhibitors of Acyl-CoA: cholesterol acyltransferase produced by *Gliocladium* sp. fo-1513 II. Structure elucidation of glisoprenins A and B. J Antibiot 45(10):1669–1676. https://doi.org/10.7164/antibiotics.45.1669

Nygren K, Dubey M, Zapparata A, Iqbal M, Tzelepis GD, Durling MB, Jensen DF, Karlsson M (2018) The mycoparasitic fungus *Clonostachys rosea* responds with both common and specific gene expression during interspecific interactions with fungal prey. Evol Appl 11(6):931–949. https://doi.org/10.1111/eva.12609

Pastor N, Palacios S, Torres A (2023) Microbial consortia containing fungal biocontrol agents, with emphasis on *Trichoderma* spp.: current applications for plant protection and effects on soil microbial communities. Eur J Plant Pathol:1–28. https://doi.org/10.1007/s10658-023-02773-1

Prasher IB, Chauhan R (2017) *Clonostachys indicus* sp. nov. from India. Kavaka 48(1):22–26

Qualhato TF, Lopes FA, Steindorff AS, Brandao RS, Jesuino RS, Ulhoa CJ (2013) Mycoparasitism studies of *Trichoderma* species against three phytopathogenic fungi: evaluation of antagonism and hydrolytic enzyme production. Biotechnol Lett 35:1461–1468. https://doi.org/10.1007/s10529-013-1225-3

Roberti R, Badiali F, Pisi A, Veronesi A, Pancaldi D, Cesari A (2006) Sensitivity of *Clonostachys rosea* and *Trichoderma* spp. as potential biocontrol agents to pesticides. J Phytopathol 154(2):100–109. https://doi.org/10.1111/j.1439-0434.2006.01069.x

Roberti R, Veronesi A, Cesari A, Cascone A, Di Berardino I, Bertini L, Caruso C (2008) Induction of PR proteins and resistance by the biocontrol agent *Clonostachys rosea* in wheat plants infected with *Fusarium culmorum*. Plant Sci 175(3):339–347. https://doi.org/10.1016/j.plantsci.2008.05.003

Rodriguez MA, Cabrera G, Gozzo F, Eberlin M, Godeas A (2011) *Clonostachys rosea* BAFC3874 as a *Sclerotinia sclerotiorum* antagonist: mechanisms involved and potential as a biocontrol agent. J Appl Microbiol 110(5):1177–1186. https://doi.org/10.1111/j.1365-2672.2011.04970.x

Rossman AY, Seifert KA, Samuels GJ, Minnis AM, Schroers HJ, Lombard L, Crous PW, Põldmaa K, Cannon PF, Summerbell RC (2013) Genera in *Bionectriaceae*, *Hypocreaceae*, and *Nectriaceae* (Hypocreales) proposed for acceptance or rejection. IMA Fungus 4(1):41–51. https://doi.org/10.5598/imafungus.2013.04.01.05

Saranraj P, Jayaprakash A (2017) Agrobeneficial entomopathogenic fungi-*Beauveria bassiana*: a review. Indo-Asian J Multidiscip Res 3(2):1051–1087

Schroers HJ, Samuels GJ, Seifert KA, Gams W (1999) Classification of the mycoparasite *Gliocladium roseum* in *Clonostachys* as *C. rosea*, its relationship to *Bionectria ochroleuca*, and notes on other *Gliocladium*-like fungi. Mycologia 91(2):365–385. https://doi.org/10.1080/00275514.1999.12061028

Shoresh M, Harman GE, Mastouri F (2010) Induced systemic resistance and plant responses to fungal biocontrol agents. Annu Rev Phytopathol 48:21–43. https://doi.org/10.1146/annurev-phyto-073009-114450

Singh SB, Herath K, Guan Z, Zink DL, Dombrowski AW, Polishook JD, Silverman KC, Lingham RB, Felock PJ, Hazuda DJ (2002) Integramides A and B, two novel non-ribosomal linear peptides containing nine C α-methyl amino acids produced by fungal fermentations that are inhibitors of HIV-1 integrase. Org Lett 4(9):1431–1434. https://doi.org/10.1021/ol025540a

Sun ZB, Sun MH, Zhou M, Li SD (2017) Transformation of the endochitinase gene Chi67-1 in *Clonostachys rosea* 67-1 increases its biocontrol activity against *Sclerotinia sclerotiorum*. AMB Express 7(1):1–9. https://doi.org/10.1186/s13568-016-0313-x

Sun ZB, Wang Q, Zhang J, Jiang WZ, Li SD, Ma GZ, Sun MH (2018) The transcription factor-encoding gene crtf is involved in *Clonostachys chloroleuca* mycoparasitism on *Sclerotinia sclerotiorum*. Microbiol Res 210:6–11. https://doi.org/10.1016/j.micres.2018.03.002

Sun ZB, Zhang J, Sun MH, Li SD (2019) Identification of genes related to chlamydospore formation in *Clonostachys rosea* 67-1. MicrobiologyOpen 8(1):e00624. https://doi.org/10.1002/mbo3.624

Sun ZB, Li SD, Ren Q, Xu JL, Lu X, Sun MH (2020) Biology and applications of *Clonostachys rosea*. J Appl Microbiol 129(3):486–495. https://doi.org/10.1111/jam.14625

Supratman U, Suzuki T, Nakamura T, Yokoyama Y, Harneti D, Maharani R, Salam S, Abdullah FF, Koseki T, Shiono Y (2021) New metabolites produced by endophyte *Clonostachys rosea* B5-2. Nat Prod Res 35(9):1525–1531. https://doi.org/10.1080/14786419.2019.1656629

Sutton JC, Li DW, Peng G, Yu H, Zhang P, Valdebenito-Sanhueza R (1997) *Gliocladium roseum* a versatile adversary of *Botrytis cinerea* in crops. Plant Dis 81(4):316–328. https://doi.org/10.1094/PDIS.1997.81.4.316

Suzuki Y, Takahashi H, Esumi Y, Arie T, Morita T, Koshino H, Uzawa J, Uramoto M, Yamaguchi I (2000) Haematocin, a new antifungal diketopiperazine produced by *Nectria haematococca* Berk, et Br.(880701a-l) causing nectria blight disease on ornamental plants. J Antibiot 53(1):45–49. https://doi.org/10.7164/antibiotics.53.45

Tamta AK, Pandey R, Sharma JR, Rai R, Barman M, Mg D, Mitra D, Mohapatra PK, Sami R, Al-Mushhin AA (2022) First record of *Clonostachys rosea* (Ascomycota: Hypocreales) entomopathogenic fungus in the mango hopper *Amritodus atkinsoni* (Hemiptera: Cicadellidae). Pathogens 11(12):1447. https://doi.org/10.3390/pathogens11121447

Thines E, Eilbert F, Anke H, Sterner O (1998) Glisoprenins C, D and E, new inhibitors of appressorium formation in *Magnaporthe grisea*, from cultures of *Gliocladium roseum* 1. Production and biological activities. J Antibiot 51(2):117–122. https://doi.org/10.7164/antibiotics.51.117

Tian T, Li SD, Sun MH (2014) Synergistic effect of dazomet soil fumigation and *Clonostachys rosea* against cucumber Fusarium wilt. Phytopathology 104(12):1314–1321. https://doi.org/10.1094/PHYTO-11-13-0297-R

Toledo A, Virla E, Humber R, Paradell S, Lastra CL (2006) First record of *Clonostachys rosea* (Ascomycota: Hypocreales) as an entomopathogenic fungus of *Oncometopia tucumana* and *Sonesimia grossa* (Hemiptera: Cicadellidae) in Argentina. J Invertebr Pathol 92(1):7–10. https://doi.org/10.1016/j.jip.2005.10.005

Torcato C, Gonçalves MF, Rodríguez-Gálvez E, Alves A (2020) *Clonostachys viticola* sp. nov., a novel species isolated from *Vitis vinifera*. Int J Syst Evol Microbiol 70(7):4321–4328. https://doi.org/10.1099/ijsem.0.004286

Tyśkiewicz R, Nowak A, Ozimek E, Jaroszuk-Ściseł J (2022) *Trichoderma*: the current status of its application in agriculture for the biocontrol of fungal phytopathogens and stimulation of plant growth. Int J Mol Sci 23(4):2329. https://doi.org/10.3390/ijms23042329

Venkatesan RM, Muthusamy K, Iruthayasamy J, Prithiviraj B, Kumaresan PV, Lakshmanan P, Perianadar IV (2023) First report of *Clonostachys rosea* as a Mycoparasite on *Sclerotinia sclerotiorum* causing head rot of cabbage in India. Plan Theory 12(1):199. https://doi.org/10.3390/plants12010199

Wang Y, Ren J, Li H, Pan Y, Liu X, Che Y, Liu G (2019) The disruption of verM activates the production of gliocladiosin A and B in *Clonostachys rogersoniana*. Org Biomol Chem 17(28):6782–6785. https://doi.org/10.1039/C9OB01102A

Wang Y, Tang DX, Luo R, Wang YB, Thanarut C, Dao VM, Yu H (2023) Phylogeny and systematics of the genus *Clonostachys*. Front Microbiol 14:1117753. https://doi.org/10.3389/fmicb.2023.1117753

Wu HQ, Sun LL, Fang L, Wang ZY, Cao CW (2018) Preparation of dry flowable formulations of *Clonostachys rosea* by spray drying and application for *Sclerotinia sclerotiorum* control. J Integr Agric 17(3):613–620. https://doi.org/10.1016/S2095-3119(17)61811-2

Yang Y, Chen Y (2004) Advances on mechanism of mycoparasites penetrating phytopathogens. J Southwest For Coll 24:70–74

Yang Z, Wu Q, Fan J, Huang J, Wu Z, Lin J, Bin S, Shu B (2021) Effects of the entomopathogenic fungus *Clonostachys rosea* on mortality rates and gene expression profiles in *Diaphorina citri* adults. J Invertebr Pathol 179:107539. https://doi.org/10.1016/j.jip.2021.107539

Zeng ZQ, Zhuang WY (2022) Three new species of *Clonostachys* (Hypocreales, Ascomycota) from China. J Fungi 8(10):1027. https://doi.org/10.3390/jof8101027

Zhai MM, Qi FM, Li J, Jiang CX, Hou Y, Shi YP, Di DL, Zhang JW, Wu QX (2016) Isolation of secondary metabolites from the soil-derived fungus *Clonostachys rosea* YRS-06, a biological control agent, and evaluation of antibacterial activity. J Agric Food Chem 64(11):2298–2306. https://doi.org/10.1021/acs.jafc.6b00556

Zhang YH, Gao HL, Ma GZ, Li SD (2004) Mycoparasitism of *Gliocladium roseum* 67-1 on *Sclerotinia sclerotiorum*. Acta Phytopathol Sin 34:211–214

Zhang L, Yang J, Niu Q, Zhao X, Ye F, Liang L, Zhang KQ (2008) Investigation on the infection mechanism of the fungus *Clonostachys rosea* against nematodes using the green fluorescent protein. Appl Microbiol Biotechnol 78:983–990. https://doi.org/10.1021/acs.jafc.6b00556

Zhao L, Groenewald J, Hernández-Restrepo M, Schroers HJ, Crous P (2023) Revising *Clonostachys* and allied genera in Bionectriaceae. Stud Mycol 62:204–265. https://doi.org/10.3114/sim.2023.105.03

Zin NA, Badaluddin NA (2020) Biological functions of *Trichoderma* spp. for agriculture applications. Ann Agric Sci 65(2):168–178. https://doi.org/10.1016/j.aoas.2020.09.003

Fungal Phytopathogens as a Source of Metabolites for Agricultural Applications

Edurne Baroja-Fernández, Abdellatif Bahaji, and Francisco José Muñoz

Abstract Current and future agricultural practices have to address a number of challenges, including increasing crop productivity and applying environmentally sensitive methodologies under conditions of climate change. In this context, the use of fungal metabolites is emerging as a strategy for ensuring agricultural and environmental sustainability while feeding the world's growing population. In this chapter, we discuss the role of fungal phytopathogens as a source of metabolites for different agricultural applications. Particular attention is given to volatile compounds and cell-free filtrates as sources of bioactive metabolites. The application of fungal phytopathogen metabolites to promote plant growth, mitigate biotic and abiotic stresses, and provide a source of natural herbicides for weed control, are all considered, as are future perspectives. A review of the literature related to this field is presented, providing a comprehensive overview of our current understanding.

Keywords Culture filtrate · Fungal metabolites · Phytopathogen fungi · Sustainable agriculture · Volatile compounds

1 Introduction

According to estimates by the United Nations Food and Agriculture Organization (FAO), the global human population is expected to reach 9.7 billion in 2050 (Giller et al. 2021). Concomitant with this projection is the need to increase the current total food production by approximately 70% (FAO 2018). Within this context,

Edurne Baroja-Fernández, Abdellatif Bahaji and Francisco José Muñoz have contributed equally to this chapter.

E. Baroja-Fernández · A. Bahaji · F. J. Muñoz (✉)
Instituto de Agrobiotecnología (IdAB), CSIC-Gobierno de Navarra, Nafarroa, Spain
e-mail: e.baroja@csic.es; abdellatif.bahaji@csic.es; francisco.munoz@csic.es

J. Poveda et al. (eds.), *Fungal Metabolites for Agricultural Applications*, Fungal Biology, https://doi.org/10.1007/978-3-031-76587-2_13

agriculture will also have to address critical problems such as climate change, abiotic stresses and plant disease (Zhao et al. 2018; Shahzad et al. 2021). To achieve these goals, appropriate strategies are required that will both increase agricultural productivity and reduce environmental impact.

Conventional agricultural practices are dependent on the extensive use of chemical fertilizers and pesticides to increase growth and yield. The indiscriminate use of these agrochemicals can have adverse effects on the soil, microflora, other organisms, the environment, and human health (Baweja et al. 2020; Tudi et al. 2021). There is therefore an urgent need for new approaches to reducing the use of fertilizers and pesticides and improving the sustainability of agriculture.

Many safe and environmentally friendly methods of increasing yields and/or protecting plants from stress and pests, while reducing the excessive application of agrochemicals, rely on the use of plant-beneficial microorganisms (PGPM) as inoculants (López-Bucio et al. 2015; Mehnaz 2016; Ahmad et al. 2018; Singh et al. 2021; Yadav 2021; Hijri 2023). However, soil inoculation with PGPM has various limitations, and it is not always possible to extrapolate the results obtained in pot/glasshouse trials to field efficacy (O'Callaghan et al. 2022; Morcillo et al. 2022). For example, PGPM efficiency depends on abiotic factors, such as soil composition, temperature and water availability (Lopes et al. 2021). PGPM inoculation can also alter native soil microbial populations through trophic competition and antagonistic/synergic interactions with the resident microorganisms (Trabelsi and Mhamdi 2013). Alternative strategies are needed to address these limitations. One of the most innovative approaches is the use of PGPM-derived metabolites as biostimulant products (Naamala and Smith 2020, 2021). Depending on the environmental conditions, beneficial microbes synthesize extensive and diverse metabolites, including carbohydrates, proteins, lipids, amino acids, siderophores, organic acids and phytohormones, which may directly or indirectly affect plant defenses, stimulate plant growth and productivity, and help improve stress tolerance (Ryu et al. 2003; Poveda and González-Andrés 2021; Reddy et al. 2021; Morcillo and Manzanera 2021; Koza et al. 2022). These microorganisms also synthesize and emit a plethora of volatile compounds (VCs) with low molecular mass, low polarity and high vapor pressure (Schulz and Dickschat 2007; Lemfack et al. 2018; Vaishnavi and Osborne 2021), which can disperse widely from their place of synthesis and migrate through the soil and air environments. These VCs could potentially play an important role as semiochemicals, participating in interactions between plants and microorganisms, both belowground and aboveground, and exerting a strong effect on microbial communities. Microbe-emitted VCs include volatile organic compounds (VOCs), which belong to different chemical classes (alcohols, aldehydes, aromatic compounds, ketones, terpenes, etc.), and volatile inorganic compounds (VICs), such as carbon monoxide, carbon dioxide, hydrogen sulfide, hydrogen, nitric oxide, nitrogen dioxide, hydrogen cyanide and ammonia. It has been demonstrated that some of the VICs act as signaling molecules that enhance growth and abiotic stress tolerance (Ameztoy et al. 2019, 2021; García-Gómez et al. 2019, 2020; Gámez-Arcas et al. 2022), and the number of volatiles identified and cataloged has increased in recent years. For example, at least 1860 unique VOCs emitted from 604 bacterial and 340 fungal species have been registered in the mVOC 3.0 database (Lemfack et al. 2018).

VCs emitted by beneficial microorganisms modulate plant growth and promote developmental changes, enhance photosynthesis, improve nutrient acquisition, induce systemic plant resistance, and mitigate biotic stresses (Ryu et al. 2003; Zhang et al. 2009; Garnica-Vergara et al. 2016; Wagh et al. 2021; Chandrasekaran et al. 2022; Srikamwang et al. 2023). However, several studies have shown that the biostimulant capacity of VCs is not necessarily restricted to beneficial microorganisms, but could also be extended to phytopathogenic microorganisms that do not normally interact mutualistically with plants. There is also evidence that plants respond differently to VCs emitted by beneficial and pathogenic microorganisms (Hernández-Calderón et al. 2018; García-Gómez et al. 2020).

Despite the benefits of microbe-derived compounds, there are still several pitfalls associated with their production, conservation and application. For example, the isolation and purification of these compounds are time-consuming processes; some metabolites are relatively crop specific and require appropriate concentrations; they can have antagonist effects on the useful soil microbiome; contradictory effects can be caused by the same compound produced by different microorganisms; and last, but not least, there is insufficient information about how plants perceive some of these compounds (Naamala and Smith 2021). Some of these challenging limitations can be overcome by using cell-free culture filtrates (CFs) of microorganisms (Pellegrini et al. 2020). CFs are liquid mixtures rich in bioactive metabolites derived from broth cultures obtained by centrifugation and subsequent filtration to eliminate all living cells. This last step is necessary to avoid the microbes spreading into the environment, especially if they are phytopathogenic microorganisms.

A number of articles have shown that the application of CFs from phytopathogenic fungi is a sustainable strategy for modulating plant growth (Sung et al. 2011; Jalander and Gachande 2012; Khokhar et al. 2013; Bhajbhuje 2015; Bhatt and Jampala 2020; Baroja-Fernández et al. 2021; Spinelli et al. 2022), mitigating the adverse effects of abiotic stresses (Khan 2011; Waqas et al. 2012; Poveda 2022), fighting bacterial and fungal phytopathogens (Pálfi et al. 2020; Tomas-Grau et al. 2020; Ávila and Poveda 2022; Paul et al. 2023), and stimulating plant-associated beneficial microbiota (Baroja-Fernández et al. 2021). Furthermore, metabolites contained in CFs from phytopathogenic fungi have huge potential as natural herbicides because of their phytotoxic activity (Evidente 2023). CFs from phytopathogenic fungi are therefore very good candidates for the formulation of new biostimulants for modern agricultural processes. This chapter accordingly provides an overview of the potential uses of metabolites from phytopathogenic fungi in sustainable agriculture, including detailed descriptions of some of those uses, and concludes with future perspectives on the use of fungal pathogens as a source of metabolites for agricultural applications.

2 Beneficial Metabolites from Fungal Phytopathogens

Phytopathogens are microorganisms that produce phytopathogenic toxins that can damage crops and reduce crop yields for human consumption. Microbial phytopathogens typically exhibit narrow host specificity, causing disease only in a single, or closely related, plant species (Li et al. 2020). However, they can also colonize a wide variety of plants asymptomatically (Michielse and Rep 2009). Fungal phytopathogenic species synthesize a broad spectrum of toxins with diverse structures, molecular weights and biosynthetic pathways, which can be classified as polyketides, phenolic compounds, terpenoids and nitrogen-containing metabolites (Xu et al. 2021a). Some toxins have been deemed virulence factors for plant invasion, playing a significant role in pathogenicity and disease development (Xu et al. 2021a). The idea that toxins produced by plant pathogens are detrimental to plants (phytotoxic) originated from the experiments of De Bary (1886), and since then they have been considered non-beneficial (poisonous) for plants. Toxins can be classified as host-specific, phytotoxic only to host plants susceptible to the pathogen, or non-specific, with the latter affecting many plants regardless of whether they are a host or non-host of the producing pathogen (Tsuge et al. 2013). In some cases, plant pathogens produce non-specific toxins that lack functional virulence determinants in non-host species, allowing the pathogen to exist as an asymptomatic endophyte (Hardoim et al. 2015; Ranathunge and Sandani 2016). In addition, some plant pathogens are known to produce a variety of toxins with biostimulant and bioprotective properties, which could be used to promote plant growth and as biocontrol agents for plant pests, diseases and weeds (Tarkowski and Vereecke 2014; Pusztahelyi et al. 2015; Peng et al. 2021; Salvatore and Andolfi 2021; Fan et al. 2023). Microbial plant pathogen toxins are therefore being explored as alternative applications in agriculture, as eco-friendly herbicides and biocides to manage weeds, pests and microbial pathogens (Pedras and Yu 2009; Rao and Ramgoapl 2010; Lou et al. 2013; Salvatore et al. 2021; Xu et al. 2021a; Peng et al. 2021; Salvatore and Andolfi 2021; Rangel et al. 2021; Wang et al. 2022; Evidente 2023; Fan et al. 2023). They have also been used to design rapid plant disease diagnostic tests (Xu et al. 2021a), and for *in vitro* selection of tolerant protoplasts, cells and calli to produce resistant plants (Švábová and Lebeda 2005; Rao and Ramgoapl 2010; Kosturova et al. 2012). Nakajima et al. (1987) and Mondal et al. (2000) have reported that some of the metabolites produced by the fungal phytopathogens *Aspergillus niger* and *Neocosposmora vasinfecta*, besides controlling plant diseases, also promote plant growth of cauliflower and lettuce seedlings and plantlets. Taking all these results into account, it is logical to believe that the application of these toxins could help improve crop yields and reduce the use of agrochemicals for weed and disease control. However, the application of toxins in the field of biological control is still at the laboratory stage, and further research is needed (Peng et al. 2021).

The release of VCs is just one of the mechanisms utilized by microbes to influence the growth and fitness of associated plants. VCs emitted by microorganisms have been shown to play a role in improving plant growth and tolerance against abiotic and biotic stresses (Kanchiswamy et al. 2015; Schmidt et al. 2016; Schalchli et al. 2016; Werner et al. 2016; Fincheira et al. 2017, 2021; Schulz-Bohm et al. 2017; Poveda 2021, 2022; Chandrasekaran et al. 2022). It has also been shown that microbial phytopathogens produce and release blends of VCs, organic and inorganic, that promote plant growth, accelerate flowering and enhance yield and resistance to (a)biotic stresses (Börjesson et al. 1990, 1992; Bitas et al. 2015; Weikl et al. 2016; Sánchez-López et al. 2016; Cordovez et al. 2017; Li and Kang 2018; García-Gómez et al. 2019; Moisan et al. 2019; Reese et al. 2020; Wang et al. 2021; Baroja-Fernández et al. 2021). The use of fungal pathogen VCs therefore appears to be a realistic strategy for improving crop yield.

All developmental and dynamic processes of plant life are directed through a complex signaling network of phytohormones regulated by the modification of endogenous phytohormone concentration gradients. Several authors have demonstrated that fungal species produce phytohormones that, when applied to plants, enhance plant growth and (a)biotic stress resistance (Hamayun et al. 2009, 2017; Contreras-Cornejo et al. 2009; Waqas et al. 2012, 2015; Khan et al. 2014; Leitão and Enguita 2016; Goyal and Kalia 2020; Ismail et al. 2021; Keswani et al. 2022). Most plant pathogenic microorganisms produce phytohormones (cytokinins (CKs), auxins, abscisic acid (ABA), gibberellins (GAs), jasmonic acid (JA), ethylene and salicylic acid (SA)) when they invade their host plant, to disturb the endogenous hormone levels of the host plant (Jaroszuk-Ściseł et al. 2019; Peng et al. 2021). They also produce and release phytohormones when grown in liquid culture (Tuomi et al. 1995; Chagué et al. 2002; Dhakal et al. 2012; Hinsch et al. 2015; Khan et al. 2015; Chanclud et al. 2016; Ismail et al. 2021; Baroja-Fernández et al. 2021). It has been demonstrated that these cultured fungal phytohormones can promote plant growth and abiotic stress resistance (Hamayun et al. 2009; Khan et al. 2015; Leitão and Enguita 2016; Ismail et al. 2021), therefore it is highly likely that phytohormones from fungal plant pathogens could be used to improve crop yields for a sustainable agriculture.

Amino acids are known to have multiple functions in plants, for example for stress tolerance, growth and development (Popko et al. 2018; Teixeira et al. 2020; Trovato et al. 2021). The application of amino acids has also been demonstrated to improve agricultural production (Colla et al. 2015; Popko et al. 2018; Pérez-Aguilar et al. 2023; Abdelkader et al. 2023) and there are, therefore, several commercial amino acid-based biostimulants. These biostimulants are mainly obtained from plant and animal hydrolyzed proteins (Calvo et al. 2014; Popko et al. 2018). However, both beneficial and pathogenic fungi secrete amino acids into culture media (Radhakrishnan et al. 2014; Baroja-Fernández et al. 2021). Radhakrishnan et al. (2014) were the first to show that CFs enriched in amino acids secreted by several *Penicillium* species enhance the shoot and root length and biomass of sesame plants. As a result, it is considered highly likely that amino acids produced by fungal pathogens, alone or in combination with other bioactive metabolites, could be used as biostimulants.

The application of CFs from beneficial fungal cultures results in an efficient technology that promotes plant growth and enhances yield and stress tolerance in a wide number of crops (Morcillo et al. 2022). Importantly, it has been shown that these effects are not restricted to CFs from PGPM but also extend to CFs from phytopathogens, which, when grown under appropriate conditions, could be used to enhance the germination, growth, yield and tolerance of several crop plants (Jalander and Gachande 2012; Khokhar et al. 2013; Bhajbhuje 2015; Khan et al. 2015; Pálfi et al. 2020; Bhatt and Jampala 2020; Ismail et al. 2021; Baroja-Fernández et al. 2021; Ávila and Poveda 2022; Spinelli et al. 2022; Morcillo et al. 2022). Compositional analyses of CFs from several fungal phytopathogens have revealed the presence of soluble sugars, amino acids, phytohormones (GAs, CKs, ABA and auxins) and VCs, indicating that they are a complex cocktail of the same types of bioactive molecules that are present in CFs from PGPM, and validating the use of CFs from fungal phytopathogens as a strategy for improving plant production in sustainable agriculture (Hamayun et al. 2009; Khan et al. 2011; Baroja-Fernández et al. 2021; Ismail et al. 2021).

Overall, these studies strongly suggest that fungal phytopathogens are a promising source of plant-beneficial bioactive metabolites for promoting plant growth, increasing yields and improving adaptation/resistance to the (a)biotic stresses of commercial crops cultured under glasshouse and field conditions (Fig. 1). The use of these metabolites, in a pure form or as a cocktail of diverse compounds (CFs or a blend of VCs), would help reduce the amount of agrochemicals required to maintain good crop production levels, and, as a consequence, reduce levels of air, soil and water pollution and improve farmers' economic viability. Table 1 summarizes the microbial phytopathogen bioactive metabolites and their beneficial effects on plants.

3 Plant Growth Promotion Metabolites from Fungal Phytopathogens

To obtain high yields, farmers regularly apply the required main macronutrients in the form of chemical fertilizers. However, excessive use of chemical fertilizers from limited natural resources is increasingly costly, both environmentally and economically. Unfortunately, a high percentage of the applied chemical fertilizers also rapidly becomes unavailable to plants, with serious environmental repercussions (Hodge et al. 2000). As part of the Farm to Fork strategy—one of the central pillars of the European Green Deal—the European Commission aims to reduce nutrient losses by at least 50%, resulting in at least 20% less fertilizer use by 2030, while ensuring no deterioration in soil fertility (Wesseler 2022). In this context, the development of strategies for improving the nutrient-use efficiency of crops will require multidisciplinary approaches in which biotechnology must play a fundamental role. One such approach involves inoculation with a reduced number of microorganisms that are known to stimulate plant growth by regulating the plant's hormonal, nutritional and water balance, solubilization of soil nutrients and nitrogen fixation (Pandian et al. 2015; Dar et al. 2017; Chen and Liao 2017; Schütz et al. 2018;

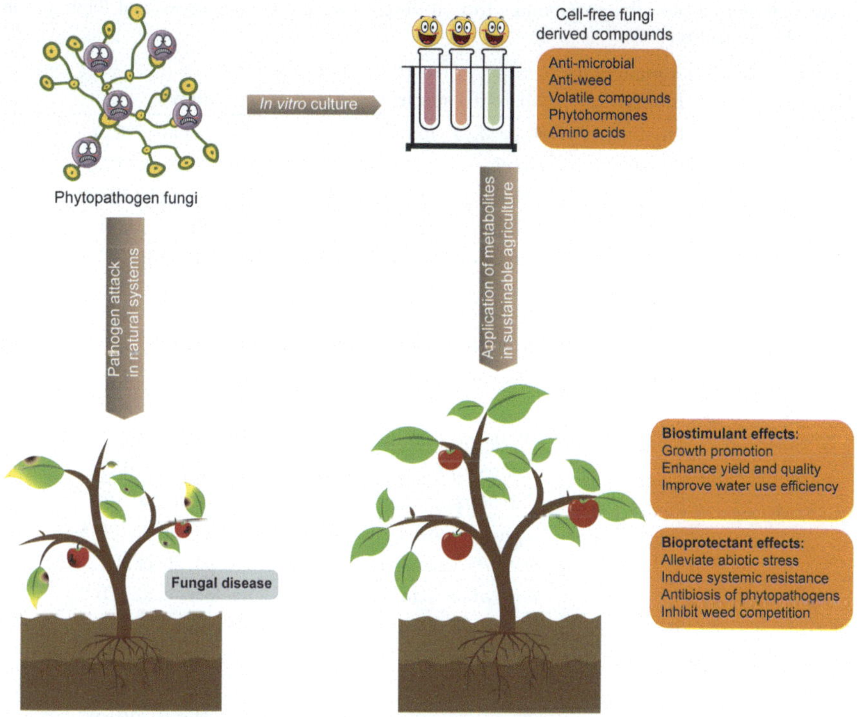

Fig. 1 Schematic representation of the use of fungal phytopathogens for the production of plant-beneficial bioactive metabolites for agronomic applications. Phytopathogen fungi present in nature, when favourable environmental conditions occur, attack plants causing several plant disease symptoms resulting in adverse effects on crop growth and yield. Fungal phytopathogens when grown *in vitro*, in specific growing media and conditions, release to the growing medium specific metabolites with potential agricultural applications. The use of cell-free derived compounds of fungal phytopathogens results in a potential source of different bioactive molecules that when applied to plants can act as biostimulants or bioprotectants increasing crop yield in a sustainable manner. (Figure was created with pictures from https://all-free-download.com)

Bargaz et al. 2018; Kafle et al. 2019; Zeffa et al. 2019; Laurent et al. 2020; Fasusi et al. 2021; Lin and Jones 2022; Faddetta et al. 2023). The application of fungal-derived metabolites is also a promising strategy for reducing dependence on agrochemicals (Vinale et al. 2012; Vinale 2014; Singh et al. 2017; García-Latorre et al. 2023).

Depending on culture conditions (media composition, growth stage and growth conditions), microbial plant pathogens can release bioactive metabolites that exert a positive effect on plant growth and metabolism (Morcillo et al. 2022). For example, Börjesson et al. (1990) showed that the culture medium was a determinant of the VCs produced by the fungal plant pathogen *Penicillium aurantiogriseum*. Conforming with this, Weikl et al. (2016) and Schenkel et al. (2018) highlighted the influence of age and growth media on the volatile profile of *Alternaria alternata* and

Table 1 Reported bioactive metabolites from different fungal phytopathogens and their role in sustainable agriculture

Metabolite	Agricultural function	Microbial specie	References
Toxins	Phytotoxic diseases	*Alternaria alternata*	Wang et al. (2022)
		Drechslera siccans	Hallock et al. (1988)
		Fimetariella rabenhorstii	Bashiri et al. (2020)
		Neofusicoccum sp.	Masi et al. (2020)
		Ustilaginoidea virens	Lu et al. (2016)
		Alternaria, Botritys, Coletotrichum, Fusarium, Gelmintosporium, Phoma, etc.	Xu et al. (2021b)
	Biocide	*Alternaria alternata*	Yagi et al. (1993)
			Shinya et al. (2006)
		Alternaria iridiaustralis	Fan et al. (2023)
		Brothryosphaeria australis	Bara et al. (2013)
		Curvularia pallescens	Elgorban et al. (2017)
		Leptosphaeria maculans	Pedras and Yu (2009)
		Pyricularia grisea	Meepagala et al. (2019)
		Verticillium dahliae	Snelders et al. (2020)
	Herbicide	*Alternaria alternata*	Abbas et al. (1995)
			Duke and Dayan (2011)
			Chen et al. (2005)
			Qiang et al. (2010)
		Alternaria iridiaustralis	Fan et al. (2023)
		Cochliobolus australiensis	Masi et al. (2017)
		Colletotrichum gloeosporioides	Xu et al. (2019, 2021b)
		Curvularia lunata	Nandhini et al. (2019)
		Fusarium oxysporum	Namasivayam et al. (2015)
		Phoma chenopodiicola	Cimmino et al. (2013)
			Evidente et al. (2015)
		Pseudopestolotiopsis theae	Gan et al. (2023)
		Pyricularia grisea	Masi et al. (2020)
		Stagonospora cirsii	Yuzikhin et al. (2007)
Volatile compounds	Growth promotion	*Alternaria alternata, Aspergillus awamori, A. brasiliensis, Botrytis aclada, Colletotrichum glosporiodes, Penicillium digitatum, P. aurantiogriseum, Verticillium dahliae*	Sánchez-López et al. (2016)
		Alternaria alternata, Penicillium aurantiogriseum	García-Gómez et al. (2019) and Baroja-Fernández et al. (2021)
		Botrytis cinerea	Kamaruzzaman et al. (2021)

(continued)

Table 1 (continued)

Metabolite	Agricultural function	Microbial specie	References
		Fusarium oxysporum	Bitas et al. (2015)
		Fusarium oxysporum, Rizoctonia solani, Sclerotinia sclerotum, Verticillium dahliae, V. longisporum	Moisan et al. (2019)
		Leptosphaeria maculans	Poveda (2022)
		Rhizoctonia solani	Cordovez et al. (2017)
		Verticillium alfalfae, V. dahliae, V. longisporum	Li et al. (2018)
	Mitigation of abiotic stress	*Fusarium oxysporum, Verticillium dahliae*	Li and Kang (2018)
		Leptosphaeria maculans	Poveda (2022)
	Mitigation of biotic stress	*Aspergillus awamori*	Fan et al. (2023)
		Botryitis cinerea	Kamaruzzaman et al. (2021)
		Fusarium oxysporum	Terra et al. (2018)
		Fusarium oxysporum, Verticillium dahliae	Li and Kang (2018)
		Fusarum verticilloides	Usseglio et al. (2017)
Phytohormones	Growth promotion	*Alternaria alternata, Fusarium tricinctum*	Khan et al. (2015)
		Alternaria sorghi	Ismail et al. (2021)
		Penicillium funiculosum	Leitão and Enguita (2016)
		Phoma herbarum, Giberella fujikuru	Hamayun et al. (2009)
	Mitigation of abiotic stress	*Fusarium oxysporum*	Hasan (2002)
		Paecilomyces formosus	Waqas et al. (2015)
		Penicillium funiculosum	Khan et al. (2011)
			Leitão and Enguita (2016)
	Mitigation of biotic stress	*Botrytis cinerea*	Chagué et al. (2002)
		Lasiodiplodia theobromae	Salvatore et al. (2020)
Culture filtrates	Growth promotion/ enhance yield	*Alternaria alternata, Alternaria solani*	Bhajbhuje (2015)
		Alternaria alternata, Fusarium tricinctum	Khan et al. (2015)
		Alternaria alternata, Penicillium aurantiogriseum	Baroja-Fernández et al. (2021)
		Alternaria sorghi	Ismail et al. (2021)
		Aspergillus fumigatus, Fusarium proliferatum	Bilal et al. (2018)
		Aspergillus niger, Rhizophus oryzae	Bhatt and Jampala (2020)
		Chaetomium globosum	Zhai et al. (2018)

(continued)

Table 1 (continued)

Metabolite	Agricultural function	Microbial specie	References
		Fusarium oxysporum, Phytium irregular, Rhizoctonia solani	Ávila and Poveda (2022)
		Fusarium solani	Parveen et al. (2019)
		Giberella fujikuroi	Khalmuratova et al. (2015)
		Giberella fujikuroi, Giberella intermedia	Khalmuratova et al. (2021)
		Penicillium aurantiogriseum	Khokhar et al. (2013)
		Pheniciliium funiculosum	Khan et al. (2011)
		Phoma herbarum	Hamayun et al. (2009)
		Sclerotium rolfsii	Singh et al. (2003)
		Shimizuomyces paradoxus	Sung et al. (2011)
	Mitigation of abiotic stress	*Leptosphaeria maculans*	Poveda (2022)
		Phoma glomerata	Waqas et al. (2012)
		Pheniciliium funiculosum	Khan et al. (2011)
	Mitigation of biotic stress	*Alternaria alternata*	Paul et al. (2022)
		Aspergillus niger, Rhizophus oryzae	Bhatt and Jampala (2020)
		Aspergillus niger	Pálfi et al. (2020)
		Aspergillus flavus, A. parasciticus	Mahapatra et al. (2015)
		Capnodium sp., Leptoxyphium sp.	Haituk et al. (2022)
		Colletotrichum acutatum	Tomas-Grau et al. (2020)
		Colletotricum fragariae	Chalfoun et al. (2011)
		Chaetomium globosum	Khan et al. (2019)
			Spinelli et al. (2022)
		Fusarium oxysporum, Phytium irregular, Rhizoctonia solani	Ávila and Poveda (2022)
		Penicillium sp.	Ali et al. (2011)
		Sclerotium rolfsii	Singh et al. (2003)

different *Fusarium oxysporum* strains. Brückner and Blechschmidt (1991) showed that high nitrogen concentrations in growing media interfere with the production of gibberellic acid in *Gibberella fujikuroi*. Bhajbhuje (2015) reported that culture age was an important determinant of *Triticum aestivum*'s response to culture filtrates. Whereas the metabolites retained in 5-day-old culture filtrates from the fungal plant pathogens *A. alternata* and *Alternaria solani* enhanced wheat germination and plantlet growth, 10–25-day-old cultures of the same microbes exerted a negative effect on germination and plantlet growth (Bhajbhuje 2015).

Over the last few years, several papers have demonstrated that VCs emitted by microbial phytopathogens, when grown under appropriate culture conditions, promote plant growth (Table 1), indicating that plant growth promotion by microbial

VCs is a widespread phenomenon, observed not only for beneficial but also for pathogenic microorganisms. Bitas et al. (2015) showed that the VCs emitted by the fungal plant pathogen *F. oxysporum* promote the growth of *Arabidopsis* and tobacco plants. Sánchez-López et al. (2016) showed that VCs emitted by phylogenetically diverse plant pathogenic fungi promote growth and flowering of various plant species, including crops. In line with these results, Cordovez et al. (2017) showed that VCs from the soil-borne phytopathogen *Rhizoctonia solani* enhance the growth and flowering of *Arabidopsis thaliana* through transcriptional changes and auxin signaling. They also showed that *R. solani* VCs primed and accelerated the development of *Arabidopsis* seedlings. Interestingly, Li et al. (2018) demonstrated that VCs produced by three of the most plant-destructive *Verticillium* species (*Verticillium dahliae*, *Verticillium alfalfae* and *Verticillium longisporum*) enhanced *A. thaliana* and *Nicothiana benthamiana* growth. Studies conducted by Moisan et al. (2019) have confirmed that VCs from *F. oxysporum*, *R. solani*, *V. dahliae* and *V. longisporum* enhanced the growth and accelerated flowering of *A. thaliana*. They also found that this phenomenon extended to plants exposed to another fungal plant pathogen, *Sclerotinia sclerotiorum*, and concluded that the effects of microbial VCs on plant growth are not related to their pathogenicity. Kamaruzzaman et al. (2021) have shown that VCs from *Botrytis cinerea* also lead to an increase in tomato growth.

Plant growth promotion by fungal VCs has frequently been attributed to VOCs with molecular masses higher than 45 Da. Studies conducted to identify the specific volatile(s) responsible for the observed growth promotion effect have found acetic acid, acetoin, 2,3-butanediol, 2-heptanone, 2-methyl-1-butanol, 3-methyl-1-butanol, sesquiterpenes, etc. (Ryu et al. 2003; Ditengou et al. 2015; Fincheira et al. 2017; Lee et al. 2019; Camarena-Pozos et al. 2019; Wu et al. 2019; Rahman et al. 2019). However, the growth promotion caused by individual VCs appears to less than that by a blend of VCs, and is dose-dependent (Cordovez et al. 2017; Kamaruzzaman et al. 2021). A mixture of synthetic VCs, at concentrations equivalent to those emitted by *F. oxysporum* grown in soil, inhibited primary root length in *Arabidopsis* plants (Schenkel et al. 2018), while small organic and inorganic volatiles emitted by fungal phytopathogens can promote photosynthesis and *Arabidopsis* plant growth (Ameztoy et al. 2019; García-Gómez et al. 2019, 2020; Gámez-Arcas et al. 2022). Collectively, these studies demonstrate that a broad number of fungal VCs can be used for improving plant growth, and that blends are better candidates than individual VCs.

Most of the studies analyzing the effects of fungal pathogen VCs on plant growth have used *Arabidopsis* plants grown *in vitro* in Murashige and Skoog medium. However, Sánchez-López et al. (2016) showed that *A. alternata* VCs promoted the growth of soil-grown *Arabidopsis*, maize and pepper plants. In agreement with this, Baroja-Fernández et al. (2021) have shown that soil application of distilled VCs from CFs of two fungal phytopathogens enhanced the yield of sweet pepper plants under glasshouse conditions, demonstrating that VCs are important determinants of a plant's response to soil-applied fungal phytopathogen-derived compounds. Additionally, it has been demonstrated that pure VCs (acetic acid, acetoin, 2,3-butanediol, 2-butanol, 2-butanone, 3-pentanol, 6-penthyl pyrone, etc.), emitted by different microbial species, including fungal phytopathogens, promote plant growth under both glasshouse and open field conditions (Ryu et al. 2003; Song and Ryu 2013; D'Alessandro et al. 2014; Velivelli et al. 2015; Kottb et al. 2015; Kong et al. 2018; Rahman et al. 2021; Li et al. 2022). These results indicate that VCs from fungal plant pathogens are indeed good candidates as new plant biostimulants for improving crop yields.

To elucidate the mode of action triggered by air-applied VCs from fungal plant pathogens, omic-technologies, *Arabidopsis* mutants and *Arabidopsis* reporter plants for hormone-inducible gene expression have all been employed. The results obtained have demonstrated that VCs from fungal pathogens promote plant development by triggering highly conserved mechanisms that disturb the endogenous hormone balance in plants (Bitas et al. 2015; Sánchez-López et al. 2016; Cordovez et al. 2017; García-Gómez et al. 2019; Kamaruzzaman et al. 2021). Similar response mechanisms have been reported in plants exposed to VCs from PGPM (Contreras-Cornejo et al. 2009; Salazar-Badillo et al. 2015; Garnica-Vergara et al. 2016; Gao et al. 2022a). Several authors have demonstrated the involvement of auxin signaling in the growth enhancement of different *Arabidopsis* ecotypes, and in *Solanum lycopersicum* and *Nicotiana benthamiana* plants after exposure to VCs emitted by the fungal phytopathogens *B. cinerea*, *F. oxysporum* and *Verticillium* species (Bitas et al. 2015; Li et al. 2018; Kamaruzzaman et al. 2021). Other authors have shown that exposure to VCs from the fungal plant pathogens *B. cinerea*, *A. alternata* and *P. aurantiogriseum* stimulates photosynthesis, promotes the accumulation of photosynthetic pigments, and induces changes in hormone (indole-3-acetic acid (IAA), CKs, GAs, ethylene and ABA) signaling and production in leaves, as a consequence of changes in the transcriptome of VC-exposed *A. thaliana* and tomato plants (Sánchez-López et al. 2016; Ameztoy et al. 2019; García-Gómez et al. 2020; Kamaruzzaman et al. 2021). García-Gómez et al. (2019) have provided evidence that VCs emitted by two different fungal plant pathogens lead to growth promotion but with distinctly different developmental changes. They also demonstrated that the main transcriptional changes in *Arabidopsis* plants exposed to VCs were the consequence of enhanced photosynthetic CO_2 fixation signaling. Ameztoy et al. (2019) showed that *A. alternata* VCs promote changes in the thiol redox-proteome of *Arabidopsis* plants, which could explain the observed plant responses. Studies characterizing the response of plant roots exposed to VCs emitted by the fungal phytopathogens *B. cinerea* and *P. aurantiogriseum* indicated changes in root metabolism and architecture, and improved nutrient and water-use efficiency through both transcriptionally and post-transcriptionally regulated mechanisms (García-Gómez et al. 2020; Kamaruzzaman et al. 2021). Interestingly, García-Gómez et al. (2020) have demonstrated that some of the mechanisms involved in plant responses to VCs emitted by plant growth promoting fungi are different to those arising from fungal phytopathogens. Furthermore, Baroja-Fernández et al. (2021) showed that pepper plant responses to soil-applied VCs did not enhance leaf chlorophyll content and photosynthesis but stimulated soil-beneficial microbiota, indicating that soil-applied VCs operate through different mechanisms than air-applied VCs. Taken together, these data show the importance of studying the specific mechanisms of action of fungal metabolites on plant growth and development.

CFs from beneficial microorganisms have been described as promising new biostimulants for enhancing the growth and yield of crops (Morcillo et al. 2022). Thus, an increasing number of papers have been published demonstrating that CFs from fungal plant pathogens promote the germination, seedling growth, plant growth and yield of several plants, including crop species (Table 1). Sung et al. (2011) showed that treating seeds with sterilized CFs from the fungal pathogen *Shimizuomyces paradoxus* resulted in improved plant height and root length of canola seedlings. Khokhar et al. (2013), studying the effect of CFs from 20 *Penicillium* species on wheat seedling growth, noted that CFs from the fungal phytopathogen *P. aurantiogriseum* resulted in the maximum fresh weight of seedlings. Bhajbhuje (2015) revealed that metabolites from 5-day-old CFs from the fungal pathogens *A. alternata* and *A. solani* could be used as a growth promoter for wheat plants. They demonstrated that seeds soaked with these CFs exhibited improved seed germination and seedling growth, and that foliar application of the same CFs in field trials enhanced the shoot and root length and dry weight of wheat plants. Parveen et al. (2019) demonstrated that CFs of the phytopathogenic fungus *Fusarium solani* increased the germination percentage of some important vegetable crops. In accordance with that, Bhatt and Jampala (2020) showed that foliar application of CFs from two pathogenic fungi, *A. niger* and *Rhizopus oryzae*, on date plants improved the quality and self-life of the fruits. Baroja-Fernández et al. (2021) have confirmed that CFs from the fungal phytopathogens *A. alternata* and *P. aurantiogriseum* promote root growth, accelerate fruit development and improve fruit composition in both glasshouse and field trials, and therefore represent a promising strategy for improving pepper yield and quality.

Compositional analyses of CFs from different fungal phytopathogens (*Phoma herbarum*, *Fusarium triticum*, *A. alternata* and *P. aurantiogriseum*) have revealed the presence of soluble sugars, amino acids, phytohormones (CKs, ABA, auxins and GAs) and VCs, indicating that they are complex cocktails of bioactive molecules (Hamayun et al. 2009; Khan et al. 2011, 2015; Baroja-Fernández et al. 2021), and a growing body of evidence has demonstrated that phytohormones secreted into culture media are important determinants of plant responses to CF treatments (Table 1). Accordingly, Hamayun et al. (2009), Khan et al. (2011) and Khalmuratova et al. (2015) have shown that GAs produced by fungal phytopathogens (*Giberella fujikuroi*, *Penicillium funiculosum* and *P. herbarum*) promote germination and seedling growth of soybean and waito C rice, the latter a known GA-deficient mutant. Khan et al. (2011, 2015) have also shown that the improved shoot and root growth, plant fresh weight and chlorophyll content of rice plants grown *in vitro* after the application of CFs from the fungal phytopathogens *P. funiculosum*, *F. triticum* and *A. alternata*, is mediated by the production of IAA by these fungi. Radhakrishnan et al. (2014) have shown that sesame seeds treated with CFs from two *Penicillium* species display increased shoot and root length and plant biomass as a result of the secretion of variable amounts of amino acids into the CFs, and proposed the possible use of these CFs as biofertilizers for improving yield. However, Baroja-Fernández et al. (2021) have emphasized that it could be the metabolites (hormones and amino acids) present in the CFs of the fungal pathogens *A. alternata* and *P. aurantiogriseum* that are involved in the improved yield observed in pepper plants treated with those CFs in glasshouse experiments. Interestingly, Ismail et al. (2021) demonstrated that CF from the fungal phytopathogen *Alternaria sorghi* contains metabolites that surpass the effect of exogenously applied hormones in increasing plant growth, photosynthetic pigments, carbohydrate content and yield of common beans in field trials. Considered together, these data reinforce the idea that the use of metabolites retained in CFs for yield promotion will also reduce the need for fertilizers and synthetic growth regulators in various crops.

However, to date, there have been no studies focused on unraveling the full potential of CFs from fungal phytopathogens in promoting plant growth and yield, and the mode of action of such formulations is still largely unknown. At best, as indicated above, some of these studies propose that the phytohormones (GAs and IAA) and amino acids released by microbes in their growing media are involved in the plant responses to fungal phytopathogen CFs. Baroja-Fernández et al. (2021) demonstrated that VCs retained in CFs are important determinants in the response of pepper plants to the soil application of fungal phytopathogen CFs. However, they also showed that soil-applied CFs promoted the activation of indigenous soil beneficial microbiota and, as a consequence, they concluded that the mechanisms involved in the plant response could be direct or mediated by changes in soil microbiota composition, or both. Further studies are needed to elucidate this.

Overall, it appears that metabolites from fungal plant pathogens represent a new, sustainable, tool for improving crop resilience and sustainable production, and can be considered a source of new microbial-derived plant biostimulants (Fig. 1). Identification of the fungal phytopathogen metabolites' mode of action on plants, based on rigorous scientific hypotheses and experimental approaches using 'omic' and genetic tools, rather than a try-and-see approach, will be crucial for developing novel and target-specific biostimulant products capable of increasing crop yields. Further investigations are also needed to select the optimum metabolite/blend of metabolites, quantity to be applied and application mode for these new biostimulants, to promote optimum plant growth and yield.

4 Mitigation of Biotic Stress

Biotic stress adversely affects plant growth and development, resulting in 20–40% reduced crop productivity worldwide, with losses caused specifically by pathogens estimated at 10–15% (Mohammad-Razdari et al. 2022). Different approaches have been applied to mitigate biotic stress, chemical control being an effective management option for many plant pests. However, the development of resistance against chemical pesticides and the large impact of their residues on human health and the environment have prompted the search for novel eco-friendly strategies to ensure crop health and production (Lamichhane et al. 2018; Fortunati et al. 2019).

Within the wide range of secondary metabolites produced by fungi, some have antimicrobial properties that could be used to protect plants from pests and diseases, and reduce crop losses caused by pathogens (Raaijmakers and Mazzola 2012; Ghorbanpour et al. 2018; Latz et al. 2018; Köhl et al. 2019; Poveda and Baptista 2021). Antimicrobial metabolites are bioactive compounds with low molecular weights, belonging to heterogenous groups such as polyketide compounds, terpenoids, non-ribosomal peptides, phenols, alkaloids, aliphatic compounds and inorganic acids. They act directly by suppressing or reducing the growth and/or proliferation of phytopathogens (Mousa and Raizada 2013; Rodrigo et al. 2022), and phytopathogenic fungi have been proposed as a good source of these antimicrobial metabolites. Griseofulvin, an antifungal polyketide metabolite compound produced by the phytopathogenic fungi *Penicillium griseofulvum* and *Penicillium lanosocoeruleum*, is the first example in the literature of antifungal secondary metabolites being used in agriculture, to prevent colonization and infection by *Botrytis* species, several *Fusarium* species and *Colletotrichum truncatum* (Dayan et al. 2009; Zhu et al. 2013; Allawi and Al-Taee 2022).

Defensins are antimicrobial peptides distributed widely across the Eukaryote domain (Pálfi et al. 2020). Defensin peptides, produced by the phytopathogenic fungus *A. niger*, show inhibitory effects on the mycelia and conidia of *B. cinerea*, *Guignardia bidwellii* and *Erysiphe necator* (Pálfi et al. 2020). These peptides represent several potential uses, such as antimicrobial effects against a wide range of phytopathogens, including viruses, bacteria and fungi, and have different modes of action, reducing the risk of resistance development against them (Paege et al. 2016; Pálfi et al. 2020). Defensins are also highly suitable for agricultural application, because they do not have any toxic effects on plants, animals and the environment (Szappanos et al. 2006; Meyer 2008). Several other secondary metabolites produced by phytopathogenic fungi with different chemical structures have also been described as antimicrobial metabolites. For example, pretrichodermamide A, isolated from the pathogenic fungus *Epiccocum nigrum*, displays antimicrobial activity towards the plant pathogenic fungus *Ustilago maydis* (Harwoko et al. 2021). Palmitic, stearic and octadecenoic acids, isolated from the plant pathogenic fungus *F. oxysporum*, have been identified as fungicides against several other species of *Fusarium* (Kılıç et al. 2021). Cuminic acid, a benzoic acid compound with antimicrobial activitiy produced by phytopathogenic fungi *Aspergillus* sp. displays inhibitory effects on the important phytopathogen *Phytophthora* sp. (Rodrigo et al. 2022). Macrosporin, anthraquinone compounds produced by phytopathogen fungi *Phoma* sp. exhibited inhibitory effects on *Fusarium graminearum* (Huang et al. 2017; Rodrigo et al. 2022).

VCs emitted by different microorganisms can protect plants against a wide range of pathogens (Morath et al. 2012; Zhang et al. 2014; Schalchli et al. 2016; Choińska et al. 2020; Joo and Hussein 2022; Mestre-Tomás et al. 2023). In particular, VCs from phytopathogenic fungi can control plant diseases by acting directly on the phytopathogen based on the presence of active antimicrobial compounds in the VC blend (Cazar et al. 2005; Wani et al. 2010; Ting et al. 2010; Singh et al. 2011; Gao et al. 2022b). Among them, 2 methyl butane, β-butyrolactone and 2-buten dinitrile, produced by species of the phytopathogenic fungus *Botrytis*, inhibit the growth of the pathogenic *F. oxysporum* (Ting et al. 2010). The mixture of sabinene, 3-methylbutan-1-ol and 2-methylpropan-1-ol, produced by species of the phytopathogen fungus *Phomopsis*, exerts antimicrobial activity against a wide range of plant pathogenic fungi, including *Pythium*, *Phytophthora*, *Sclerotinia*, *Rhizoctonia*, *Fusarium*, *Botrytis*, *Verticillium* and *Colletotrichum* (Singh et al. 2011). Butyrolactone, produced by the phytopathogenic fungus *Aspegillus terreus*, has an inhibitory effect on the pathogen *B. cinerea* (Cazar et al. 2005).

Some phytopathogenic fungi can produce VCs with nematicidal activity, such as 1-Octen-3-ol produced by *F. oxysporum* (Bitas et al. 2015), which affects the juvenile (J_2) stage and egg hatching of the root knot nematode *Meloidogyne incognita* (Veronico et al. 2023). The compound 2-methylbutyl acetate produced by *F. oxysporum* also suppresses the J_2 stage of *M. incognita* and reduces the number of eggs hatching (Terra et al. 2018).

Secondary metabolites produced by fungi, as well as their antimicrobial activity, can induce systemic resistance responses against a wide range of crop pathogens (Kim et al. 2007; Mousa and Raizada 2013; Pieterse et al. 2014; Pascholati et al. 2019; Poveda et al. 2020; Bolívar-Anillo et al. 2020; Fontana et al. 2021; Xu et al. 2021b; Rodrigo et al. 2022). These secondary metabolites are referred to as biotic elicitors, and have great potential in plant disease control applications (Patel et al. 2020; Sák et al. 2021; Guo and Cheng 2022). Based on the signaling pathway induced in plants by elicitors, systemic resistance can be divided into systemic acquired resistance (SAR), which is induced by non-pathogenic microorganisms and biotrophic pathogens and has SA as the main signaling agent, and induced systemic resistance (ISR), which is induced by rhizospheric growth-promoting microorganisms or necrotrophic pathogens and has JA and ethylene as the main signaling agents (Pieterse et al. 2014; Kamle et al. 2020). The mechanisms implicated by these different recognition and response pathways may invoke crosstalk between the three different signal transduction pathways, JA, ethylene and SA.

Fungal elicitors are a group of substances obtained from phytopathogenic fungi and can be classified into two main categories based on their chemical composition: saccharide and protein elicitors (Guo and Cheng 2022). Chitin and chitosan are saccharide elicitors that stimulate the plant immune system against pathogens (Miya et al. 2007; Iriti and Faoro 2009; Fesel and Zuccaro 2016; Geoghegan and Gurr 2017; Pusztahelyi 2018). Glucan is a saccharide elicitor that may enhance resistance in plants by promoting the accumulation of phytoalexins and promoting their activity against pathogenic infection (Yamaguchi et al. 2000; Bowman and Free 2006; Fesel and Zuccaro 2016; Sun et al. 2019; Ruiz-Herrera and Ortiz-Castellanos 2019; Rebaque et al. 2021). Necrosis- and ethylene-inducing peptide 1 (NEP1)-like proteins are protein elicitors purified from CFs of the phytopathogen *F. oxysporum*, and activate plant immune responses (Böhm et al. 2014; Albert et al. 2015;

Seidl and Van Den Ackerveken 2019). They have been shown to be widely distributed among plant pathogenic fungi, including *B. cinerea*, *Boytritis elliptica*, *Colletotrichum higginsianum*, *V. dahliae* and *Magnaporthe oryzae* (Böhm et al. 2014; Albert et al. 2015; Van Den Ackerveken 2017; Seidl and Van Den Ackerveken 2019; Lenarčič et al. 2019). Harpin protein elicitors (Hrip1) are heat-stable, cysteine-free and glycine-rich proteins that can increase plant resistance by eliciting a variety of immune responses (Wei et al. 1992). It has been reported that Hrip1 from the phytopathogenic fungus *Alternaria tenuissima* is able to induce pathogenesis-related (PR) gene expression and SAR in tobacco and *Arabidopsis* plants (Kulye et al. 2012; Miao et al. 2019).

An alternative method for protecting plants from pests and diseases is the application of CFs from different phytopathogenic fungi. These CFs contain specific molecules, such as proteins, amino acids, oligosaccharides, oligopeptides and toxins, that induce systemic immune responses in plants (Huet et al. 1992; da Silva et al. 2016; Rodrigues et al. 2020). Tomas-Grau et al. (2020) observed that CFs obtained from the phytopathogenic fungus *Colletotrichum acutatum* protected strawberry plants against the grey mold *B. cinerea*, inducing both local acquired resistance (LAR) and SAR. Ávila and Poveda (2022) showed that the application of CFs from the phytopathogen *F. oxysporum* to the rhizosphere of *A. thaliana* plants induced gene expression of the SA and JA pathways, in both radicular and aerial tissue, indicating activation of the plant's immune response. Paul et al. (2023) reported that CFs from the phytopathogenic *A. alternata* were able to induce effective immune responses in *Catharanthus roseus* and reduce disease intensity during leaf blight caused by *A. alternata*.

Based on the studies mentioned, phytopathogenic fungi could provide a promising source of bioactive metabolites that protect plants from pests and diseases, improve adaptation/tolerance to biotic stresses, and increase yields of valuable crops in glasshouse and open field conditions (Fig. 1). The use of these bioactive compounds, in a pure form or as a cocktail of diverse compounds (fungal elicitors, CFs or a blend of VCs), would provide an eco-friendly alternative to agrochemicals, thus reducing the impact of agriculture on the environment and human and animal health.

5 Mitigation of Abiotic Stress

Plants are exposed to several abiotic stresses such as drought, salinity, high temperatures, flooding, heavy metal toxicity, UV-B radiation and ozone which have negative effects on crop growth, productivity and quality (Gull et al. 2019). Abiotic stresses can strongly reduce crop performance, with crop yield losses ranging from 50% to 70% (Francini and Sebastiani 2019). This situation will worsen due to climate change and the pressures of an ever-growing human population (Kopecká et al. 2023). Increases in temperature and water scarcity are prognosticated to occur as an outcome of climate change and thus will pose a huge humanity challenge to ensuring food security (FAO 2018).

A safe and environmentally friendly practice to protect plants from abiotic stress is based on the application of metabolites produced by microorganisms as biostimulants. VCs emitted by beneficial microorganisms have been used to mitigate abiotic stress (Jalali et al. 2017; Li and Kang 2018; Poveda 2021). Jalali et al. (2017) reported that VCs emitted by different species of *Trichoderma* induce salt tolerance in *A. thaliana* plants by reducing the H_2O_2 accumulation. According to Cappellari and Banchio (2020), acetoin emitted by *Bacillus amyloliquefaciens* increased tolerance to salinity stress of *Mentha piperita.* Furthermore, 2R,3R-butanediol emitted by *Pseudomonas chlororaphis* was determinant in inducing tolerance to drought in *A. thaliana* (Cho et al. 2008). Recently, Baroja-Fernández et al. (2021) demonstrated that these two VCs are also produced by the phytopathogen fungi *A. alternata* and *P. aurantiogriseum*, therefore it is highly likely that VCs from fungal plant pathogens could be used to improve plant protection against abiotic stress. In this line, Li and Kang (2018) have shown that VCs produced by phytopathogen fungi *V. dahlia* increased salt tolerance in *A. thaliana* through auxin signaling. García-Gómez et al. (2020) have also reported that VCs emitted by phytopathogen fungi *P. aurantiogriseum* enhanced plant tolerance to drought stress in *Arabidopsis* plants due to the increase of water use efficiency in leaves and the down-regulated expression of aquaporins in roots. A recent study conducted by Poveda (2022) showed that VCs from phytopathogen fungi *Leptosphaeria maculans* alleviate the negative effects of drought and salinity stresses on tomato calli, as a consequence of the induction of abiotic stress tolerance genes.

Another possible application of fungal metabolites to mitigate adverse abiotic stress effects on crop production is through the application of fungi CFs. Khan et al. (2011) have reported that CFs from phytopathogenic fungi *P. funiculosum* ameliorated the adverse effects of salinity stress in soybean plants. The authors proposed that the presence of GAs and IAA in the CFs mitigated the negative effects of salinity stress by improving soybean growth and metabolism. Furthermore, Waqas et al. (2012) have shown that treatment with CFs of phytopathogen fungi *P. glomerata* and *Penicillium* sp. significantly mitigated negative effects of drought and salinity stress in cucumber plants due to the altered level of JA, elevated SA and reduced synthesis of ABA. Similar results have been reported by Poveda (2022) who observed that CFs from phytopathogen *L. maculans* mitigated the detrimental effects of abiotic stresses on tomato calli and, proposed that it could be related to the increased expression of ABA-dependent abiotic stress genes and the accumulation of antioxidants in calli tissue.

Taken together, these studies show that metabolites from phytopathogen fungi trigger plants' defense mechanisms to improve tolerance to different abiotic stresses, being an efficient eco-friendly approach to mitigate abiotic stress in a wide range of crops under a changing climate scenario (Fig. 1).

6 Weed Control by Fungal Pathogen-Derived Metabolites

Weeds have long been recognized as one of the most serious pest groups for farmers, causing major crop losses (Cimmino et al. 2015). Weeds compete with crop plants for water and nutrients, drastically reducing crop yields and endangering global food security. Weeds could reduce the future global food supply by 20–40%. In addition, weeds can cause severe damage to important infrastructures such as railways, highways, pipelines and irrigation canals (Kim et al. 2012; Schoenenberger and D'Andrea 2012).

Weed management is therefore an important agricultural practice. Farmers use a variety of methods for weed control, including mechanical and agronomic strategies (Derrouch et al. 2020; Monteiro and Santos 2022). In recent decades, because of their low efficacy, traditional methods have been overshadowed by the extensive and heavy use of synthetic herbicides (Cimmino et al. 2015; Cordeau et al. 2016). Although synthetic herbicides do provide an effective strategy for weed control, excessive and inappropriate use has generated large levels of environmental pollution, as well as posing serious risks to human and animal health (Myers et al. 2016; Mostafalou and Abdollahi 2017; Martins-Gomes et al. 2022). Some of the most toxic herbicides have already been prohibited, and some of the most widely used herbicides (glyphosates) are in the process of being banned. However, the overuse of a small number of herbicides with similar modes of action has also led to the selection of resistant weeds (Perotti et al. 2020; Torra et al. 2021; Hulme 2022). Herbicide resistance is increasing, and the situation is getting worse not only because of the restriction of pesticide use but also because of climate change (Montull and Torra 2023). The International Survey of Herbicide-Resistant Weeds (www.weedscience.org) has reported at least 522 unique cases (species × site of action) of herbicide-resistant weeds globally, including 269 species (154 dicots and 115 monocots) (Heap 2014).

In this context, the search for new herbicides, based on the use of natural, environmentally friendly, compounds, is of paramount importance. In the 1990s, fungal toxins emerged as metabolites with potential use as herbicides (phytotoxins) (Evidente 2023). Phytotoxins are secondary metabolites belonging to different classes of natural compounds that play an important role in the development of disease symptoms in agricultural and forest plants, by targeting the host cell or interfering with the host's immune system (Cimmino et al. 2015; Agudo-Jurado et al. 2023). Phytotoxins are good candidates to be used as herbicides because they are characterized by high target selectivity, low persistence in the environment, reduced soil and water contamination, and provide new mechanisms of action to the herbicide arsenal (Duke and Dayan 2011; Cimmino et al. 2015; Vurro et al. 2018).

Evidente (2023) reviewed the occurrence of fungal metabolites with herbicidal activity in papers published from 1980 to 2022. He found 203 compounds from phytopathogenic fungi that could be exploited as putative bioherbicides, and the list of compounds exhibiting phytotoxicity activity is constantly expanding. Mikami et al. (1971) isolated the first compound (tenuazonic acid) with phytotoxic qualities from the fungus *Alternaria longipes*. This compound produced necrotic spots on tobacco leaves and inhibited the growth of shoots and roots of rice seedlings (Mikami et al. 1971; Iwasaki et al. 1972). Since then, the number of papers on phytotoxins and their possible use as herbicides has increased exponentially over the years.

Phytopathogenic fungi are a source of substances with pathogenic capacity against specific plants. Inspired by the fact that phytotoxins produced by fungi are a good source of natural and safe bioherbicides, several authors and companies have instigated searches for these compounds. Gan et al. (2023) characterized eight compounds from the fungus *Pseudopestalotiopsis theae*, isolated from leaves of *Illigera celebica*. Four of them showed phytotoxic effects against seed germination and seedling growth of *Setaria viridis*, and were proposed as eco-friendly herbicides. Species from the genus *Colletotrichum*, a pathogenic fungus that produces anthracnose disease in a wide range of hosts, have been used to identify and assess phytotoxic metabolites for bioherbicide application (Masi et al. 2018; Xu et al. 2019, 2021a). Xu et al. (2019) isolated the phytotoxin dirhamnolipid from cultures of the fungus *Colletotrichum gloeosporioides* BWH-1. This compound showed bioherbicide activity when used both alone and as an adjuvant to commercial herbicides. Masi et al. (2018) isolated three phytotoxins from *C. gloeosporioides*, but only one showed potential herbicidal activity against *Ambrosia artemisiifolia*, a widespread invasive weed responsible for serious allergies in humans.

Phytotoxins are produced, in most cases, by phytopathogenic fungi. So, to prevent the spread of these microbes in nature, the use of cultures with live microorganisms as bioherbicides is not a feasible option. An alternative could be the use of cultures containing dead microbes (Duke 2023), but the regulation of these products is unclear. For example, the United States Environmental Protection Agency (US EPA) (https://www.epa.gov) provides the following definition for microbial pesticides: "Microbial pesticides consist of a microorganism (e.g., a bacterium, fungus, virus or protozoan) as the active ingredient". The US EPA does not discriminate between microorganisms that are alive or dead, making it difficult to register them as pesticides. For this reason, most published scientific work on phytotoxins (Cimmino et al. 2013; Evidente et al. 2015; Sica et al. 2016; Masi et al. 2018; Poluektova et al. 2018; Xu et al. 2019; Zorrilla et al. 2023) and virtually all microbial bioherbicides currently available on the market (table 1 in Evidente (2023) and table 1 in Duke (2023)) use metabolites purified from liquid or solid cultures of microorganisms.

Despite the extensive research performed on fungal metabolites and the large number of fungal phytotoxins characterized, several technological, biotechnological and physiological factors hinder the commercial development of these products (Vurro et al. 2012; Sica et al. 2016). One of those limitations is the relatively low levels of phytotoxins present in fungal culture filtrates. Production of phytotoxins at pilot or commercial levels requires selecting the best production method, optimizing the nutrient media and fermentation conditions, and designing a simple purification method (Vurro et al. 2012; Cimmino et al. 2015). For these reasons, a small number of secondary metabolites have been studied for scale-up in bioreactors

(Junker et al. 2009). Another challenge is the formulation of a natural herbicide with the best surfactants and adjuvants for altering their physicochemical properties to allow slow and gradual release of the active ingredient in the field (Gitsopoulos et al. 2014; Poluektova et al. 2018). In this respect, the use of nanoparticles offers several advantages, such as enhanced effectiveness of bioherbicide formulations in the field, environmental protection, and improved shelf life (Amna et al. 2019; Kremer 2019; Pérez-de-Luque 2023; Campos et al. 2023). For example, Namasivayam et al. (2015) extracted metabolites with bioherbicidal properties from a liquid culture of *F. oxysporum* and coated chitosan nanoparticles with them. These nanoparticles showed a phytotoxic effect against *Ninidam theenjan*, highlighting their potential.

In the search for new, effective and environmentally friendly alternatives to chemical herbicides, phytotoxins have emerged as a strong candidate (Fig. 1). However, much developmental work still needs to be carried out before they can be widely used as herbicides.

7 Conclusion and Future Perspectives

All information presented in this chapter indicates the great potential of the use of fungal phytopathogen-metabolites for improving crop resilience and sustainable production in climatic changing conditions. We expect to start a new era in which fungal pathogen-metabolites might be an excellent option to reduce costly and unsuitable agrochemicals and water consumption. However, despite the extensive research performed on fungal metabolites, most of the studies have been conducted at laboratory scale and, as a consequence the technology based on its application in the field is still starting and many technical and physiological aspects should be optimized before it can be widely used in open filed. Therefore, further studies will be necessary. First, all metabolites, obtained from phytopathogenic fungi, should be characterized to unveil their possible risk against humans, animals and the environment before being used in field applications. These will require collaborative efforts of different expertise, and technological advancements for selecting the non-toxic metabolites at non-toxic concentrations before manufacturing. Second, the production of the metabolite/blend of metabolites should be optimize. The growth conditions, medium composition and oxygen availability, are aspects to be taken into account for increasing the metabolite(s) yield, and for scaling up the production of the metabolites from lab scale to 100 L-fermentors. Third, studies conducted to select the appropriate application mode (seed coating, foliar application or soil application) and dose for each type of metabolite, according to their chemical-physical properties, will be necessary before product commercialization. Fourth, efficient formulations are required for the delivery of fungal phytopathogen metabolites in the field, providing them with stability and allowing for their progressive release. Nowadays, apart from liquid, solid and encapsulation formulations, nanoencapsulation is emerging as a new efficient technology. Fifth, identification of the fungal phytopathogens metabolites' mode of action on plants based on reasonable scientific hypotheses and some experimental approaches using 'omic' and genetic tools, rather than a try-and-see approach, may be crucial for developing novel and target-specific biostimulant/bioprotectan products capable of increasing crop yields. And last but not least, the economic aspects of these treatments should be addressed. These new products should be economically competitive in comparison with agrochemicals. The yield increase obtained after its field application should compensate its cost, improving farmers´ crop incomes.

We can conclude that, despite having great potential for improving crop production the technology based on fungal phytopathogen-metabolites application in the field is still starting and many technical-economical aspects should be optimized to achieve greater efficacy of the products before their effective implementation in agriculture.

Acknowledgments This work was supported by the Departamento de Universidad, Innovación y Transformación Digital (Gobierno de Navarra) grants PC036-037 BIOMEF and PC148-149 REPABIO.

Bibliography

Abbas HK, Duke SO, Paul RN et al (1995) AAL-toxin, a potent natural herbicide which disrupts sphingolipid metabolism of plants. Pestic Sci 43:181–187. https://doi.org/10.1002/ps.2780430302

Abdelkader M, Voronina L, Baratova L et al (2023) Biostimulants-based amino acids augment physio-biochemical responses and promote salinity tolerance of lettuce plants (*Lactuca sativa* L.). Horticulturae 9:807. https://doi.org/10.3390/horticulturae9070807

Agudo-Jurado FJ, Reveglia P, Rubiales D et al (2023) Status of phytotoxins isolated from necrotrophic fungi causing diseases on grain legumes. Int J Mol Sci 24:5116. https://doi.org/10.3390/ijms24065116

Ahmad M, Pataczek L, Hilger TH et al (2018) Perspectives of microbial inoculation for sustainable development and environmental management. Front Microbiol 9:2992. https://doi.org/10.3389/fmicb.2018.02992

Albert I, Böhm H, Albert M et al (2015) An RLP23–SOBIR1–BAK1 complex mediates NLP-triggered immunity. Nat Plants 1:15140. https://doi.org/10.1038/nplants.2015.140

Ali SZ, Sandhya V, Grover M et al (2011) Effect of inoculation with a thermotolerant plant growth promoting *Pseudomonas putida* strain AKMP7 on growth of wheat (*Triticum* spp.) under heat stress. J Plant Interact 6:239–246. https://doi.org/10.1080/17429145.2010.545147

Allawi MYA, Al-Taee WSQ (2022) Investigation of a new local isolate of *Penicillium lanosocoeruleum* that produces the antifungal griseofulvin. Pak J Med Health Sci 16:380–386. https://doi.org/10.53350/pjmhs22164380

Ameztoy K, Baslam M, Sánchez-López ÁM et al (2019) Plant responses to fungal volatiles involve global posttranslational thiol redox proteome changes that affect photosynthesis. Plant Cell Environ 42:2627–2644. https://doi.org/10.1111/pce.13601

Ameztoy K, Sánchez-López ÁM, Muñoz FJ et al (2021) Proteostatic regulation of MEP and shikimate pathways by redox-activated photosynthesis signaling in plants exposed to small fungal volatiles. Front Plant Sci 12:637976. https://doi.org/10.3389/fpls.2021.637976

Amna, Alharby HF, Hakeem KR, Qureshi MI (2019) Weed control through herbicide-loaded nanoparticles. In: Husen A, Iqbal M (eds) Nanomaterials and plant potential. Springer, Cham, pp 507–527

Ávila AC, Poveda J (2022) Induction of immune response in *Arabidopsis thaliana* treated with phytopathogen filtrates. Biol Life Sci Forum 11:85. https://doi.org/10.3390/IECPS2021-11974

Bara RA, Zerfaß I, Lai D et al (2013) New natural product from *Botryosphaeria australis*, an endophyte from mangrove *Avicennia marina*. Squalen Bull Mar Fish Postharvest Biotechnol 8:139. https://doi.org/10.15578/squalen.v8i3.86

Bargaz A, Lyamlouli K, Chtouki M et al (2018) Soil microbial resources for improving fertilizers efficiency in an integrated plant nutrient management system. Front Microbiol 9:1606. https://doi.org/10.3389/fmicb.2018.01606

Baroja-Fernández E, Almagro G, Sánchez-López ÁM et al (2021) Enhanced yield of pepper plants promoted by soil application of volatiles from cell-free fungal culture filtrates is associated with activation of the beneficial soil microbiota. Front Plant Sci 12:752653. https://doi.org/10.3389/fpls.2021.752653

Bashiri S, Abdollahzadeh J, Di Lecce R et al (2020) Rabenchromenone and rabenzophenone, phytotoxic tetrasubstituted chromenone and hexasubstituted benzophenone constituents produced by the oak-decline-associated fungus *Fimetariella rabenhorstii*. J Nat Prod 83:447–452. https://doi.org/10.1021/acs.jnatprod.9b01017

Baweja P, Kumar S, Kumar G (2020) Fertilizers and pesticides: their impact on soil health and environment. In: Giri B, Varma A (eds) Soil health. Springer, Cham, pp 265–285

Bhajbhuje M (2015) Response of metabolites from culture filtrates of *Alternaria* species against *Triticum aestivum* L. Int J Life Sci 3:55–62

Bhatt K, Jampala SSM (2020) Influence of pre-harvest foliar spray of fungal culture filtrates on post-harvest biology of date fruit harvested at Khalal stage. Postharvest Biol Technol 166:111220. https://doi.org/10.1016/j.postharvbio.2020.111220

Bilal L, Asaf S, Hamayun M et al (2018) Plant growth promoting endophytic fungi *Asprgillus fumigatus* TS1 and *Fusarium proliferatum* BRL1 produce gibberellins and regulates plant endogenous hormones. Symbiosis 76:117–127. https://doi.org/10.1007/s13199-018-0545-4

Bitas V, McCartney N, Li N et al (2015) *Fusarium oxysporum* volatiles enhance plant growth via affecting auxin transport and signaling. Front Microbiol 6. https://doi.org/10.3389/fmicb.2015.01248

Böhm H, Albert I, Oome S et al (2014) A conserved peptide pattern from a widespread microbial virulence factor triggers pattern-induced immunity in Arabidopsis. PLoS Pathog 10:e1004491. https://doi.org/10.1371/journal.ppat.1004491

Bolívar-Anillo H, Garrido C, Collado I (2020) Endophytic microorganisms for biocontrol of the phytopathogenic fungus *Botrytis cinerea*. Phytochem Rev 19. https://doi.org/10.1007/s11101-019-09603-5

Börjesson T, Stöllman U, Schnürer J (1990) Volatile metabolites and other indicators of *Penicillium aurantiogriseum* growth on different substrates. Appl Environ Microbiol 56:3705–3710. https://doi.org/10.1128/aem.56.12.3705-3710.1990

Börjesson T, Stöllman U, Schnürer J (1992) Volatile metabolites produced by six fungal species compared with other indicators of fungal growth on cereal grains. Appl Environ Microbiol 58:2599–2605. https://doi.org/10.1128/aem.58.8.2599-2605.1992

Bowman SM, Free SJ (2006) The structure and synthesis of the fungal cell wall. BioEssays 28:799–808. https://doi.org/10.1002/bies.20441

Brückner B, Blechschmidt D (1991) Nitrogen regulation of gibberellin biosynthesis in *Gibberella fujikuroi*. Appl Microbiol Biotechnol 35:646–650. https://doi.org/10.1007/BF00169631

Calvo P, Nelson L, Kloepper JW (2014) Agricultural uses of plant biostimulants. Plant Soil 383:3–41. https://doi.org/10.1007/s11104-014-2131-8

Camarena-Pozos DA, Flores-Núñez VM, López MG et al (2019) Smells from the desert: microbial volatiles that affect plant growth and development of native and non-native plant species. Plant Cell Environ 42:1368–1380. https://doi.org/10.1111/pce.13476

Campos EVR, Ratko J, Bidyarani N et al (2023) Nature-based herbicides and micro-/nanotechnology fostering sustainable agriculture. ACS Sustain Chem Eng 11:9900–9917. https://doi.org/10.1021/acssuschemeng.3c02282

Cappellari LDR, Banchio E (2020) Microbial volatile organic compounds produced by *Bacillus amyloliquefaciens* GB03 ameliorate the effects of salt stress in *Mentha piperita* principally through acetoin emission. J Plant Growth Regul 39:764–775. https://doi.org/10.1007/s00344-019-10020-3

Cazar ME, Schmeda-Hirschmann G, Astudillo L (2005) Antimicrobial butyrolactone i derivatives from the ecuadorian soil fungus *Aspergillus terreus* thorn. var terreus. World J Microbiol Biotechnol 21:1067–1075. https://doi.org/10.1007/s11274-004-8150-5

Chagué V, Elad Y, Barakat R et al (2002) Ethylene biosynthesis in *Botrytis cinerea*. FEMS Microbiol Ecol 40:143–149. https://doi.org/10.1111/j.1574-6941.2002.tb00946.x

Chalfoun NR, Castagnaro AP, Díaz Ricci JC (2011) Induced resistance activated by a culture filtrate derived from an avirulent pathogen as a mechanism of biological control of anthracnose in strawberry. Biol Control 58:319–329. https://doi.org/10.1016/j.biocontrol.2011.05.007

Chanclud E, Kisiala A, Emery NRJ et al (2016) Cytokinin production by the rice blast fungus is a pivotal requirement for full virulence. PLoS Pathog 12:e1005457. https://doi.org/10.1371/journal.ppat.1005457

Chandrasekaran M, Paramasivan M, Sahayarayan JJ (2022) Microbial volatile organic compounds: an alternative for chemical fertilizers in sustainable agriculture development. Microorganisms 11:42. https://doi.org/10.3390/microorganisms11010042

Chen L, Liao H (2017) Engineering crop nutrient efficiency for sustainable agriculture: engineering crop nutrient efficiency. J Integr Plant Biol 59:710–735. https://doi.org/10.1111/jipb.12559

Chen S, Dai X, Qiang S, Tang Y (2005) Effect of a nonhost-selective toxin from *Alternaria alternata* on chloroplast-electron transfer activity in *Eupatorium adenophorum*. Plant Pathol 54:671–677. https://doi.org/10.1111/j.1365-3059.2005.01249.x

Cho SM, Kang BR, Han SH et al (2008) 2R,3R-butanediol, a bacterial volatile produced by *Pseudomonas chlororaphis* O6, is involved in induction of systemic tolerance to drought in *Arabidopsis thaliana*. Mol Plant Microbe Interact 21:1067–1075. https://doi.org/10.1094/MPMI-21-8-1067

Choińska R, Piasecka-Jóźwiak K, Chabłowska B et al (2020) Biocontrol ability and volatile organic compounds production as a putative mode of action of yeast strains isolated from organic grapes and rye grains. Antonie Van Leeuwenhoek 113:1135–1146. https://doi.org/10.1007/s10482-020-01420-7

Cimmino A, Andolfi A, Zonno MC et al (2013) Chenopodolans A–C: phytotoxic furopyrans produced by *Phoma chenopodiicola*, a fungal pathogen of *Chenopodium album*. Phytochemistry 96:208–213. https://doi.org/10.1016/j.phytochem.2013.10.007

Cimmino A, Masi M, Evidente M et al (2015) Fungal phytotoxins with potential herbicidal activity: chemical and biological characterization. Nat Prod Rep 32:1629–1653. https://doi.org/10.1039/C5NP00081E

Colla G, Nardi S, Cardarelli M et al (2015) Protein hydrolysates as biostimulants in horticulture. Sci Hortic 196:28–38. https://doi.org/10.1016/j.scienta.2015.08.037

Contreras-Cornejo HA, Macías-Rodríguez L, Cortés-Penagos C, López-Bucio J (2009) *Trichoderma virens*, a plant beneficial fungus, enhances biomass production and promotes lateral root growth through an auxin-dependent mechanism in Arabidopsis. Plant Physiol 149:1579–1592. https://doi.org/10.1104/pp.108.130369

Cordeau S, Triolet M, Wayman S et al (2016) Bioherbicides: dead in the water? A review of the existing products for integrated weed management. Crop Prot 87:44–49. https://doi.org/10.1016/j.cropro.2016.04.016

Cordovez V, Mommer L, Moisan K et al (2017) Plant phenotypic and transcriptional changes induced by volatiles from the fungal root pathogen *Rhizoctonia solani*. Front Plant Sci 8:1262. https://doi.org/10.3389/fpls.2017.01262

D'Alessandro M, Erb M, Ton J et al (2014) Volatiles produced by soil-borne endophytic bacteria increase plant pathogen resistance and affect tritrophic interactions: bacterial volatiles and multitrophic interactions. Plant Cell Environ 37:813–826. https://doi.org/10.1111/pce.12220

da Silva A, Freitas K, Batista J et al (2016) Induction of defense mechanisms from filtrates of saprophytic fungi against early blight disease in tomato. Afr J Microbiol Res 10:1849–1859. https://doi.org/10.5897/AJMR2016.8106

Dar TA, Uddin M, Ali A et al (2017) Understanding the dynamics of phosphorus starvation and plant growth. In: Naeem M, Ansari AA, Gill SS (eds) Essential plant nutrients: uptake, use efficiency, and management. Springer International Publishing, Cham, pp 147–154

Dayan FE, Cantrell CL, Duke SO (2009) Natural products in crop protection. Bioorg Med Chem 17:4022–4034. https://doi.org/10.1016/j.bmc.2009.01.046

De Bary A (1886) Über einige Sclerotinien und Sclerotinienkrankheiten. Bot Zeit 44:377–474

Derrouch D, Chauvel B, Felten E, Dessaint F (2020) Weed management in the transition to conservation agriculture: farmers' response. Agronomy 10:843. https://doi.org/10.3390/agronomy10060843

Dhakal R, Bajpai VK, Baek K-H (2012) Production of gaba (gamma – aminobutyric acid) by microorganisms: a review. Braz J Microbiol 43:1230–1241. https://doi.org/10.1590/S1517-83822012000400001

Ditengou FA, Müller A, Rosenkranz M et al (2015) Volatile signalling by sesquiterpenes from ectomycorrhizal fungi reprogrammes root architecture. Nat Commun 6:6279. https://doi.org/10.1038/ncomms7279

Duke SO (2023) Why are there no widely successful microbial bioherbicides for weed management in crops? Pest Manag Sci. https://doi.org/10.1002/ps.7595

Duke SO, Dayan FE (2011) Modes of action of microbially-produced phytotoxins. Toxins 3:1038–1064. https://doi.org/10.3390/toxins3081038

Elgorban AM, El-Samawaty AE-RM, Abd-Elkader OH et al (2017) Bioengineered silver nanoparticles using *Curvularia pallescens* and its fungicidal activity against *Cladosporium fulvum*. Saudi J Biol Sci 24:1522–1528. https://doi.org/10.1016/j.sjbs.2016.09.019

Evidente A (2023) Specialized metabolites produced by phytotopatogen fungi to control weeds and parasite plants. Microorganisms 11:843. https://doi.org/10.3390/microorganisms11040843

Evidente M, Cimmino A, Zonno MC et al (2015) Phytotoxins produced by *Phoma chenopodicola*, a fungal pathogen of *Chenopodium album*. Phytochemistry 117:482–488. https://doi.org/10.1016/j.phytochem.2015.07.008

Faddetta T, Polito G, Abbate L et al (2023) Bioactive metabolite survey of actinobacteria showing plant growth promoting traits to develop novel biofertilizers. Metabolites 13:374. https://doi.org/10.3390/metabo13030374

Fan J, Guo F, Zhao C et al (2023) Secondary metabolites with herbicidal and antifungal activities from marine-derived fungus *Alternaria iridiaustralis*. J Fungi 9:716. https://doi.org/10.3390/jof9070716

FAO (2018) The future of food and agriculture—alternative pathways to 2050. Summary version. Rome. 60pp. Licence: CC BY-NC-SA 3.0 IGO

Fasusi OA, Cruz C, Babalola OO (2021) Agricultural sustainability: microbial biofertilizers in rhizosphere management. Agriculture 11:163. https://doi.org/10.3390/agriculture11020163

Fesel PH, Zuccaro A (2016) β-glucan: crucial component of the fungal cell wall and elusive MAMP in plants. Fungal Genet Biol 90:53–60. https://doi.org/10.1016/j.fgb.2015.12.004

Fincheira P, Parra L, Mutis A et al (2017) Volatiles emitted by *Bacillus* sp. BCT9 act as growth modulating agents on *Lactuca sativa* seedlings. Microbiol Res 203:47–56. https://doi.org/10.1016/j.micres.2017.06.007

Fincheira P, Quiroz A, Tortella G et al (2021) Current advances in plant-microbe communication via volatile organic compounds as an innovative strategy to improve plant growth. Microbiol Res 247:126726. https://doi.org/10.1016/j.micres.2021.126726

Fontana DC, De Paula S, Torres AG et al (2021) Endophytic fungi: biological control and induced resistance to phytopathogens and abiotic stresses. Pathogens 10:570. https://doi.org/10.3390/pathogens10050570

Fortunati E, Mazzaglia A, Balestra GM (2019) Sustainable control strategies for plant protection and food packaging sectors by natural substances and novel nanotechnological approaches. J Sci Food Agric 99:986–1000. https://doi.org/10.1002/jsfa.9341

Francini A, Sebastiani L (2019) Abiotic stress effects on performance of horticultural crops. Horticulturae 5:67. https://doi.org/10.3390/horticulturae5040067

Gámez-Arcas S, Baroja-Fernández E, García-Gómez P et al (2022) Action mechanisms of small microbial volatile compounds in plants. J Exp Bot 73:498–510. https://doi.org/10.1093/jxb/erab463

Gan D, Liu J-Q, Yang Y-J et al (2023) Phytotoxic meroterpenoids with herbicidal activities from the phytopathogenic fungus *Pseudopestalotiopsis theae*. Phytochemistry 206:113522. https://doi.org/10.1016/j.phytochem.2022.113522

Gao Y, Feng J, Wu J et al (2022a) Transcriptome analysis of the growth-promoting effect of volatile organic compounds produced by *Microbacterium aurantiacum* GX14001 on tobacco (*Nicotiana benthamiana*). BMC Plant Biol 22:208. https://doi.org/10.1186/s12870-022-03591-z

Gao Y, Ren H, He S et al (2022b) Antifungal activity of the volatile organic compounds produced by *Ceratocystis fimbriata* strains WSJK-1 and Mby. Front Microbiol 13:1034939. https://doi.org/10.3389/fmicb.2022.1034939

García-Gómez P, Almagro G, Sánchez-López ÁM et al (2019) Volatile compounds other than CO_2 emitted by different microorganisms promote distinct posttranscriptionally regulated responses in plants. Plant Cell Environ 42:1729–1746. https://doi.org/10.1111/pce.13490

García-Gómez P, Bahaji A, Gámez-Arcas S et al (2020) Volatiles from the fungal phytopathogen *Penicillium aurantiogriseum* modulate root metabolism and architecture through proteome resetting. Plant Cell Environ 43:2551–2570. https://doi.org/10.1111/pce.13817

García-Latorre C, Rodrigo S, Santamaría O (2023) Potential of fungal endophytes isolated from pasture species in spanish dehesas to produce enzymes under salt conditions. Microorganisms 11:908. https://doi.org/10.3390/microorganisms11040908

Garnica-Vergara A, Barrera-Ortiz S, Muñoz-Parra E et al (2016) The volatile 6-pentyl-2H-pyran-2-one from *Trichoderma atroviride* regulates *Arabidopsis thaliana* root morphogenesis via auxin signaling and *ETHYLENE INSENSITIVE 2* functioning. New Phytol 209:1496–1512. https://doi.org/10.1111/nph.13725

Geoghegan IA, Gurr SJ (2017) Investigating chitin deacetylation and chitosan hydrolysis during vegetative growth in *Magnaporthe oryzae*. Cell Microbiol 19:e12743. https://doi.org/10.1111/cmi.12743

Ghorbanpour M, Omidvari M, Abbaszadeh-Dahaji P et al (2018) Mechanisms underlying the protective effects of beneficial fungi against plant diseases. Biol Control 117:147–157. https://doi.org/10.1016/j.biocontrol.2017.11.006

Giller KE, Delaune T, Silva JV et al (2021) The future of farming: who will produce our food? Food Secur 13:1073–1099. https://doi.org/10.1007/s12571-021-01184-6

Gitsopoulos TK, Damalas CA, Georgoulas I (2014) Improving diquat efficacy on grasses by adding adjuvants to the spray solution before use. Planta Daninha 32:355–360. https://doi.org/10.1590/S0100-83582014000200013

Goyal A, Kalia A (2020) Fungal phytohormones: plant growth-regulating substances and their applications in crop productivity. In: Yadav AN, Mishra S, Kour D et al (eds) Agriculturally important fungi for sustainable agriculture: volume 1: perspective for diversity and crop productivity. Springer, Cham, pp 143–169

Gull A, Ahmad Lone A, Ul Islam Wani N (2019) Biotic and abiotic stresses in plants. In: Bosco De Oliveira A (ed) Abiotic and biotic stress in plants. IntechOpen, London

Guo J, Cheng Y (2022) Advances in fungal elicitor-triggered plant immunity. Int J Mol Sci 23:12003. https://doi.org/10.3390/ijms231912003

Haituk S, Withee P, Sangta J et al (2022) Production of non-volatile metabolites from sooty molds and their bio-functionalities. Processes 10:1–19. https://doi.org/10.3390/pr10020329

Hallock YF, Clardy J, Kenfield DS, Strobel G (1988) De-O-methyldiaporthin, a phytotoxin from *Drechslera siccans*. Phytochemistry 27:3123–3125. https://doi.org/10.1016/0031-9422(88)80012-8

Hamayun M, Khan S, Khan A et al (2009) *Phoma herbarum* as a new gibberellin-producing and plant growth-promoting fungus. J Microbiol Biotechnol 19:1244–1249. https://doi.org/10.4014/jmb.0901.0030

Hamayun M, Hussain A, Khan SA et al (2017) Gibberellins producing endophytic fungus *Porostereum spadiceum* agh786 rescues growth of salt affected soybean. Front Microbiol 8:686. https://doi.org/10.3389/fmicb.2017.00686

Hardoim PR, Van Overbeek LS, Berg G et al (2015) The hidden world within plants: ecological and evolutionary considerations for defining functioning of microbial endophytes. Microbiol Mol Biol Rev 79:293–320. https://doi.org/10.1128/MMBR.00050-14

Harwoko H, Daletos G, Stuhldreier F et al (2021) Dithiodiketopiperazine derivatives from endophytic fungi *Trichoderma harzianum* and *Epicoccum nigrum*. Nat Prod Res 35:257–265. https://doi.org/10.1080/14786419.2019.1627348

Hasan HAH (2002) Gibberellin and auxin production by plant root-fungi and their biosynthesis under salinity-calcium interaction. Plant Soil Environ 48:101–106. https://doi.org/10.17221/4207-PSE

Heap I (2014) Herbicide resistant weeds. In: Pimentel D, Peshin R (eds) Integrated pest management: pesticide problems, vol 3. Springer, pp 281–301

Hernández-Calderón E, Aviles-Garcia ME, Castulo-Rubio DY et al (2018) Volatile compounds from beneficial or pathogenic bacteria differentially regulate root exudation, transcription of iron transporters, and defense signaling pathways in *Sorghum bicolor*. Plant Mol Biol 96:291–304. https://doi.org/10.1007/s11103-017-0694-5

Hijri M (2023) Microbial-based plant biostimulants. Microorganisms 11:686. https://doi.org/10.3390/microorganisms11030686

Hinsch J, Vrabka J, Oeser B et al (2015) *De novo* biosynthesis of cytokinins in the biotrophic fungus *Claviceps purpurea*. Environ Microbiol 17:2935–2951. https://doi.org/10.1111/1462-2920.12838

Hodge A, Robinson D, Fitter A (2000) Are microorganisms more effective than plants at competing for nitrogen? Trends Plant Sci 5:304–308. https://doi.org/10.1016/S1360-1385(00)01656-3

Huang S, Xu J, Li F et al (2017) Identification and antifungal activity of metabolites from the mangrove fungus *Phoma* sp. 128. Chem Nat Compd 53:237–240. https://doi.org/10.1007/s10600-017-1961-z

Huet et al (1992) Structures of elicitin isoforms secreted by *Phytophthora drechsleri*.

Hulme PE (2022) Global drivers of herbicide-resistant weed richness in major cereal crops worldwide. Pest Manag Sci 78:1824–1832. https://doi.org/10.1002/ps.6800

Iriti M, Faoro F (2009) Chitosan as a MAMP, searching for a PRR. Plant Signal Behav 4:66–68. https://doi.org/10.4161/psb.4.1.7408

Ismail MA, Amin MA, Eid AM et al (2021) Comparative study between exogenously applied plant growth hormones versus metabolites of microbial endophytes as plant growth-promoting for *Phaseolus vulgaris* l. Cells 10:1059. https://doi.org/10.3390/cells10051059

Iwasaki S, Muro H, Nozoe S et al (1972) Isolation of 3,4-dihydro-3,4,8-trihydroxy-1(2H)-naphthalenone and tenuazonic acid from *Pyricularia oryzae* cavara. Tetrahedron Lett 13:13–16. https://doi.org/10.1016/S0040-4039(01)84225-8

Jalali F, Zafari D, Salari H (2017) Volatile organic compounds of some *Trichoderma* spp. increase growth and induce salt tolerance in *Arabidopsis thaliana*. Fungal Ecol 29:67–75. https://doi.org/10.1016/j.funeco.2017.06.007

Jalander V, Gachande BD (2012) Effect of fungal metabolites of some rhizosphere fungi on seed germination and seedling growth of some pulses and cereals. Sci Res Rep 2:265–267

Jaroszuk-Ściseł J, Tyśkiewicz R, Nowak A et al (2019) Phytohormones (auxin, gibberellin) and acc deaminase *in vitro* synthesized by the mycoparasitic *Trichoderma* DEMTkZ3A0 strain and changes in the level of auxin and plant resistance markers in wheat seedlings inoculated with this strain conidia. Int J Mol Sci 20. https://doi.org/10.3390/ijms20194923

Joo JH, Hussein KA (2022) Biological control and plant growth promotion properties of volatile organic compound-producing antagonistic *Trichoderma* spp. Front Plant Sci 13:897668. https://doi.org/10.3389/fpls.2022.897668

Junker B, Walker A, Hesse M et al (2009) Pilot-scale process development and scale up for antifungal production. Bioprocess Biosyst Eng 32:443–458. https://doi.org/10.1007/s00449-008-0264-y

Kafle A, Cope K, Raths R et al (2019) Harnessing soil microbes to improve plant phosphate efficiency in cropping systems. Agronomy 9:127. https://doi.org/10.3390/agronomy9030127

Kamaruzzaman M, Wang Z, Wu M et al (2021) Promotion of tomato growth by the volatiles produced by the hypovirulent strain QT5-19 of the plant gray mold fungus *Botrytis cinerea*. Microbiol Res 247:126731. https://doi.org/10.1016/j.micres.2021.126731

Kamle M, Borah R, Bora H et al (2020) Systemic acquired resistance (SAR) and induced systemic resistance (ISR): role and mechanism of action against phytopathogens. In: Hesham AE-L, Upadhyay RS, Sharma GD et al (eds) Fungal Biotechnology and Bioengineering. Springer, Cham, pp 457–470

Kanchiswamy CN, Malnoy M, Maffei ME (2015) Bioprospecting bacterial and fungal volatiles for sustainable agriculture. Trends Plant Sci 20:206–211. https://doi.org/10.1016/j.tplants.2015.01.004

Keswani C, Singh SP, García-Estrada C et al (2022) Biosynthesis and beneficial effects of microbial gibberellins on crops for sustainable agriculture. J Appl Microbiol 132:1597–1615. https://doi.org/10.1111/jam.15348

Khalmuratova I, Kim H, Nam Y-J et al (2015) Diversity and plant growth promoting capacity of endophytic fungi associated with halophytic plants from the west coast of Korea. Mycobiology 43:373–383. https://doi.org/10.5941/MYCO.2015.43.4.373

Khalmuratova I, Choi D-H, Kim J-G, Lee I-S (2021) Endophytic fungi of salt-tolerant plants: diversity and ability to promote plant growth. J Microbiol Biotechnol 31:1526–1532. https://doi.org/10.4014/jmb.2106.06007

Khan AL (2011) Salinity stress resistance offered by endophytic fungal interaction between *Penicillium minioluteum* lhl09 and *Glycine max*. L. J Microbiol Biotechnol 21:893–902. https://doi.org/10.4014/jmb.1103.03012

Khan AL, Hamayun M, Kim Y-H et al (2011) Ameliorative symbiosis of endophyte (*Penicillium funiculosum* LHL06) under salt stress elevated plant growth of *Glycine max* L. Plant Physiol Biochem 49:852–861. https://doi.org/10.1016/j.plaphy.2011.03.005

Khan AL, Waqas M, Hussain J et al (2014) Endophytes *Aspergillus caespitosus* LK12 and *Phoma* sp. LK13 of *Moringa peregrina* produce gibberellins and improve rice plant growth. J Plant Interact 9:731–737. https://doi.org/10.1080/17429145.2014.917384

Khan AR, Ullah I, Waqas M et al (2015) Plant growth-promoting potential of endophytic fungi isolated from *Solanum nigrum* leaves. World J Microbiol Biotechnol 31:1461–1466. https://doi.org/10.1007/s11274-015-1888-0

Khan B, Yan W, Wei S et al (2019) Nematicidal metabolites from endophytic fungus *Chaetomium globosum* YSC5. FEMS Microbiol Lett 366:fnz169. https://doi.org/10.1093/femsle/fnz169

Khokhar I, Haider MS, Mukhtar I et al (2013) Effect of *Penicillium* species culture filtrate on seedling growth of wheat. Int Res J Agric Sci Soil Sci 3:24–29

Kılıç G, Gonca T, Bozdeveci A et al (2021) Antimicrobial, cytotoxic, antiviral effects, and apectroscopic characterization of metabolites produced by *Fusarium oxysporum* YP9B. Rec Nat Prod 15:547–567. https://doi.org/10.25135/rnp.208.20.06.1674

Kim H-Y, Choi GJ, Lee HB et al (2007) Some fungal endophytes from vegetable crops and their anti-oomycete activities against tomato late blight. Lett Appl Microbiol 44:332–337. https://doi.org/10.1111/j.1472-765X.2006.02093.x

Kim MH, Sung CY, Li M-H, Chu K-H (2012) Bioretention for stormwater quality improvement in Texas: removal effectiveness of *Escherichia coli*. Sep Purif Technol 84:120–124. https://doi.org/10.1016/j.seppur.2011.04.025

Köhl J, Kolnaar R, Ravensberg WJ (2019) Mode of action of microbial biological control agents against plant diseases: relevance beyond efficacy. Front Plant Sci 10:845. https://doi.org/10.3389/fpls.2019.00845

Kong HG, Shin TS, Kim TH, Ryu C-M (2018) Stereoisomers of the bacterial volatile compound 2,3-butanediol differently elicit systemic defense responses of pepper against multiple viruses in the field. Front Plant Sci 9:90. https://doi.org/10.3389/fpls.2018.00090

Kopecká R, Kameniarová M, Černý M et al (2023) Abiotic stress in crop production. Int J Mol Sci 24:6603. https://doi.org/10.3390/ijms24076603

Kosturova G, Rodeva R, Tasheva K et al (2012) Effect of crude culture filtrates of the pathogenic fungus *Phoma medicaginis* on *in vitro* cultures of pea. AgroLife Sci J 1:126–131

Kottb M, Gigolashvili T, Großkinsky DK, Piechulla B (2015) *Trichoderma* volatiles effecting *Arabidopsis*: from inhibition to protection against phytopathogenic fungi. Front Microbiol 6. https://doi.org/10.3389/fmicb.2015.00995

Koza N, Adedayo A, Babalola O, Kappo A (2022) Microorganisms in plant growth and development: roles in abiotic stress tolerance and secondary metabolites secretion. Microorganisms 10:1528. https://doi.org/10.3390/microorganisms10081528

Kremer RJ (2019) Bioherbicides and nanotechnology: current status and future trends. In: Nanobiopesticides today and future perspectives. Elsevier, London, pp 353–366

Kulye M, Liu H, Zhang Y et al (2012) Hrip1, a novel protein elicitor from necrotrophic fungus, *Alternaria tenuissima*, elicits cell death, expression of defence-related genes and systemic acquired resistance in tobacco: Hrip1, a novel protein elicitor, induces systemic acquired resistance in tobacco. Plant Cell Environ 35:2104–2120. https://doi.org/10.1111/j.1365-3040.2012.02539.x

Lamichhane JR, Osdaghi E, Behlau F et al (2018) Thirteen decades of antimicrobial copper compounds applied in agriculture. A review. Agron Sustain Dev 38:28. https://doi.org/10.1007/s13593-018-0503-9

Latz MAC, Jensen B, Collinge DB, Jørgensen HJL (2018) Endophytic fungi as biocontrol agents: elucidating mechanisms in disease suppression. Plant Ecol Divers 11:555–567. https://doi.org/10.1080/17550874.2018.1534146

Laurent E-A, Ahmed N, Durieu C et al (2020) Marine and fungal biostimulants improve grain yield, nitrogen absorption and allocation in durum wheat plants. J Agric Sci 158:279–287. https://doi.org/10.1017/S0021859620000660

Lee S, Behringer G, Hung R, Bennett J (2019) Effects of fungal volatile organic compounds on *Arabidopsis thaliana* growth and gene expression. Fungal Ecol 37:1–9. https://doi.org/10.1016/j.funeco.2018.08.004

Leitão AL, Enguita FJ (2016) Gibberellins in *Penicillium* strains: challenges for endophyte-plant host interactions under salinity stress. Microbiol Res 183:8–18. https://doi.org/10.1016/j.micres.2015.11.004

Lemfack MC, Gohlke B-O, Toguem SMT et al (2018) mVOC 2.0: a database of microbial volatiles. Nucleic Acids Res 46:D1261–D1265. https://doi.org/10.1093/nar/gkx1016

Lenarčič T, Pirc K, Hodnik V et al (2019) Molecular basis for functional diversity among microbial Nep1-like proteins. PLoS Pathog 15:e1007951. https://doi.org/10.1371/journal.ppat.1007951

Li N, Kang S (2018) Do volatile compounds produced by *Fusarium oxysporum* and *Verticillium dahliae* affect stress tolerance in plants? Mycology 9:166–175. https://doi.org/10.1080/21501203.2018.1448009

Li N, Wang W, Bitas V et al (2018) Volatile compounds emitted by diverse *Verticillium* species enhance plant growth by manipulating auxin signaling. Mol Plant-Microbe Interact 31:1021–1031. https://doi.org/10.1094/MPMI-11-17-0263-R

Li J, Cornelissen B, Rep M (2020) Host-specificity factors in plant pathogenic fungi. Fungal Genet Biol 144:103447. https://doi.org/10.1016/j.fgb.2020.103447

Li J, Wang J, Liu H et al (2022) Application of microbial inoculants significantly enhances crop productivity: a meta-analysis of studies from 2010 to 2020. J Sustain Agric Environ 1:216–225. https://doi.org/10.1002/sae2.12028

Lin Y, Jones ML (2022) Evaluating the growth-promoting effects of microbial biostimulants on greenhouse floriculture crops. HortScience 57:97–109. https://doi.org/10.21273/HORTSCI16149-21

Lopes MJDS, Dias-Filho MB, Gurgel ESC (2021) Successful plant growth-promoting microbes: inoculation methods and abiotic factors. Front Sustain Food Syst 5:606454. https://doi.org/10.3389/fsufs.2021.606454

López-Bucio J, Pelagio-Flores R, Herrera-Estrella A (2015) *Trichoderma* as biostimulant: exploiting the multilevel properties of a plant beneficial fungus. Sci Hortic 196:109–123. https://doi.org/10.1016/j.scienta.2015.08.043

Lou J, Fu L, Peng Y, Zhou L (2013) Metabolites from *Alternaria* fungi and their bioactivities. Molecules 18:5891–5935. https://doi.org/10.3390/molecules18055891

Lu Y, Li S, Shao M et al (2016) Isolation, identification, derivatization and phytotoxic activity of secondary metabolites produced by *Cladosporium oxysporum* DH14, a locust-associated fungus. J Integr Agric 15:832–839. https://doi.org/10.1016/S2095-3119(15)61145-5

Mahapatra R, Jampala SSM, Patel DR (2015) Induction of systemic acquired resistance in *Zea mays* L. by *Aspergillus flavus* and *A. parasiticus* derived elicitors. Arch Phytopathol Plant Protect 48:120–134. https://doi.org/10.1080/03235408.2014.884523

Martins-Gomes C, Silva TL, Andreani T, Silva AM (2022) Glyphosate vs. glyphosate-based herbicides exposure: a review on their toxicity. J Xenobiot 12:21–40. https://doi.org/10.3390/jox12010003

Masi M, Meyer S, Pescitelli G et al (2017) Phytotoxic activity against *Bromus tectorum* for secondary metabolites of a seed-pathogenic *Fusarium* strain belonging to the *F. tricinctum* species complex. Nat Prod Res 31:2768–2777. https://doi.org/10.1080/14786419.2017.1297445

Masi M, Zonno MC, Cimmino A et al (2018) On the metabolites produced by *Colletotrichum gloeosporioides* a fungus proposed for the *Ambrosia artemisiifolia* biocontrol; spectroscopic data and absolute configuration assignment of colletochlorin a. Nat Prod Res 32:1537–1547. https://doi.org/10.1080/14786419.2017.1385020

Masi M, Santoro E, Clement S et al (2020) Further secondary metabolites produced by the fungus *Pyricularia grisea* isolated from buffelgrass (*Cenchrus ciliaris*). Chirality 32:1234–1242. https://doi.org/10.1002/chir.23270

Meepagala KM, Clausen BM, Johnson RD et al (2019) A phytotoxic and antifungal metabolite (Pyrichalasin H) from a fungus infecting *Brachiaria eruciformis* (signal grass). J Agric Chem Environ 08:115–128. https://doi.org/10.4236/jacen.2019.83010

Mehnaz S (2016) An overview of globally available bioformulations. In: Arora NK, Mehnaz S, Balestrini R (eds) Bioformulations: for sustainable agriculture. Springer, New Delhi, pp 267–281

Mestre-Tomás J, Esgueva-Vilà D, Fuster-Alonso A et al (2023) Chitosan modulates volatile organic compound emission from the biocontrol fungus *Pochonia chlamydosporia*. Molecules 28:4053. https://doi.org/10.3390/molecules28104053

Meyer V (2008) A small protein that fights fungi: AFP as a new promising antifungal agent of biotechnological value. Appl Microbiol Biotechnol 78:17–28. https://doi.org/10.1007/s00253-007-1291-3

Miao X, Qu H, Han Y et al (2019) The protein elicitor Hrip1 enhances resistance to insects and early bolting and flowering in *Arabidopsis thaliana*. PLoS One 14:e0216082. https://doi.org/10.1371/journal.pone.0216082

Michielse CB, Rep M (2009) Pathogen profile update: *Fusarium oxysporum*. Mol Plant Pathol 10:311–324. https://doi.org/10.1111/j.1364-3703.2009.00538.x

Mikami Y, Nishijima Y, Iimura H et al (1971) Chemical studies on brown-spot disease of tobacco plants: part I. Tenuazonic acid as a vivotoxin of *Alternaria longipes*. Agric Biol Chem 35:611–618. https://doi.org/10.1080/00021369.1971.10859948

Miya A, Albert P, Shinya T et al (2007) CERK1, a LysM receptor kinase, is essential for chitin elicitor signaling in *Arabidopsis*. Proc Natl Acad Sci 104:19613–19618. https://doi.org/10.1073/pnas.0705147104

Mohammad-Razdari A, Rousseau D, Bakhshipour A et al (2022) Recent advances in E-monitoring of plant diseases. Biosens Bioelectron 201:113953. https://doi.org/10.1016/j.bios.2021.113953
Moisan K, Cordovez V, Van De Zande EM et al (2019) Volatiles of pathogenic and non-pathogenic soil-borne fungi affect plant development and resistance to insects. Oecologia 190:589–604. https://doi.org/10.1007/s00442-019-04433-w
Mondal G, Dureja P, Sen B (2000) Fungal metabolites from *Aspergillus niger* AN27 related to plant growth promotion. Indian J Exp Biol 38:84–87
Monteiro A, Santos S (2022) Sustainable approach to weed management: the role of precision weed management. Agronomy 12:118. https://doi.org/10.3390/agronomy12010118
Montull JM, Torra J (2023) Herbicide resistance is increasing in Spain: concomitant management and prevention. Plan Theory 12:469. https://doi.org/10.3390/plants12030469
Morath SU, Hung R, Bennett JW (2012) Fungal volatile organic compounds: a review with emphasis on their biotechnological potential. Fungal Biol Rev 26:73–83. https://doi.org/10.1016/j.fbr.2012.07.001
Morcillo R, Manzanera M (2021) The effects of plant-associated bacterial exopolysaccharides on plant abiotic stress tolerance. Metabolites 11:337. https://doi.org/10.3390/metabo11060337
Morcillo RJL, Baroja-Fernández E, López-Serrano L et al (2022) Cell-free microbial culture filtrates as candidate biostimulants to enhance plant growth and yield and activate soil- and plant-associated beneficial microbiota. Front Plant Sci 13:1040515. https://doi.org/10.3389/fpls.2022.1040515
Mostafalou S, Abdollahi M (2017) Pesticides: an update of human exposure and toxicity. Arch Toxicol 91:549–599. https://doi.org/10.1007/s00204-016-1849-x
Mousa WK, Raizada MN (2013) The diversity of anti-microbial secondary metabolites produced by fungal endophytes: an interdisciplinary perspective. Front Microbiol 4. https://doi.org/10.3389/fmicb.2013.00065
Myers JP, Antoniou MN, Blumberg B et al (2016) Concerns over use of glyphosate-based herbicides and risks associated with exposures: a consensus statement. Environ Health 15:19. https://doi.org/10.1186/s12940-016-0117-0
Naamala J, Smith DL (2020) Relevance of plant growth promoting microorganisms and their derived compounds, in the face of climate change. Agronomy 10:1179. https://doi.org/10.3390/agronomy10081179
Naamala J, Smith DL (2021) Microbial derived compounds, a step toward enhancing microbial inoculants technology for sustainable agriculture. Front Microbiol 12:634807. https://doi.org/10.3389/fmicb.2021.634807
Nakajima H, Nishimura K, Hamasaki T et al (1987) Structure of Neovasinone, a new alpha-pyrone plant growth regulator produced by the fungus, *Neocosmospora vasinfecta* E.F. Smith. Agric Biol Chem 51:1221–1224
Namasivayam S, Tony B, Bharani R, Raj F (2015) Herbicidal activity of soil isolate of *Fusarium oxysporum* free and chitosan nanoparticles coated metabolites against economic important weed. Asian J Microbiol Biotechnol Environ Sci 17:1015–1020
Nandhini C, Ganesh P, Yoganathan K, Kumar D (2019) Efficacy of *Colletotrichum gloeosporioides*, potential fungi for bio control of *Echinochloa crus-galli* (Barnyard grass). J Drug Deliv Ther 9:72–75. https://doi.org/10.22270/jddt.v9i6-s.3751
O'Callaghan M, Ballard RA, Wright D (2022) Soil microbial inoculants for sustainable agriculture: limitations and opportunities. Soil Use Manag 38:1340–1369. https://doi.org/10.1111/sum.12811
Paege N, Jung S, Schäpe P et al (2016) A transcriptome meta-analysis proposes novel biological roles for the antifungal protein AnAFP in *Aspergillus niger*. PLoS One 11:e0165755. https://doi.org/10.1371/journal.pone.0165755
Pálfi X, Karácsony Z, Csikós A et al (2020) The potential use of the culture filtrate of an *Aspergillus niger* strain in the management of fungal diseases of grapevine. J Cent Eur Agric 21:839–850. https://doi.org/10.5513/JCEA01/21.4.2942

Pandian K, Arunachalam P, Govindaraj M (2015) Implications and ways to enhance nutrient use efficiency under changing climate. In: Crop improvement in the era of climate change. IK Publishers, Delhi, pp 115–142

Parveen S, Wani AH, Bhat MY (2019) Effect of culture filtrates of pathogenic and antagonistic fungi on seed germination of some economically important vegetables. Braz J Biol Sci 6:133–139. https://doi.org/10.21472/bjbs.061212

Pascholati S, De Souza V, Cardoso Filh J (2019) Indução de resistência por *Trichoderma*. In: Meyer M, Mazaro S, da Silva J (eds) Trichoderma: uso na agricultura. Embrapa, Brasília, pp 235–254

Patel ZM, Mahapatra R, Jampala SSM (2020) Chapter 11 – Role of fungal elicitors in plant defense mechanism. In: Sharma V, Salwan R, Al-Ani LKT (eds) Molecular aspects of plant beneficial microbes in agriculture. Academic, London, pp 143–158

Paul A, Shamim N, Acharya K, Chakraborty N (2022) Twin effects of fungal elicitor produced from *Alternaria alternata* on induced systemic resistance and secondary metabolites production in *Catharanthus roseus* (L.) G. Don. Plant Stress 6:100124. https://doi.org/10.1016/j.stress.2022.100124

Paul A, Sarkar A, Acharya K, Chakraborty N (2023) Fungal elicitor-mediated induction of innate immunity in *Catharanthus roseus* against leaf blight disease caused by *Alternaria alternata*. J Plant Growth Regul 42:491–501. https://doi.org/10.1007/s00344-021-10569-y

Pedras MSC, Yu Y (2009) Phytotoxins, elicitors and other secondary metabolites from phytopathogenic "blackleg" fungi: structure, phytotoxicity and biosynthesis. Nat Prod Commun 4:1934578X0900400. https://doi.org/10.1177/1934578X0900400927

Pellegrini M, Pagnani G, Bernardi M et al (2020) Cell-free supernatants of plant growth-promoting bacteria: a review of their use as biostimulant and microbial biocontrol agents in sustainable agriculture. Sustain Switz 12:1–22. https://doi.org/10.3390/su12239917

Peng Y, Li SJ, Yan J et al (2021) Research progress on phytopathogenic fungi and their role as biocontrol agents. Front Microbiol 12:670135. https://doi.org/10.3389/fmicb.2021.670135

Pérez-Aguilar H, Lacruz-Asaro M, Arán-Ais F (2023) Evaluation of biostimulants based on recovered protein hydrolysates from animal by-products as plant growth enhancers. J Plant Sci Phytopathol 7:042–047. https://doi.org/10.29328/journal.jpsp.1001104

Pérez-de-Luque A (2023) Can nanotechnology improve the application of bioherbicides? Pest Manag Sci. https://doi.org/10.1002/ps.7526

Perotti VE, Larran AS, Palmieri VE et al (2020) Herbicide resistant weeds: a call to integrate conventional agricultural practices, molecular biology knowledge and new technologies. Plant Sci 290:110255. https://doi.org/10.1016/j.plantsci.2019.110255

Pieterse CMJ, Zamioudis C, Berendsen RL et al (2014) Induced systemic resistance by beneficial microbes. Annu Rev Phytopathol 52:347–375. https://doi.org/10.1146/annurev-phyto-082712-102340

Poluektova E, Tokarev Y, Sokornova S et al (2018) Curvulin and Phaeosphaeride A from *Paraphoma* sp. VIZR 1.46 isolated from *Cirsium arvense* as potential herbicides. Molecules 23:2795. https://doi.org/10.3390/molecules23112795

Popko M, Michalak I, Wilk R et al (2018) Effect of the new plant growth biostimulants based on amino acids on yield and grain quality of winter wheat. Molecules 23:470. https://doi.org/10.3390/molecules23020470

Poveda J (2021) Beneficial effects of microbial volatile organic compounds (MVOCs) in plants. Appl Soil Ecol 168:104118. https://doi.org/10.1016/j.apsoil.2021.104118

Poveda J (2022) Effect of volatile and non-volatile metabolites from *Leptosphaeria maculans* on tomato calli under abiotic stresses. Plant Stress 3:100054. https://doi.org/10.1016/j.stress.2021.100054

Poveda J, Baptista P (2021) Filamentous fungi as biocontrol agents in olive (*Olea europaea* L.) diseases: Mycorrhizal and endophytic fungi. Crop Prot 146:105672. https://doi.org/10.1016/j.cropro.2021.105672

Poveda J, González-Andrés F (2021) *Bacillus* as a source of phytohormones for use in agriculture. Appl Microbiol Biotechnol 105:8629–8645. https://doi.org/10.1007/s00253-021-11492-8

Poveda J, Abril-Urias P, Escobar C (2020) Biological control of plant-parasitic nematodes by filamentous fungi inducers of resistance: *Trichoderma*, mycorrhizal and endophytic fungi. Front Microbiol 11:992. https://doi.org/10.3389/fmicb.2020.00992

Pusztahelyi T (2018) Chitin and chitin-related compounds in plant–fungal interactions. Mycology 9:189–201. https://doi.org/10.1080/21501203.2018.1473299

Pusztahelyi T, Holb IJ, Pócsi I (2015) Secondary metabolites in fungus-plant interactions. Front Plant Sci 6. https://doi.org/10.3389/fpls.2015.00573

Qiang S, Wang L, Wei R et al (2010) Bioassay of the herbicidal activity of aac-toxin produced by *Alternaria alternata* isolated from *Ageratina adenophora*. Weed Technol 24:197–201. https://doi.org/10.1614/WT-D-09-00016.1

Raaijmakers JM, Mazzola M (2012) Diversity and natural functions of antibiotics produced by beneficial and plant pathogenic bacteria. Annu Rev Phytopathol 50:403–424. https://doi.org/10.1146/annurev-phyto-081211-172908

Radhakrishnan R, Kang S-M, Baek I-Y, Lee I-J (2014) Characterization of plant growth-promoting traits of *Penicillium* species against the effects of high soil salinity and root disease. J Plant Interact 9:754–762. https://doi.org/10.1080/17429145.2014.930524

Rahman MM, Mostofa MG, Rahman MA et al (2019) Acetic acid: a cost-effective agent for mitigation of seawater-induced salt toxicity in mung bean. Sci Rep 9:15186. https://doi.org/10.1038/s41598-019-51178-w

Rahman M, Mostofa MG, Keya SS et al (2021) Acetic acid improves drought acclimation in soybean: an integrative response of photosynthesis, osmoregulation, mineral uptake and antioxidant defense. Physiol Plant 172:334–350. https://doi.org/10.1111/ppl.13191

Ranathunge NP, Sandani HBP (2016) Deceptive behaviour of *Colletotrichum truncatum*: strategic survival as an asymptomatic endophyte on non-host species. J Plant Prot Res 56:157–162. https://doi.org/10.1515/jppr-2016-0026

Rangel LI, Hamilton O, Jonge R, Bolton MD (2021) Fungal social influencers: secondary metabolites as a platform for shaping the plant-associated community. Plant J 108:632–645. https://doi.org/10.1111/tpj.15490

Rao S, Ramgoapl S (2010) Effect of *Alternaria helianthi* culture filtrate on callus and regeneration of plantlets from tolerant callus in sunflower (*Helianthus annuus* L.). Indian. J Biotechnol 9(1):187–191

Rebaque D, Del Hierro I, López G et al (2021) Cell wall-derived mixed-linked β-1,3/1,4-glucans trigger immune responses and disease resistance in plants. Plant J 106:601–615. https://doi.org/10.1111/tpj.15185

Reddy S, Sinha A, Osborne WJ (2021) Microbial secondary metabolites: recent developments and technological challenges. In: Volatiles and Metabolites of Microbes. Elsevier, London, pp 1–22

Reese KL, Rasley A, Avila JR et al (2020) Metabolic profiling of volatile organic compounds (VOCs) emitted by the pathogens *Francisella tularensis* and *Bacillus anthracis* in liquid culture. Sci Rep 10:9333. https://doi.org/10.1038/s41598-020-66136-0

Rodrigo S, García-Latorre C, Santamaria O (2022) Metabolites produced by fungi against fungal phytopathogens: review, implementation and perspectives. Plan Theory 11:81. https://doi.org/10.3390/plants11010081

Rodrigues M, da Silva A, Batista J et al (2020) Induction of defense mechanisms in tomato plants by saprobic fungi filtrates against early blight disease. Rev Caatinga 33:671–678. https://doi.org/10.1590/1983-21252020v33n310rc

Ruiz-Herrera J, Ortiz-Castellanos L (2019) Cell wall glucans of fungi. A review. Cell Surf 5:100022. https://doi.org/10.1016/j.tcsw.2019.100022

Ryu C-M, Farag MA, Hu C-H et al (2003) Bacterial volatiles promote growth in *Arabidopsis*. Proc Natl Acad Sci 100:4927–4932. https://doi.org/10.1073/pnas.0730845100

Sák M, Dokupilová I, Kaňuková Š et al (2021) Biotic and abiotic elicitors of stilbenes production in *Vitis vinifera* l. cell culture. Plan Theory 10:490. https://doi.org/10.3390/plants10030490

Salazar-Badillo FB, Sánchez-Rangel D, Becerra-Flora A et al (2015) *Arabidopsis thaliana* polyamine content is modified by the interaction with different *Trichoderma* species. Plant Physiol Biochem 95:49–56. https://doi.org/10.1016/j.plaphy.2015.07.003

Salvatore MM, Andolfi A (2021) Phytopathogenic fungi and toxicity. Toxins 13:689. https://doi.org/10.3390/toxins13100689

Salvatore MM, Alves A, Andolfi A (2020) Secondary metabolites of *Lasiodiplodia theobromae*: distribution, chemical diversity, bioactivity, and implications of their occurrence. Toxins 12:457. https://doi.org/10.3390/toxins12070457

Salvatore MM, Alves A, Andolfi A (2021) Secondary metabolites produced by *Neofusicoccum* species associated with plants: a review. Agriculture 11:149. https://doi.org/10.3390/agriculture11020149

Sánchez-López ÁM, Baslam M, De Diego N et al (2016) Volatile compounds emitted by diverse phytopathogenic microorganisms promote plant growth and flowering through cytokinin action. Plant Cell Environ 39:2592–2608. https://doi.org/10.1111/pce.12759

Schalchli H, Tortella GR, Rubilar O et al (2016) Fungal volatiles: an environmentally friendly tool to control pathogenic microorganisms in plants. Crit Rev Biotechnol 36:144–152. https://doi.org/10.3109/07388551.2014.946466

Schenkel D, Maciá-Vicente JG, Bissell A, Splivallo R (2018) Fungi indirectly affect plant root architecture by modulating soil volatile organic compounds. Front Microbiol 9:1847. https://doi.org/10.3389/fmicb.2018.01847

Schmidt R, Etalo DW, de Jager V et al (2016) Microbial small talk: volatiles in fungal–bacterial interactions. Front Microbiol 6:1495

Schoenenberger N, D'Andrea L (2012) Surveying the occurrence of subspontaneous glyphosate-tolerant genetically engineered *Brassica napus* L. (Brassicaceae) along Swiss railways. Environ Sci Eur 24:23. https://doi.org/10.1186/2190-4715-24-23

Schulz S, Dickschat JS (2007) Bacterial volatiles: the smell of small organisms. Nat Prod Rep 24:814–842. https://doi.org/10.1039/b507392h

Schulz-Bohm K, Martín-Sánchez L, Garbeva P (2017) Microbial volatiles: small molecules with an important role in intra- and inter-kingdom interactions. Front Microbiol 8:2484. https://doi.org/10.3389/fmicb.2017.02484

Schütz L, Gattinger A, Meier M et al (2018) Improving crop yield and nutrient use efficiency via biofertilization—a global meta-analysis. Front Plant Sci 8:2204. https://doi.org/10.3389/fpls.2017.02204

Seidl MF, Van Den Ackerveken G (2019) Activity and phylogenetics of the broadly occurring family of microbial Nep1-like proteins. Annu Rev Phytopathol 57:367–386. https://doi.org/10.1146/annurev-phyto-082718-100054

Shahzad A, Ullah S, Dar AA et al (2021) Nexus on climate change: agriculture and possible solution to cope future climate change stresses. Environ Sci Pollut Res 28:14211–14232. https://doi.org/10.1007/s11356-021-12649-8

Shinya T, Ménard R, Kozone I et al (2006) Novel β-1,3-, 1,6-oligoglucan elicitor from *Alternaria alternata* 102 for defense responses in tobacco: novel β-glucan elicitor from *A. alternata* 102. FEBS J 273:2421–2431. https://doi.org/10.1111/j.1742-4658.2006.05249.x

Sica VP, Figueroa M, Raja HA et al (2016) Optimizing production and evaluating biosynthesis *in situ* of a herbicidal compound, mevalocidin, from *Coniolariella* sp. J Ind Microbiol Biotechnol 43:1149–1157. https://doi.org/10.1007/s10295-016-1782-2

Singh UP, Sarma BK, Singh DP (2003) Effect of plant growth-promoting rhizobacteria and culture filtrate of *Sclerotium rolfsii* on phenolic and salicylic acid contents in chickpea (*Cicer arietinum*). Curr Microbiol 46:131–140. https://doi.org/10.1007/s00284-002-3834-2

Singh SK, Strobel GA, Knighton B et al (2011) An endophytic *Phomopsis* sp. possessing bioactivity and fuel potential with its volatile organic compounds. Microb Ecol 61:729–739. https://doi.org/10.1007/s00248-011-9818-7

Singh R, Kumar M, Mittal A, Mehta PK (2017) Microbial metabolites in nutrition, healthcare and agriculture. 3 Biotech 7:15. https://doi.org/10.1007/s13205-016-0586-4

Singh M, Bhasin S, Madan N et al (2021) Bioinoculants for agricultural sustainability. In: Soni R, Suyal DC, Bhargava P, Goel R (eds) Microbiological activity for soil and plant health management. Springer, Singapore, pp 629–641

Snelders NC, Rovenich H, Petti GC et al (2020) A plant pathogen utilizes effector proteins for microbiome manipulation. Microbiology. https://doi.org/10.1101/2020.01.30.926725

Song G, Ryu C-M (2013) Two volatile organic compounds trigger plant self-defense against a bacterial pathogen and a sucking insect in cucumber under open field conditions. Int J Mol Sci 14:9803–9819. https://doi.org/10.3390/ijms14059803

Spinelli V, Brasili E, Sciubba F et al (2022) Biostimulant effects of *Chaetomium globosum* and *Minimedusa polyspora* culture filtrates on *Cichorium intybus* plant: growth performance and metabolomic traits. Front Plant Sci 13:879076. https://doi.org/10.3389/fpls.2022.879076

Srikamwang C, Onsa NE, Sunanta P et al (2023) Role of microbial volatile organic compounds in promoting plant growth and disease resistance in horticultural production. Plant Signal Behav 2227440. https://doi.org/10.1080/15592324.2023.2227440

Sun C, Jin L, Cai Y et al (2019) (1→3)-β-D-glucan from yeast cell wall: characteristic and potential application in controlling postharvest disease of pear. Postharvest Biol Technol 154:105–114. https://doi.org/10.1016/j.postharvbio.2019.04.021

Sung G-H, Shrestha B, Park K-B et al (2011) Enhancing effect of *Shimizuomyces paradoxus* on seed germination and seedling growth of canola, plant growth of cucumber, and harvest of tomato. Mycobiology 39:7–11. https://doi.org/10.4489/MYCO.2011.39.1.007

Švábová L, Lebeda A (2005) *In vitro* selection for improved plant resistance to toxin-producing pathogens. J Phytopathol 153:52–64. https://doi.org/10.1111/j.1439-0434.2004.00928.x

Szappanos H, Szigeti GP, Pál B et al (2006) The antifungal protein AFP secreted by *Aspergillus giganteus* does not cause detrimental effects on certain mammalian cells. Peptides 27:1717–1725. https://doi.org/10.1016/j.peptides.2006.01.009

Tarkowski P, Vereecke D (2014) Threats and opportunities of plant pathogenic bacteria. Biotechnol Adv 32:215–229. https://doi.org/10.1016/j.biotechadv.2013.11.001

Teixeira WF, Soares LH, Fagan EB et al (2020) Amino acids as stress reducers in soybean plant growth under different water-deficit conditions. J Plant Growth Regul 39:905–919. https://doi.org/10.1007/s00344-019-10032-z

Terra WC, Campos VP, Martins SJ et al (2018) Volatile organic molecules from *Fusarium oxysporum* strain 21 with nematicidal activity against *Meloidogyne incognita*. Crop Prot 106:125–131. https://doi.org/10.1016/j.cropro.2017.12.022

Ting A, Mah S, Tee C (2010) Identification of volatile metabolites from fungal endophytes with biocontrol potential towards *Fusarium oxysporum* f. sp. *cubense* race 4. Am J Agric Biol Sci 5:177–182. https://doi.org/10.3844/ajabssp.2010.177.182

Tomas-Grau RH, Hael-Conrad V, Requena-Serra FJ et al (2020) Biological control of strawberry grey mold disease caused by *Botrytis cinerea* mediated by *Colletotrichum acutatum* extracts. BioControl 65:461–473. https://doi.org/10.1007/s10526-020-10003-4

Torra J, Osuna MD, Merotto A, Vila-Aiub M (2021) Editorial: multiple herbicide-resistant weeds and non-target site resistance mechanisms: a global challenge for food production. Front Plant Sci 12:763212. https://doi.org/10.3389/fpls.2021.763212

Trabelsi D, Mhamdi R (2013) Microbial inoculants and their impact on soil microbial communities: a review. Biomed Res Int 2013:1–11. https://doi.org/10.1155/2013/863240

Trovato M, Funck D, Forlani G et al (2021) Editorial: amino acids in plants: regulation and functions in development and stress defense. Front Plant Sci 12:772810. https://doi.org/10.3389/fpls.2021.772810

Tsuge T, Harimoto Y, Akimitsu K et al (2013) Host-selective toxins produced by the plant pathogenic fungus *Alternaria alternata*. FEMS Microbiol Rev 37:44–66. https://doi.org/10.1111/j.1574-6976.2012.00350.x

Tudi M, Daniel Ruan H, Wang L et al (2021) Agriculture development, pesticide application and its impact on the environment. Int J Environ Res Public Health 18:1112. https://doi.org/10.3390/ijerph18031112

Tuomi T, Laakso S, Rosenqvist H (1995) Plant hormones in fungi and bacteria from malting barley. J Inst Brew 101:351–357. https://doi.org/10.1002/j.2050-0416.1995.tb00872.x

Usseglio VL, Pizzolitto RP, Rodriguez C et al (2017) Volatile organic compounds from the interaction between *Fusarium verticillioides* and maize kernels as a natural repellents of *Sitophilus zeamais*. J Stored Prod Res 73:109–114. https://doi.org/10.1016/j.jspr.2017.08.001

Vaishnavi J, Osborne WJ (2021) Microbial volatiles: small molecules with an important role in intra- and interbacterial genus interactions-quorum sensing. Volatiles and Metabolites of Microbes. Elsevier, In, pp 35–50

Van Den Ackerveken G (2017) How plants differ in toxin-sensitivity. Science 358:1383–1384. https://doi.org/10.1126/science.aar4188

Velivelli SLS, Kromann P, Lojan P et al (2015) Identification of mVOCs from andean rhizobacteria and field evaluation of bacterial and mycorrhizal inoculants on growth of potato in its center of origin. Microb Ecol 69:652–667. https://doi.org/10.1007/s00248-014-0514-2

Veronico P, Sasanelli N, Troccoli A et al (2023) Evaluation of fungal volatile organic compounds for control the plant parasitic nematode *Meloidogyne incognita*. Plan Theory 12:1935. https://doi.org/10.3390/plants12101935

Vinale F (2014) Biopesticides and biofertilizers based on fungal secondary metabolites. J Biofertilizers Biopestic 5. https://doi.org/10.4172/2155-6202.1000e119

Vinale F, Sivasithamparam K, Ghisalberti EL et al (2012) *Trichoderma* secondary metabolites that affect plant metabolism. Nat Prod Commun 7:1934578X1200701. https://doi.org/10.1177/1934578X1200701133

Vurro M, Andolfi A, Boari A et al (2012) Optimization of the production of herbicidal toxins by the fungus *Ascochyta caulina*. Biol Control 60:192–198. https://doi.org/10.1016/j.biocontrol.2011.10.005

Vurro M, Boari A, Casella F, Zonno MC (2018) Fungal phytotoxins in sustainable weed management. Curr Med Chem 25:268–286. https://doi.org/10.2174/0929867324666170426152331

Wagh MS, Osborne WJ, Sivarajan S (2021) Volatile organic compounds for enhancement of plant growth through plant growth promoting rhizobacteria. In: Volatiles and metabolites of microbes. Elsevier, London, pp 325–347

Wang Y, Zhao A, Morcillo RJL et al (2021) A bacterial effector protein uncovers a plant metabolic pathway involved in tolerance to bacterial wilt disease. Mol Plant 14:1281–1296. https://doi.org/10.1016/j.molp.2021.04.014

Wang H, Guo Y, Luo Z et al (2022) Recent advances in *Alternaria* phytotoxins: a review of their occurrence, structure, bioactivity, and biosynthesis. J Fungi 8:168. https://doi.org/10.3390/jof8020168

Wani MA, Sanjana K, Kumar DM, Lal DK (2010) GC – MS analysis reveals production of 2 – phenylethanol from *Aspergillus niger* endophytic in rose. J Basic Microbiol 50:110–114. https://doi.org/10.1002/jobm.200900295

Waqas M, Khan AL, Kamran M et al (2012) Endophytic fungi produce gibberellins and indoleacetic acid and promotes host-plant growth during stress. Molecules 17:10754–10773. https://doi.org/10.3390/molecules170910754

Waqas M, Khan AL, Shahzad R et al (2015) Mutualistic fungal endophytes produce phytohormones and organic acids that promote japonica rice plant growth under prolonged heat stress. J Zhejiang Univ Sci B 16:1011–1018. https://doi.org/10.1631/jzus.B1500081

Wei Z-M, Laby RJ, Zumoff CH et al (1992) Harpin, elicitor of the hypersensitive response produced by the plant pathogen *Erwinia amylovora*. Science 257:85–88. https://doi.org/10.1126/science.1621099

Weikl F, Ghirardo A, Schnitzler J-P, Pritsch K (2016) Sesquiterpene emissions from *Alternaria alternata* and *Fusarium oxysporum*: effects of age, nutrient availability, and co-cultivation. Sci Rep 6:22152. https://doi.org/10.1038/srep22152

Werner S, Polle A, Brinkmann N (2016) Belowground communication: impacts of volatile organic compounds (VOCs) from soil fungi on other soil-inhabiting organisms. Appl Microbiol Biotechnol 100:8651–8665. https://doi.org/10.1007/s00253-016-7792-1

Wesseler J (2022) The EU's farm-to-fork strategy: an assessment from the perspective of agricultural economics. Appl Econ Perspect Policy 44:1826–1843. https://doi.org/10.1002/aepp.13239

Wu Y, Zhou J, Li C, Ma Y (2019) Antifungal and plant growth promotion activity of volatile organic compounds produced by *Bacillus amyloliquefaciens*. MicrobiologyOpen 8:e00813. https://doi.org/10.1002/mbo3.813

Xu Z, Shi M, Tian Y et al (2019) Dirhamnolipid produced by the pathogenic fungus *Colletotrichum gloeosporioides* BWH-1 and its herbicidal activity. Molecules 24:2969. https://doi.org/10.3390/molecules24162969

Xu D, Xue M, Shen Z et al (2021a) Phytotoxic secondary metabolites from fungi. Toxins 13:261. https://doi.org/10.3390/toxins13040261

Xu K, Li X-Q, Zhao D-L, Zhang P (2021b) Antifungal secondary metabolites produced by the fungal endophytes: chemical diversity and potential use in the development of biopesticides. Front Microbiol 12:689527. https://doi.org/10.3389/fmicb.2021.689527

Yadav AN (2021) Beneficial plant-microbe interactions for agricultural sustainability. J Appl Biol Biotechnol 9:i–iv. https://doi.org/10.7324/JABB.2021.91ed

Yagi A, Okamura N, Haraguchi H et al (1993) Antimicrobial tetrahydroanthraquinones from a strain of *Alternaria solani*. Int J Plant Biochem 33:87–91. https://doi.org/10.1016/0031-9422(93)85401-C

Yamaguchi T, Yamada A, Hong N et al (2000) Differences in the recognition of glucan elicitor signals between rice and soybean: β-glucan fragments from the rice blast disease fungus *Pyricularia oryzae* that elicit phytoalexin biosynthesis in suspension-cultured rice cells. Plant Cell 12:817–826

Yuzikhin O, Mitina G, Berestetskiy A (2007) Herbicidal potential of stagonolide, a new phytotoxic nonenolide from *Stagonospora cirsii*. J Agric Food Chem 55:7707–7711. https://doi.org/10.1021/jf070742c

Zeffa DM, Perini LJ, Silva MB et al (2019) *Azospirillum brasilense* promotes increases in growth and nitrogen use efficiency of maize genotypes. PLoS One 14:e0215332. https://doi.org/10.1371/journal.pone.0215332

Zhai X, Luo D, Li X et al (2018) Endophyte *Chaetomium globosum* d38 promotes bioactive constituents accumulation and root production in *Salvia miltiorrhiza*. Front Microbiol 8:2694. https://doi.org/10.3389/fmicb.2017.02694

Zhang H, Sun Y, Xie X et al (2009) A soil bacterium regulates plant acquisition of iron via deficiency-inducible mechanisms. Plant J 58:568–577. https://doi.org/10.1111/j.1365-313X.2009.03803.x

Zhang Y, Askim JR, Zhong W et al (2014) Identification of pathogenic fungi with an optoelectronic nose. Analyst 139:1922–1928. https://doi.org/10.1039/C3AN02112B

Zhao X, Cui H, Wang Y et al (2018) Development strategies and prospects of nano-based smart pesticide formulation. J Agric Food Chem 66:6504–6512. https://doi.org/10.1021/acs.jafc.7b02004

Zhu Y-J, Pan Z-Z, Yu X-J et al (2013) Inhibitory activities of griseofulvin and its analogues against phytopathogenic fungi. Asian J Chem 25:8788–8792. https://doi.org/10.14233/ajchem.2013.15664

Zorrilla JG, Masi M, Clement S et al (2023) Production of (10S,11S)-(—)-epi-pyriculol and its HPLC quantification in liquid cultures of *Pyricularia grisea*, a potential mycoherbicide for the control of buffelgrass (*Cenchrus ciliaris*). J Fungi 9:316. https://doi.org/10.3390/jof9030316

The manufacturer's authorised representative in the EU is Springer Nature Customer Service Centre GmbH, Europaplatz 3, 69115 Heidelberg, Germany. If you have any concerns regarding our products, please contact ProductSafety@springernature.com

Printed and bound by CPI Group (UK) Ltd, Croydon, CR0 4YY
30/09/2026
02225600-0006